DATA ANALYTICS AND SMART TECHNOLOGIES IN SOLID WASTE MANAGEMENT

Advancing Sustainability

DATA ANALYTICS AND SMART TECHNOLOGIES IN SOLID WASTE MANAGEMENT

Advancing Sustainability

Moharana Choudhury
Sushobhan Majumdar, PhD
Avnish Chauhan, PhD
Editors

APPLE ACADEMIC PRESS

First edition published 2026

Apple Academic Press Inc.
1265 Goldenrod Circle, NE,
Palm Bay, FL 32905 USA

760 Laurentian Drive, Unit 19,
Burlington, ON L7N 0A4, Canada

CRC Press
2385 NW Executive Center Drive,
Suite 320, Boca Raton FL 33431

4 Park Square, Milton Park,
Abingdon, Oxon, OX14 4RN, UK

Library and Archives Canada Cataloguing in Publication

CIP data on file with Canada Library and Archives

Library of Congress Cataloging-in-Publication Data

CIP data on file with US Library of Congress

ISBN: 978-1-77964-082-6 (hbk)
ISBN: 978-1-77964-083-3 (ebk)
DOI: 10.1201/9781779640833

About the Editors

Moharana Choudhury
Research Coordinator, Environmental Research and Management Division, Voice of Environment (VoE), Guwahati, Assam, India

Moharana Choudhury is an Indian researcher and environmentalist. He has worked on various scientific projects in various government departments and has more than 14 years of research and working experience in environmental research and services. He began his career as a research intern at the National Institute of Disaster Management in New Delhi, followed by the North Eastern Space Application Center (Department of Space, Government of India/ISRO Unit) as a Junior Research Fellow. Mr. Choudhury has authored or edited more than 15 books with internationally reputed publishing houses and more than 45 articles in national and internationally reputed journals and proceedings and more than 60 book chapters. He has written many articles and scientific reports in popular magazines. He has been working as an editor of various prestigious international publishing houses worldwide, such as Elsevier, Royal Society of Chemistry (RSC), Frontiers, Springer, Wiley, CRC Press/Routledge group, Apple Academic Press, Cambridge Scholar Publishing, etc. Mr. Choudhury is working on various environmental conservation activities in Assam and throughout country. He has worked on several initiatives to integrate environmentally sustainable projects for iconic heritage places, wetlands, and ecotourism spots in Assam, India, especially for cleanliness and beating single-use plastic pollution missions. He contributes to various environmental research, management, social media mass awareness, and research studies. He is also well known in India for his efforts to implement environmental education and safeguard the privileges of environmental degree holders in India. He has completed his education as an MSc, PGDGIS, PGDEM, and PGCGARD from Gurukul Kangri Deemed to be University, Haridwar, Uttarakhand, North Eastern Hill University (NEHU), Shillong Meghalaya, Tezpur University, Assam, and National Institute of Rural Development (NIRDPR) Hyderabad, India, respectively

Dr. Sushobhan Majumdar
Assistant Professor, Department of Geography, Pandit Raghunath Murmu Smriti Mahavidyalaya (affiliated to Bankura University), India

Sushobhan Majumdar, PhD, is an Assistant Professor in the Department of Geography at Pandit Raghunath Murmu Smriti Mahavidyalaya (affiliated to Bankura University), India. Previously he was employed by the Government of West Bengal as a Group A Officer in the Minority Affairs and Minority Education Department. He is presently engaged in research on the design and development of environmental geographies. Urban geography, social geography, regional development and planning, remote sensing and GIS, and geography of transport are among his main areas of study. Additionally, he is working on different development projects for the Government of West Bengal Socio Economic Planning Unit of the Kolkata Metropolitan Development Authority (KMDA). He has more than 50 research papers published in a variety of reputable journals, proceedings, and edited volumes both nationally and internationally. He has to date published four edited volumes for Springer Nature and Routledge Group on urban-related issues. He is now editor of various renowned publishing houses in the world, including Elsevier, Royal Society of Chemistry, Springer, Routledge, Wiley, etc.

Dr. Avnish Chauhan
Associate Professor, Department of Environmental Science, Graphic Era Hill University, Dehradun, Uttarakhand, India

Avnish Chauhan, PhD, is working as Associate Professor in Department of Environmental Science, Graphic Era Hill University, Dehradun, Uttarakhand, India. Dr. Chauhan has more than 15 years of teaching experience. He has published more than 56 national and international research papers and edited and written about 33 text and reference books. He is NET and PhD and was awarded with a gold medal by the Indian Academy of Environmental

Sciences for his contribution. He has been awarded with a Best Environmentalist Award 2020 by the Deccan Environment Research Organization, Karnataka. He is an appointed reviewer for Springer, Taylor & Francis, and other reputed publishers. He is presently engaged in research on solid waste and environmental pollution. Air pollution, water pollution, vermiculture, soil pollution, impact of air pollutants on vegetation, and bioremediation are among his main areas of study.

Contents

Contributors .... *xiii*

Foreword 1 by Ernesto D. R. Santibanez Gonzalez .... *xix*

Foreword 2 by Mika Sillanpää .... *xxi*

Acknowledgment .... *xxiii*

Preface .... *xxv*

Introduction .... *xxix*

PART I: Managing Our Waste: An Introduction to Solid Waste Management **1**

1. Unveiling the Diversity: Distinct Categories of Various Solid Waste Categories: Sustainable Approach for Management **3**
Kiran Bala, Sakshi Verma, Anshul, Arushi Dogra, Chandan Mahata, and Varsha Srivastava

2. Does Waste Matter? The Critical Importance of Waste Management: A Case Study of South Asia **21**
Qi He, Wanying Liao, Jiajun Xu, and Hongtao Wang

3. A Case on Unseen Consequences of Waste: Environmental and Health Impacts of Inadequate Management **37**
Vijai, Worakamol Wisetsri, and Mohd Asif Shah

4. A Global Perspective on Progressive Paths about the Evolution of Solid-Waste Management **51**
Vijai, Worakamol Wisetsri, and Sridevi

PART II: Waste Generation, Composition, and Effective Segregation: A Comprehensive Guide **73**

5. Waste Generation Dynamics, Influencing Factors, and Trends in Western Europe **75**
Ashish Sharma, Hemlata Sahal, Rakesh Choudhary, and Mariana Martinez Serratos

6. **Unveiling Waste Trends: Composition and Generation Patterns in Southeast Asia 91**
Bancha Yingngam

7. **Unblocking Resources: Techniques and Technologies for Material Recovery in Solid Waste Management 107**
Kondle Ravi, Kushagra Sharma, Bolagam Ravikumar, Vimal Choudhary, Mohd Asif Shah, and Adnan Lakdawala

8. **A Comprehensive Study on Assessing the Significance and Impact of Waste Segregation at the Source 121**
Avnish Chauhan, Chandan Mahata, Samuele Barrili, and Deep Gupta

PART III: Responsible Waste Management: Disposal, Treatment, and Recycling Strategies 137

9. **Managing Landfills: Design, Operation, and Environmental Consideration: A Case Study of Developed Countries 139**
Murad Muhammad, Sana Ullah, and Muqadas Batool

10. **Anaerobic Digestion and Composting: A Sustainable Solution for Organic Waste Management 149**
Ratul Pegu, Sarmistha Paul, Bishal Paul, Akash Chetry, and Elkana Hezron

11. **Insights into Hazardous and Special Wastes: A Case Study of Latin America 165**
Prangya Rath, Farhana Rahman, Harshita Jain, Laxmi Kant Bhardwaj, Pankaj Kanti Jodder, Bhawana Sharma, Moharana Choudhury, Cíntia Soares, and Natan Padoin

12. **Burning Bright: Incineration Processes, Energy Recovery, and Emission Control 179**
Kiran Bala, Anshul, Sakshi Verma, Arushi Dogra, Pankaj Kanti Jodder, and Abhishek Nalluri

PART IV: Shaping the Future: Policies, Regulations, and Outlook in Environmental Management 197

13. **Revolutionizing Waste Management: Cutting-Edge Technologies Leading the Charge Toward Sustainable Futures 199**
Ratan Chowdhury, Gisandu K. Malunguja, and Ambarish Bhuyan

14. Innovative Approaches to Waste Management for Sustainable Practices Through Data-Driven Solutions 213
Hem Chandra Joshi, Abbas Shieh, Anshul Jaiswal, Lalitha Shanmugasundaram, Mariana Flores Garcia, and Sushobhan Majumdar

15. Smart Waste Management: Technologies for Monitoring and Effective Control 231
Shipra Mohan Tripathi and Ravi Kumar Gangwar

16. Emerging Waste Contaminants Such as Macroplastics/ Microplastics: Strategies to Combat Plastic Pollution 251
Kiran Bala, Arushi Dogra, Farhana Rahman, Sakshi Verma, Anshul, Pushp Lata, and Sangeeta Jaiswal

17. Waste Management in a Changing Climate: Building Resilience for the Future 271
Pujyasmita Nayak, Suman Kumar Das, Samikshya Mishra, and Amit Kumar Pandit

18. Current Trends in Waste Management: Case Studies from the African Continent 287
Rakesh Pant, Khushi Sah, Anushka Dabral, Harjeet Tomer, Sheetal, Shivangi Singh, Amit Gupta, and Dinaol Zewude

19. Current Trends in Waste Management: Case Studies from the Oceania Subcontinent 303
Tsenbeni N. Lotha, Latonglila Jamir, Prabhat Kumar Pathak, and Tuani Lidiawati Simangunsong

20. Current Trends in Waste Management: Case Studies from the Gulf Countries 317
Atin Kumar, Nitish Karn, Nandini Arya, Mohamed A. E. Abdel Rahman, and Antonio Scopa

Index *333*

Contributors

Mohamed A. E. Abdel Rahman
Division of Environmental Studies and Land Use, National Authority for Remote Sensing and Space Sciences (NARSS), Cairo, Egypt

Anshul
Institute of Environmental Studies, Kurukshetra University, Kurukshetra, Haryana, India

Nandini Arya
School of Agriculture, Uttaranchal University, Dehradun, Uttarakhand, India

Kiran Bala
Department of Zoology, Deshbandhu College, University of Delhi, Delhi, India

Samuele Barrili
M4W—Marketing for Waste, Italy

Muqadas Batool
Department of Botany, Postgraduate College Layyah, Bahauddin Zakariya University, Multan, Pakistan

Laxmi Kant Bhardwaj
Amity Institute of Environmental Toxicology, Safety & Management (AIETSM), Amity University, Noida, Uttar Pradesh, India

Ambarish Bhuyan
Department of Botany, D.H.S. Kanoi College, Dibrugarh, Assam, India

Avnish Chauhan
Department of Environmental Science, Graphic Era Hill University, Dehradun, Uttarakhand, India

Akash Chetry
Department of Environmental Sciences, Tezpur University, Tezpur, Assam, India

Rakesh Choudhary
Department of Civil Engineering, National Institute of Technology Delhi, New Delhi, Delhi, India

Vimal Choudhary
Department of Horticulture, School of Agriculture, Lovely Professional University, Phagwara, Punjab, India

Moharana Choudhury
Research Coordinator, Environmental Research and Management Division, Voice of Environment (VoE), Guwahati, Assam, India

Ratan Chowdhury
Department of Botany, Rangapara College, Rangapara, Assam, India

Anushka Dabral
UCL School of Pharmacy, University College London, Gower Street, London, UK

Suman Kumar Das
University of Brescia, Brescia, Italy

Arushi Dogra
Faculty of Life Sciences and Biotechnology, South Asian University, New Delhi, Delhi, India

Ravi Kumar Gangwar
Department of Soil Science, Institute of Environmental Sciences, Hungarian University of Agriculture and Life Sciences, Gödöllő, Hungary

Mariana Flores Garcia
Faculty of Architecture, Universidad La Salle Bajío, León de los Aldama, Mexico

Amit Gupta
Department of Zoology, University of Jammu, Baba Saheb Ambedkar Road, Jammu Tawi, Jammu and Kashmir, India

Deep Gupta
School of Smart Agriculture, COER University Roorkee, Haridwar, Uttarakhand, India

Qi He
Institute of South Asian Studies, Sichuan University, Chengdu, Sichuan, China

Elkana Hezron
Department of Sustainable Agriculture, Biodiversity and Ecosystem Management, School of Life Sciences and Bioengineering, The Nelson Mandela African Institution of Science and Technology, Arusha, Tanzania

Harshita Jain
Amity Institute of Environmental Sciences (AIES), Amity University, Noida, Uttar Pradesh, India

Anshul Jaiswal
Department of Botany, Banaras Hindu University, Varanasi, Uttar Pradesh, India

Sangeeta Jaiswal
Department of Internal Medicine, Division of Gastroenterology, University of Michigan, Ann Arbor, MI, USA

Latonglila Jamir
Department of Environmental Science, Nagaland University, Lumami, Nagaland, India

Pankaj Kanti Jodder
Department of Geography and Earth Science, University of North Carolina at Charlotte, Charlotte, NC, USA

Hem Chandra Joshi
Faculty of Natural Resource Management, Far Western University, Bhimdatta, Nepal

Nitish Karn
School of Agriculture, Uttaranchal University, Dehradun, Uttarakhand, India

Atin Kumar
School of Agriculture, Uttaranchal University, Dehradun, Uttarakhand, India

Adnan Lakdawala
George Washington University, Washington, DC, USA

Pushp Lata
Department of Zoology, University of Delhi, New Delhi, India

Wanying Liao
Faculty of Arts and Social Sciences, Universiti Malaya, Kuala Lumpur, Malaysia

Tsenbeni N. Lotha
Department of Environmental Science, Nagaland University, Lumami, Nagaland, India

Chandan Mahata
Department of Agriculture and Biological Engineering, University of Illinois Urbana-Champaign, Urbana, IL, USA; Department of Biological Systems Engineering, Virginia Tech, Blacksburg, VA, USA

Sushobhan Majumdar
Department of Geography, Pandit Raghunath Murmu Smriti Mahavidyalaya, Bankura, West Bengal, India

Gisandu K. Malunguja
Department of Natural Sciences, Mbeya University of Science and Technology, Mbeya, Tanzania

Samikshya Mishra
Department of Environmental Sciences, Sambalpur University, Sambalpur, Odisha, India

Murad Muhammad
State Key Laboratory of Desert and Oasis Ecology, Key Laboratory of Ecological Safety and Sustainable Development in Arid Lands, Xinjiang Institute of Ecology and Geography, Chinese Academy of Sciences, Urumqi, PR China; Xinjiang Key Laboratory of Biodiversity Conservation and Application in Arid Lands, Xinjiang Institute of Ecology and Geography, Chinese Academy of Sciences, Urumqi, China; University of Chinese Academy of Sciences, Beijing, PR China

Abhishek Nalluri
Huazhong University of Science and Technology, Wuhan, China

Pujyasmita Nayak
India Meteorological Department, Bhubaneswar, Odisha, India

Natan Padoin
Department of Chemical and Food Engineering, Federal University of Santa Catarina, Florianópolis, Santa Catarina, Brazil

Amit Kumar Pandit
National Institute of Aerospace, Hampton, VA, USA

Rakesh Pant
Department of Microbiology, Graphic Era (Deemed to Be) University, Dehradun, Uttarakhand, India; Department of Quality, Eureka Forbes Ltd Lal Tappar, Dehradun, Uttarakhand, India

Prabhat Kumar Pathak
Department of Biochemistry and Molecular Biology, Oklahoma State University, Stillwater, OK, USA

Bishal Paul
Department of Environmental Sciences, Tezpur University, Tezpur, Assam, India

Sarmistha Paul
Department of Botany, Visva Bharati University, Santiniketan, Bolpur, Birbhum, West Bengal, India

Ratul Pegu
Department of Environmental Sciences, Tezpur University, Tezpur, Assam, India

Farhana Rahman
University of Technology Sydney (UTS), Ultimo, Sydney, NSW, Australia

Prangya Rath
Amity Institute of Environmental Sciences (AIES), Amity University, Noida, Uttar Pradesh, India

Kondle Ravi
Department of Horticulture, School of Agriculture, Lovely Professional University, Phagwara, Punjab, India

Bolagam Ravikumar
Department of Horticulture, School of Agriculture, Lovely Professional University, Phagwara, Punjab, India

Khushi Sah
Department of Microbiology, Graphic Era (Deemed to Be) University, Dehradun, Uttarakhand, India

Hemlata Sahal
Department of Environmental Science, Maharaja Ganga Singh University, Bikaner, Rajasthan, India

Antonio Scopa
Scuola di ScienzeAgrarie, Forestali, Alimentari ed Ambientali (SAFE) Universitàdegli Studi della Basilicata Viale dell'Ateneo Lucano, Potenza, Italy

Mariana Martinez Serratos
School of Engineering, Wentworth Institute of Technology, Boston, MA, USA

Mohd Asif Shah
Department of Economics, Kebri Dehar University, Kebri Dehar, Ethiopia

Lalitha Shanmugasundaram
George Washington University, Washington, DC, USA

Ashish Sharma
WRI India Ross Centre for Sustainable Cities and Transport, Mumbai, Maharashtra, India; Sustainability Alternatives Ltd., London, United Kingdom

Bhawana Sharma
Department of Bioscience and Biotechnology, Banasthali Vidyapith, Vanasthali, Rajasthan, India

Kushagra Sharma
Department of Horticulture, Jawaharlal Nehru Krishi Vishwa Vidyalaya, Jabalpur, Madhya Pradesh, India

Sheetal
Department of Microbiology, Graphic Era (Deemed to Be) University, Dehradun, Uttarakhand, India

Abbas Shieh
Department of Planning, Policy, and Design, Azad University, Australia

Tuani Lidiawati Simangunsong
Department of Engineering Department, University of Surabaya, Surabaya, Indonesia; Center for Environmental and Renewable Energy Studies, University of Surabaya, Surabaya, Indonesia

Shivangi Singh
Department of Microbiology, Graphic Era (Deemed to Be) University, Dehradun, Uttarakhand, India

Cíntia Soares
Laboratory of Materials and Scientific Computing (LabMAC), Chemical and Food Engineering Department, Federal University of Santa Catarina, Florianópolis, Santa Catarina, Brazil

Sridevi
School of Commerce, Vel Tech Rangarajan Dr. Sagunthala R&D Institute of Science and Technology, Chennai, Tamilnadu, India

Varsha Srivastava
Research Unit of Sustainable Chemistry, Faculty of Technology, University of Oulu, Oulu, Finland

Harjeet Tomer
Department of Microbiology, Graphic Era (Deemed to Be) University, Dehradun, Uttarakhand, India

Shipra Mohan Tripathi
Department of Biotechnology, Utkarsh School of Management and Technology (Affiliated to MJP Rohilkhand University), Bareilly, Uttar Pradesh, India

Sana Ullah
Department of Agricultural Leadership, Education & Communication College of Agricultural and Environmental Sciences University of Georgia, Athens, GA, USA

Sakshi Verma
Department of Nutrition Biology, Central University of Haryana, Mahendergarh, Haryana, India

Vijai
School of Commerce, Vel Tech Rangarajan Dr. Sagunthala R&D Institute of Science and Technology, Chennai, Tamilnadu, India

Hongtao Wang
College of Urban and Environmental Sciences, Central China Normal University, Wuhan, Hubei, China

Worakamol Wisetsri
Department of Manufacturing and Service Industry Management, Faculty of Business and Industrial Development, King Mongkut's University of Technology North Bangkok, Bangkok, Thailand

Jiajun Xu
Institute for Advanced Studies, Universiti Malaya, Kuala Lumpur, Malaysia

Bancha Yingngam
Department of Pharmaceutical Chemistry and Technology, Faculty of Pharmaceutical Sciences, Ubon Ratchathani University, Ubon Ratchathani, Thailand

Dinaol Zewude
Department of Biology, School of Applied Natural Sciences, Adama Science and Technology University, Adama, Oromia, Ethiopia

Foreword 1

We are living in an era where the sustainability of the planet is no longer an option but a necessity; solid waste management is at the forefront of our environmental challenges. This edited volume, *Advancing Sustainability: Data Analytics and Smart Technologies in Solid Waste Management,* is a timely and invaluable contribution to this critical field. This book fuses the knowledge and experience of leading specialists, providing a comprehensive exploration of how data analytics and smart technologies can revolutionize solid waste management and pave the way for a sustainable future.

The book is carefully structured into four sections, each addressing a fundamental aspect of solid waste management. The first section, "Managing Our Waste: An Introduction to Solid Waste Management," sets the stage with an in-depth look at the various categories of solid waste and its management. These initial chapters explore the diversity of waste types, the importance of proper waste management, and the often-overlooked health and environmental impacts of improper practices. The global perspective offered highlights the progressive paths taken by different regions in their evolution of solid waste management.

In the second section, "Waste Generation, Composition and Effective Segregation: A Comprehensive Guide", the focus is on the dynamics of waste generation, the factors influencing these trends and the importance of effective segregation. The chapters in this section examine waste generation patterns in different regions and present advanced techniques for material recovery, highlighting the role of waste segregation at source to improve efficiency.

The third section, "Responsible waste management: disposal, treatment and recycling strategies", addresses the practical aspects of waste management, covering sustainable solutions and real-world insights through case studies. Explores the benefits of incineration processes, including energy recovery and emissions control.

Finally, the fourth section, "Shaping the Future: Policies, Regulations and Perspectives in Environmental Management," examines how cutting-edge technologies and innovative approaches can lead to more sustainable practices. This section discusses data-driven solutions, smart waste management technologies and strategies to combat emerging pollutants such as plastic and microplastics, and presents current trends in waste management through various case studies.

The authors have managed to publish a book that can be an essential resource for policymakers, professionals, researchers and students. By integrating global perspectives, innovative methodologies and practical strategies, this book not only points the way forward but also calls for the collective action necessary to build a sustainable and resilient future for all.

—Ernesto D. R. Santibanez Gonzalez, PhD
Professor and Scientist;
Academic Visitor, London Business School, UK;
Associate Editor, Journal of Cleaner Production

Foreword 2

This comprehensive volume explores the multifaceted domain of solid waste management through modern data analytics and innovative technologies, focusing on enhancing sustainability. The book is well organized into four major vital sections, each offering valuable insights and practical approaches to the complex challenges of waste management.

The first subsection is "Managing Our Waste: an Introduction to Solid Waste Management." This introductory section provides a foundational understanding of solid waste management. It delves into the historical evolution, current practices, and emerging trends that shape how communities and industries handle waste. It also highlights the importance of adopting sustainable practices to mitigate environmental impact.

The following section is "Waste Generation, Composition, and Effective Segregation: A Comprehensive Guide." This section offers a detailed examination of waste generation patterns and the composition of waste streams. It presents methodologies for adequate waste segregation, emphasizing the role of accurate data and intelligent technologies in optimizing the separation process. This guide aims to equip readers with the knowledge needed to enhance efficiency in waste sorting and reduction at the source.

The following section is "Responsible Waste Management: Disposal, Treatment, and Recycling Strategies". In this section, the book explores various strategies for responsible waste management, including disposal, treatment, and recycling. It discusses state-of-the-art technologies and best practices for minimizing waste, recovering valuable resources, and reducing the environmental footprint of waste disposal. This part of the book is crucial for understanding the waste lifecycle and the importance of closing the loop in waste management. The last section is "Shaping the Future: Policies, Regulations, and Outlook in Environmental Management." This final section addresses the broader context of environmental management, focusing on policies and regulations that govern waste management practices. It provides an outlook on future trends and innovations that promise to transform the sector. This section underscores the need for comprehensive regulatory frameworks and forward-thinking policies to foster a sustainable future.

I trust that you will find *Data Analytics and Smart Technologies in Solid Waste Management* an invaluable resource in advancing sustainability in waste management. This book's insights and practical strategies are designed to support professionals, policymakers, and researchers in making informed decisions and driving positive change in their communities.

—Mika Sillanpää, PhD
Professor, University of Johannesburg, South Africa

Acknowledgment

The editorial team would like to express their deepest gratitude to the esteemed publisher for giving us the opportunity to edit this book volume, *Data Analytics and Smart Technologies in Solid Waste Management: Advancing Sustainability.*

Second, the editorial team is thankful to all the esteemed contributors worldwide for their significant contributions as authors.

The editors thank subject-expert reviewers for their critical and significant efforts to improve the chapter quality and spontaneous constructive suggestions for this book.

Additionally, we appreciate the assistance of the esteemed publication team. Their spontaneous guidance and cooperation have been invaluable in navigating the complexities of the publishing process.

Finally, the editorial team would like to express gratitude to their parents, family members, and friends for their continuous support, motivation, and valuable suggestions throughout the book editing period.

This book is dedicated to all striving to create a more sustainable future through innovative technologies and data-driven solutions in the environment and solid waste management field.

Thank you.

Editors

Preface

The global challenge of managing solid waste sustainably has become increasingly pressing with rapid urbanization, industrial growth, and population expansion. Traditional methods of waste management practices, often rely on landfilling and incineration, are proving inadequate in addressing the environmental, economic, and social impacts of escalating waste volumes. These outdated practices contribute significantly to greenhouse gas emissions, resource depletion, and public health concerns, necessitating a paradigm shift toward more sustainable and efficient waste management solutions.

In this era marked by unprecedented technological advancement and heightened environmental awareness, this book, *Data Analytics and Smart Technologies in Solid Waste Management: Advancing Sustainability,* offers a crucial and timely exploration of how modern innovations and technologies can transform waste management practices. This book bridges the gap between cutting-edge data analytics and sustainable environmental practices, presenting a comprehensive examination of how smart technologies can enhance the efficiency, effectiveness, and ecological impact of solid waste management systems. As the world grapples with mounting waste generation and its adverse effects on ecosystems and human health, this volume underscores the urgent need for innovative solutions that leverage big data, machine learning, IoT, and other technological advancements. Through a series of insightful chapters authored by leading experts in this particular field, readers will gain a deep understanding of the current landscape, emerging trends, and future directions of waste management practices.

By highlighting real-world case studies and practical applications, this book not only addresses theoretical aspects but also provides actionable insights for researchers, practitioners, decision makers, and policymakers. Our hope is that this work will inspire and empower stakeholders across the globe to adopt sustainable practices, ultimately contributing to a cleaner, healthier planet for future generations. As we embark on this journey through the evolving world of solid waste management, we invite you to discover the transformative potential of data analytics and smart technologies in advancing sustainability.

This edited book provides a good and generalized background of the topic that quickly gives the reader an appreciation of the wide range of applications for these technologies.

Chapter 1 explores the diverse categories of solid waste, aiming to develop a practical and inclusive approach to waste management. It identifies, classifies, and understands each type, addressing their unique challenges and opportunities, and laying the foundation for sustainable solutions.

Chapter 2 presents a comprehensive investigation into the pressing issue of solid waste management in the South Asian region. This study delves into the environmental, social, and public health implications of inadequate waste management practices and focuses on the unique challenges faced by countries in South Asia.

Chapter 3 provides a comprehensive examination of the hidden but significant repercussions of poor waste management practices on a global scale. This study investigates the environmental and health consequences arising from inefficient waste disposal methods, drawing insights from various case studies worldwide.

Chapter 4 offers a comprehensive exploration of the development and transformation of solid waste management practices on a global scale. This study takes a broad view of the evolution of waste management techniques, policies, and strategies from various regions worldwide.

Chapter 5 offers an in-depth examination of the patterns and drivers of waste generation in the Western European region. This study provides a comprehensive analysis of the factors contributing to waste production and explores the evolving trends in waste generation over time.

Chapter 6 provides an in-depth analysis of the waste generation dynamics and composition in the Southeast Asian region. This study aims to uncover and understand the trends, factors, and characteristics associated with waste production in this geographically and culturally diverse part of the world.

Chapter 7 is a comprehensive exploration of innovative methods and technologies designed to recover valuable materials from solid waste streams. This study highlights the importance of resource recovery in sustainable waste management practices and investigates the techniques and technologies that can effectively extract valuable resources from waste materials.

Chapter 8 presents a detailed investigation into the practice of waste segregation at its origin, typically within households or businesses. This study aims to assess the importance and consequences of separating waste into distinct categories before disposal.

Chapter 9 provides a comprehensive examination of landfill management practices in advanced economies. This study offers a detailed analysis of the planning, operation, and environmental considerations associated with landfill sites, with a focus on developed nations.

Chapter 10 offers a comprehensive exploration of environmentally friendly methods for managing organic waste. This study focuses on two sustainable approaches, anaerobic digestion and composting, as effective techniques for processing and reducing organic waste.

Chapter 11 provides an in-depth investigation into the identification and management of hazardous and special wastes in the context of Latin American countries. This study focuses on the challenges and strategies related to the proper handling of waste materials that pose unique risks to the environment and public health.

Chapter 12 presents a detailed examination of waste incineration as a waste management method. This study focuses on the processes involved in incineration, the recovery of energy from waste, and the measures for controlling emissions and environmental impact.

Chapter 13 offers a comprehensive exploration of cutting-edge technologies and innovative approaches in the field of waste management. This study focuses on how technological advancements are transforming waste management practices and contributing to sustainability.

Chapter 14 explores modern waste management techniques utilizing data analytics. It emphasizes the integration of advanced technologies, such as AI and IoT, to optimize waste collection, recycling, and disposal processes. The research highlights the benefits of data-driven strategies in improving efficiency, reducing environmental impact, and promoting sustainable waste management practices.

Chapter 15 provides an in-depth exploration of advanced technologies and systems designed to revolutionize waste management processes. This study focuses on the development and implementation of smart solutions for monitoring and improving waste collection and disposal.

Chapter 16 offers a comprehensive examination of the challenges posed by single-use plastics and explores strategies and approaches to mitigate their environmental impact.

Chapter 17 provides a comprehensive exploration of the intersection between waste management practices and the challenges posed by a changing climate. This study focuses on strategies and approaches to enhance the resilience of waste management systems in the face of climate change impacts.

Chapter 18 examines contemporary waste management practices across various African countries. Through detailed case studies, it identifies prevailing trends, challenges, and innovative solutions being implemented. The research provides insights into regional disparities, governmental policies, and community-driven initiatives aimed at improving waste management and promoting sustainability in the continent.

Chapter 19 explores the region's strategies, challenges, and innovations in handling waste. The research highlights the impact of policy frameworks, technological advancements, and community involvement on improving waste management and fostering sustainable practices in the Oceania subcontinent.

Chapter 20 explores the latest waste management practices in Gulf countries through detailed case studies. It examines regional approaches, challenges, and advancements in waste handling. The research emphasizes the role of government policies, technological innovations, and sustainable practices in enhancing waste management systems in the Gulf region.

As we stand at the crossroads of environmental responsibility and technological advancement, this book underscores the critical importance of leveraging data analytics and smart technologies to pave the way for a more sustainable approach to managing solid waste. The journey toward sustainability is complex and challenging, but with the right tools and knowledge, it is a journey that holds immense promise for our planet and future generations.

Introduction

The global challenge of managing solid waste sustainably has become increasingly pressing with rapid urbanization, industrial growth, and population expansion. Traditional methods of waste management practices, often rely on landfilling and incineration, are proving inadequate in addressing the environmental, economic, and social impacts of escalating waste volumes. These outdated practices contribute significantly to greenhouse gas emissions, resource depletion, and public health concerns, necessitating a paradigm shift toward more sustainable and efficient waste management solutions.

In this era marked by unprecedented technological advancement and heightened environmental awareness, this book, *Data Analytics and Smart Technologies in Solid Waste Management: Advancing Sustainability,* offers a crucial and timely exploration of how modern innovations and technologies can transform waste management practices. This book bridges the gap between cutting-edge data analytics and sustainable environmental practices, presenting a comprehensive examination of how smart technologies can enhance the efficiency, effectiveness, and ecological impact of solid waste management systems. As the world grapples with mounting waste generation and its adverse effects on ecosystems and human health, this volume underscores the urgent need for innovative solutions that leverage big data, machine learning, IoT, and other technological advancements. Through a series of insightful chapters authored by leading experts in this particular field, readers will gain a deep understanding of the current landscape, emerging trends, and future directions of waste management practices.

By highlighting real-world case studies and practical applications, this book not only addresses theoretical aspects but also provides actionable insights for researchers, practitioners, decision makers, and policymakers. Our hope is that this work will inspire and empower stakeholders across the globe to adopt sustainable practices, ultimately contributing to a cleaner, healthier planet for future generations. As we embark on this journey through the evolving world of solid waste management, we invite

PART I

Managing Our Waste: An Introduction to Solid Waste Management

CHAPTER 1

Unveiling the Diversity: Distinct Categories of Various Solid Waste Categories: Sustainable Approach for Management

KIRAN BALA[1*], SAKSHI VERMA[2], ANSHUL[3], ARUSHI DOGRA[4], CHANDAN MAHATA[5,6], and VARSHA SRIVASTAVA[7]

[1]*Department of Zoology, Deshbandhu College, University of Delhi, Delhi, India*

[2]*Department of Nutrition Biology, Central University of Haryana, Mahendergarh, Haryana, India*

[3]*Institute of Environmental Studies, Kurukshetra University, Kurukshetra, Haryana, India*

[4]*Faculty of Life Sciences and Biotechnology, South Asian University, New Delhi, Delhi, India*

[5]*Department of Agriculture and Biological Engineering, University of Illinois Urbana-Champaign, Urbana, IL, USA*

[6]*Department of Biological Systems Engineering, Virginia Tech, Blacksburg, VA, USA*

[7]*Research Unit of Sustainable Chemistry, Faculty of Technology, University of Oulu, Oulu, Finland*

*Corresponding author

Data Analytics and Smart Technologies in Solid Waste Management: Advancing Sustainability. Moharana Choudhury, Sushobhan Majumdar, and Avnish Chauhan (Eds.)

DOI: 10.1201/9781779640833-2

ABSTRACT

This chapter discusses the multiple aspects of solid waste management. Considering the uncontrolled increase in waste production globally, creating effective and sustainable planning policies motivates a comprehensive understanding of various waste categories. The chapter explores innovative strategies and significant sustainable technologies that support efficient recycling processes, encourage material reuse, and reduce waste production at its source. It also explores the potential for converting waste into energy. This chapter comprehensively explains the complex issues associated with each waste category, from the rapidly expanding e-waste issue triggered by technological advancements to the risks that hazardous waste poses to human health and the environment. In addition to providing information, it aims to inspire and educate people about the importance of adopting ethical waste management techniques that support the global goal of a healthier and cleaner environment.

1.1 INTRODUCTION

Solid waste is a critical global concern, and in recent times, India is facing this serious issue as well. The massive population increase of more than 1.44 billion in India generates a staggering almost 62 million tons of solid waste annually, making India the world's third-largest waste generator [1]. India has been modernizing in every way possible over the years. Rapid urbanization, population growth, and changes in the pattern of consumption have magnified the scale of solid waste management [2]. Numerous environmental issues are the result of ineffective solid waste management. Ecosystems and public health are harmed by air, water, and soil pollution, which is a result of improper disposal procedures and inadequate waste treatment [3]. The accumulation of waste in waste disposal facilities and landfills contributes to environmental degradation by releasing greenhouse gases and contaminating groundwater [4]. Furthermore, incorrect garbage disposal promotes the development of infections, such as respiratory and vector-borne disorders [5]. The effects impact not only the environment but also society, the economy, and people's general well-being and standard of living [1]. Management of solid waste sustainably is important for achieving the development goals sustainably and for addressing challenges related to large-scale trash disposal as well as health and environmental concerns. A poorly handled municipal solid waste (MSW) creates risks for the local population. The development of alternative waste management techniques is crucial due to the consequences related to the environment as well as finance in concern with the appropriate environmentally responsible disposal of contemporary trash. Composting is a low-cost, sustainable method for managing MSW; however, only 6–7% of MSW was recycled using this technique [6].

Everywhere on the planet, different kinds of solid waste are produced. The primary factors that may be influencing these are as follows:

1. The level of emergency (waste volume and composition may vary with time; during COVID-19, for example, waste production from medical waste grew dramatically).
2. The geographical region (the more developed countries have better waste management practices in comparison to some less developed countries).
3. Seasonal variations (it can also affect the type of food available, and waste management can be a major issue in the rainy season).
4. Material status and sociocultural customs of the impacted population.

Waste production is typically low and mostly biodegradable in rural populations since food rations consist of dry, unpackaged products. Categories of solid waste, along with their origin, have been summarized in Table 1.1. Uprooted urban individuals are more likely to generate larger volumes of nonbiodegradable waste, especially when packed food rations are provided [7].

According to the European Commission, the initial steps in solid waste management are illustrated in the Flowchart (Figure 1.1).

1.2 BROAD CATEGORIES OF SOLID WASTE

Solid waste can be categorized in several ways based on different criteria; here, we are exploring categories of solid waste based on the sources and their types [8].

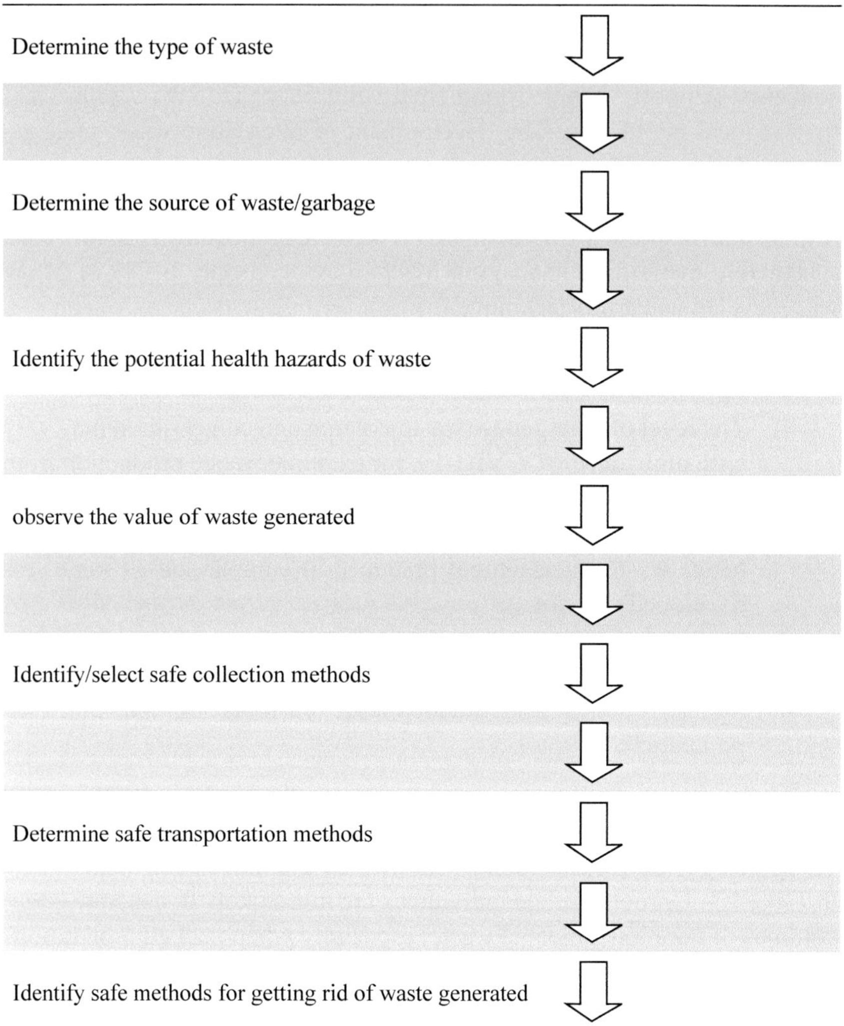

FIGURE 1.1 Flowchart showing initial steps in the management of solid waste.

1.2.1 CATEGORY OF SOLID WASTE BASED ON SOURCES

We deal with solid waste every day, often without realizing how much variety there is in where it comes from. Think of the trash from our homes—leftover food, packaging, or broken items—that is residential waste. Then there is what piles up in markets, restaurants, or shops—commercial waste like glass, metal scraps, and food remnants. When we walk through parks, along streets, or past open areas, the litter we see there falls under open area waste. Schools, offices, and government buildings generate their share too, mostly paper, plastic, and a lot of discarded packaging—what we call institutional waste. Out on farms, leftover plant material and discarded tools make up agricultural waste. The debris from buildings going up or coming down, and leftovers from factory processes? That is industrial waste. And finally, all the mess that comes from city-level activities—street sweeping, gardening, and construction—that is bundled into municipal solid waste. Recognizing where each type comes from helps us manage it smarter—and maybe think twice before tossing something carelessly.

TABLE 1.1 Category of Solid Waste Based on Sources

Type of Solid Waste	Sources	References
1. Municipal solid waste	Waste materials that are produced as a result of various municipal activities, including street cleaning, landscaping, construction, demolition, etc.	[2]
2. Commercial solid waste	Trash produced by businesses, such as food leftovers, glassware, ashes, metals, etc., the company includes commercial stores, restaurants, and marketplaces	[9]
3. Open areas waste	These contain waste from streets, lanes, parks, open spaces, playgrounds, the coasts, leisure zones, etc.	[8]
4. Institutional wastes	Mostly made up of paper, plastic, and glassware, educational institutions, management, and public belongings, such as offices, colleges, and schools	[10]
5. Agricultural wastes	Solid wastes from fields, orchards, vineyards, farms, etc.	[2]
6. Residential wastes	Wastes produced by typical home activities, apartments, etc.	[9]
7. Industrial wastes	Process wastes, ashes, construction and demolition wastes, and other wastes resulting from continuous industrial activity	[8]

1.2.2 CATEGORY OF SOLID WASTE BASED ON TYPES

Solid waste comes in many forms, each telling a different story of how we live, work, and consume. Ashes and residue, for instance, are the powdery byproducts left after burning firewood or charcoal—common in households that still rely on traditional cooking methods. Street waste gathers up whatever is scattered along our roadsides and empty lots: bits of cardboard, vegetable peels, and stray plastic wrappers—all signs of a community going about its day. Then there is the biodegradable versus nonbiodegradable divide: food scraps and garden waste break down naturally, while materials like glass, metal, and plastic linger, recyclable but stubborn. On the rural side, farm waste arises from everything from milking cows to harvesting crops. More troubling are hazardous wastes—chemicals or discarded products from industries and institutions that can harm both us and the planet if not handled properly. And after all the water from our homes and cities is cleaned, sewage waste is what is left behind—sludge that needs careful burial to keep harmful microbes in check. Each waste type demands a thoughtful approach, not just for disposal but for the ripple effects it can have on health and the environment (Table 1.2).

TABLE 1.2 Category of Solid Waste Based on Types

Type of Solid Waste	Sources	References
Ashes and residue	The residues or leftovers of the remaining substances from the burning of charcoal, and other combustible material like wood used for cooking or heating in house establishments. Since ash residues are fine, powdery substances that are inorganic, they are valuable in landfills	[8, 10]
Street waste	Type of waste collected from the street, footpath, and vacant plots mainly consists of vegetable peels, cardboard, plastics etc.	[2]
Biodegradable, nonbiodegradable	Waste consists mainly of organic materials produced by various domestic and commercial processes, known as biodegradable waste. Wastes that are not biodegradable are made up of recyclable and inorganic elements, including metal, glass, plastic, and cans	[8]
Farm wastes	Wastes are produced by a variety of agricultural operations, such as planting, harvesting, producing milk, raising animals for slaughtering, and running feedlots	[9]
Hazardous wastes	Hazardous wastes are defined as wastes of industrial, institutional, or consumer origin that pose a risk to the environment and human health either now or in the future	[11]
Sewage wastes	After sewage is treated, the solid residue is left behind. Since they were made by treating organic sludge that was extracted from both raw and processed sewage, it requires dumping and burying quickly to keep putrescible organic material from harboring pathogens	[11]

1.3 SUSTAINABLE WASTE MANAGEMENT APPROACH

Many waste management techniques used today are mostly centered on the challenging issue of efficiently handling the increasing amount of waste produced in metropolitan areas. The most used strategies by municipal systems are collection and disposal strategies, including the organized collection of waste from households and enterprises using designated bins, including scheduled pickups, and having proper disposal sites [12]. On the other hand, relying solely on conventional landfills results in significant environmental issues and demands alternative approaches. Initiatives aimed at recycling waste are crucial in minimizing the detrimental effects of waste on the environment continuously.

1.3.1 INNOVATIVE STRATEGIES FOR WASTE REDUCTION

Pollution of the air, land, and water is linked to solid waste generation [13]. Monitoring waste generation is the initial step in any waste management strategy, regardless of location or country [14].

1.3.1.1 SOURCE SEGREGATION

Source separation involves the segregation of different types of waste at the point of generation. The primary objective is to categorize waste into recyclable materials, organic waste, and nonrecyclables.

The most commonly, easily, and conveniently adopted route for waste disposal is a landfill [14, 15]. Developed countries have adopted an improved kind of landfill known as "sanitary landfilling" (diagram shown in Figure 1.1). The method mentioned here fully adheres to the standard protocols of the United States Environmental Protection Agency (USEPA) [16]. Comparing sanitary landfills to conventional landfills, they are safer for the environment because they are constructed with some architectural upgrades that consist of four layers. A typical sanitary landfill is explained in Figure 1.2 [17]:

1. Dense clay lining makes up the bottom layer, which is covered in high-density polythene sheets to prevent toxic effluents from seeping through.

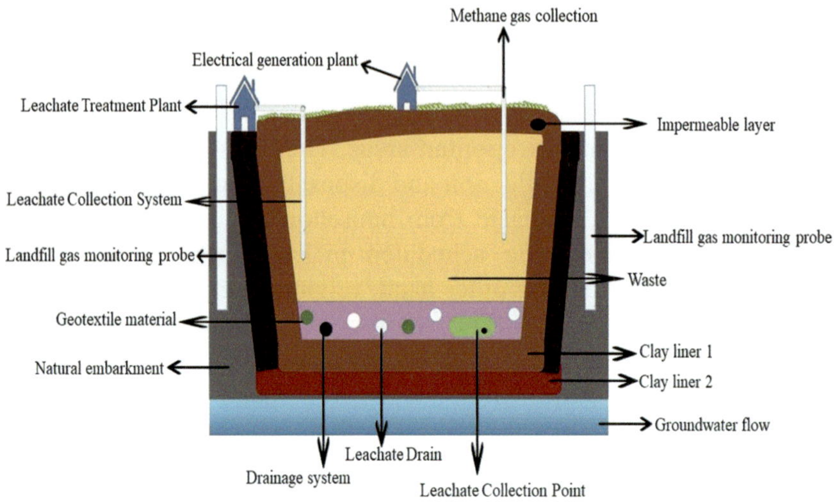

FIGURE 1.2 Modern sanitary landfill.

2. Appropriate drainage pipes are included in the second layer to direct the flow of incoming poisonous fluids from the dump toward a treatment facility.
3. The third layer is used to collect landfill gases and move them for further processing and usage.
4. The highest or fourth layer contains the largest reservoir for solid waste. Finally, a sanitary landfill's mouth is covered with dirt to keep an eye on the release of harmful bacteria and an unpleasant stench [17].

Over time, it has been noted that toxic heavy metals, CH_4 and N_2O are significantly extruded from common landfills [18]. Since skilled labor is limited in developing nations, this is the main barrier to the implementation of such a practice [19]. Strict environmental regulations severely restrict the use of other waste management techniques, such as pyrolysis, incineration, and gasification, even in highly industrialized nations [20]. However, in developing or low-income nations, incineration is seen as an economical way to eliminate solid waste method, regardless of its pollution factor [21]. The national economy's stagnation and a lack of environmental awareness are the main causes of degradation in the solid waste treatment infrastructure [22].

1.3.2 THE THREE RS: REDUCE, REUSE, AND RECYCLE

The main contribution of the 3R principles is that they help us to move toward more sustainable living; there is analysis and study [21, 23]. It demonstrates that by emphasizing the 3R principle more, we may minimize the amount of garbage we produce. The work that is provided for recycling purposes consists of gathering, classifying, and processing recyclable materials [23, 24]. As the reduce, reuse, and recycle principle needs less energy it helps to improve the air quality index by creating less air pollution, water pollution, and soil pollution by decreasing the waste that goes to landfills [25]. The role of each 3R principle is explained here in little detail.

1.3.2.1 REDUCE

The term "reduce" refers to minimizing the quantity of waste that humans produce. It emphasizes how crucial it is to use resources more wisely, consume less, and pay attention to one's consumption patterns. Reducing packaging is one way to reduce waste; other strategies include avoiding single-use items, choosing products with little packaging, and consuming resources with awareness. Through these initiatives, the decreased principle helps to save natural resources, lower pollution, and create a more sustainable future while also easing the load on landfills and incinerators [26].

1.3.2.2 REUSE

Reuse involves finding ways to use products and materials again rather than disposing of them after a single use. This can include fixing things, giving them to a different individual so they can be used for something else, or repairing them. Reuse eliminates the impact of consumption on the environment, conserves resources, and extends the life of goods and materials [27, 28]. The USEPA states that a product's environmental impact can be greatly decreased by reusing it for an additional year. Furthermore, studies have shown that promoting reuse initiatives can lead to economic benefits by creating job opportunities and supporting local economies. Solid waste management systems can successfully reduce trash generation and help create a more sustainable future by encouraging consumers, businesses, and their surroundings to adopt reuse practices (USEPA).

1.3.2.3 RECYCLE

It is the process of gathering, processing, and repurposing waste materials to create new goods or raw materials for new product creation. Recycling reduces the quantity of waste delivered to landfills and incinerators, saves energy, and protects natural resources [26]. Paper, plastics, glass, metals, and certain kinds of technological waste are items that are frequently recycled. When recycling is done instead of creating new materials from virgin resources, the EPA claims that recycling conserves natural resources, lowers greenhouse gas emissions, and saves energy [27, 28]. Additionally, recycling supports recycling markets and businesses, which promotes job creation and economic progress [23, 24]. Solid waste management systems have the potential to efficiently handle waste streams and simultaneously support sustainable development and environmental protection by endorsing recycling activities and allocating funds for recycling infrastructure (USEPA).

1.4 WASTE-TO-ENERGY CONVERSION

Waste-to-energy conversion technologies are newly evolving methods that transform solid waste into usable energy, effectively addressing both waste management and energy generation [29]. These techniques encompass various approaches.

Solid wastes have a significant quantity of stored energy that can be converted into heat, gas, oil, or electricity using waste management and transformation technologies. Thermochemical and biochemical conversion are the two types of waste-to-energy conversion that are discussed below [30].

Thermochemical conversion: This method produces energy by utilizing the heat breakdown of waste materials. Some of the techniques used in thermochemical conversion are described below:

1.4.1 INCINERATION

This technique creates heat energy and ash residue by directly burning solid waste in an oxygen-rich environment. According to the literature, 65–80% of heat energy is derived from incineration [31]. A running flow chart showing the management of solid waste by the incineration process has been presented in Figure 1.3. The main steps involved in the incineration process are follows:

1. Combustion
2. Boiler/steam turbine
3. Exhaust gas cleaning
4. Dry or wet scrubbers
5. Activated carbon injection
6. Bag house filter
7. Selective noncatalytic reduction

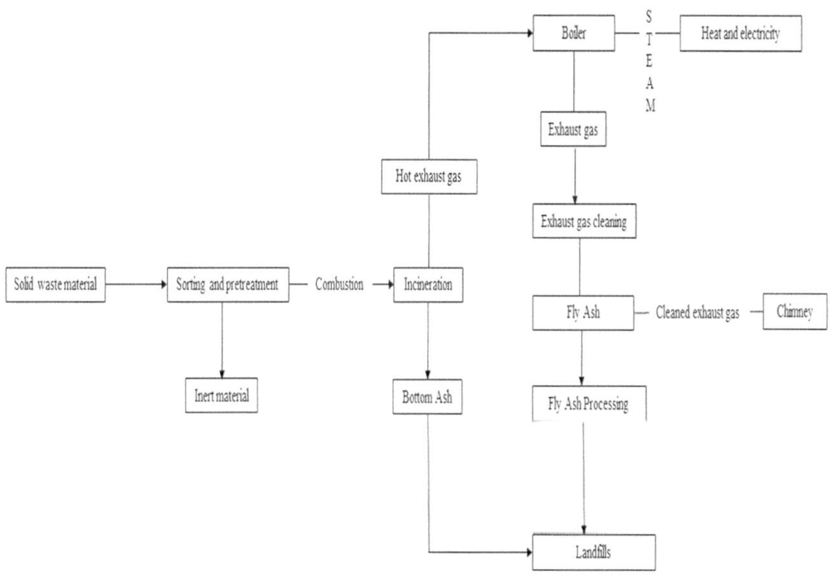

FIGURE 1.3 Running flowchart showing the management of solid waste by incineration process [30].

1.4.1.1 ADVANTAGES OF INCINERATORS

Incineration is safest from the hygienic point of vicw, it requires very little space and noise, and also it is odorless. This technique can reduce trace volume to a very large extent, and the cost of transportation is very low in incineration. A typical plant of incineration with a 300-t/day capacity can generate 3 MW of power [32].

1.4.1.2 DISADVANTAGES OF INCINERATORS

The expense of incineration is high as compared with commonly used landfills, it can create air pollution as the burning process is involved in this, particulate production is expected, this technique can produce SO_x and NO_x emission, dioxin production, and incinerators can only be operated by the high skilled person because of handling requirements [33].

1.4.2 PYROLYSIS

It is a more economically friendly alternative to incineration, which is more environmentally friendly as well. This is the thermal decomposition of organic matter by applying a relatively high amount of heat (ranging from 500 to 800°C) in an oxygen-limited environment with vapor ranging from 3 to 1500 s [34]. Although pyrolysis is the initial stage of gasification and combustion, it differs from combustion, which requires oxygen above every other requirement, in the sense that it takes place without oxygen [35]. Because of its high calorific value (23 MJ/kg), the oil produced during the pyrolysis process can be used as an additive to regular diesel to increase its performance [36, 37].

1.4.2.1 PROCESS OF PYROLYSIS

Organic molecules are broken down into their gaseous components, a solid carbon and ash residue, and a liquid called pyrolytic oil (sometimes called bio-oil) during the pyrolysis process. Organic pollutants are broken down into smaller molecular weight compounds during destruction [38]. When heat is present, organic compounds that can break down or degrade can be treated using pyrolysis. Among these, some examples are dioxins, polychlorinated biphenyls, and polycyclic aromatic hydrocarbons. It does not help eliminate or destroy inorganic materials, such as metals [34]. Figure 1.4 describes a typical pyrolysis process.

1.4.3 GASIFICATION

It is a type of solid waste management that is described as the supply of all gasification agents enabling the conversion of a solid or liquid carbon-based substance (feedstock) into a combustible gaseous product (combustible gas) at a very high temperature (>600°C) [39, 40]. Gasification is

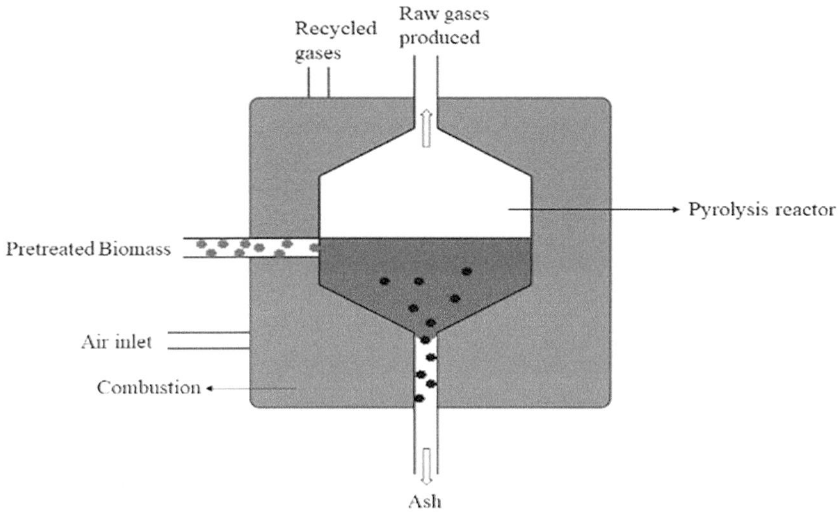

FIGURE 1.4 Process of pyrolysis.

divided into four categories based on the oxidation medium: steam, plasma, oxygen-enriched air, pure oxygen, and partial oxidation with air [41]. An explanatory representation of the gasification process is given in Figure 1.5.

The steps involved in pyrolysis are as follows:

1. Vaporization
2. Pyrolysis/devolatilization
3. The secondary crackling of tars
4. Gasification

1.4.4 BIOCHEMICAL CONVERSION

It is a kind of conversion process that is mostly preferred for solid wastes that contain a decayable substance in a higher fraction and have moisture content, as these facilitate microbial activity [14]. This method depends on the assistance of microbes or enzymes to denature solid waste. Fermentation and anaerobic digestion are the two main methods of biochemical conversion. Some of the techniques used in biochemical conversion are described below.

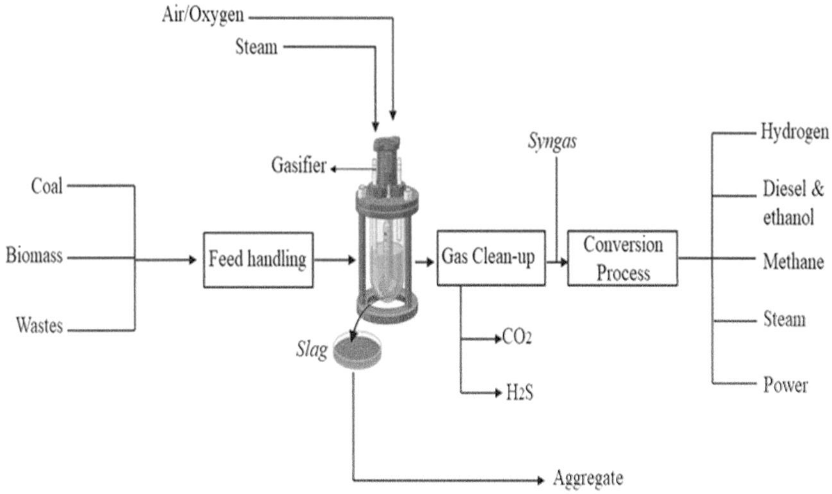

FIGURE 1.5 An explanative representation of the gasification process.

1.4.4.1 ANAEROBIC DIGESTION

Another name for this process is bio-methanation. It makes use of a biogas reactor to produce sludge and biogas ($CH_4 + CO_2$) by feeding organic solid waste into an oxygen-deficient environment [42].

1.4.4.2 BIOETHANOL PRODUCTION

A newly developed technique to manage solid waste is the production of bioethanol, especially the kind of waste that is generated in the kitchen as well as waste from vegetables or crop products, while generating renewable energy. The whole process is completed in three main steps pretreatment, hydrolysis, and fermentation [43, 44]. Moreover, the waste left after ethanol production can be further processed into valuable products like biochar, which can be used for adsorbing contaminants from wastewater, thus contributing to a circular economy and a more sustainable environment [43].

1.5 CONCLUSION

The rate at which solid waste is being generated is increasing at an alarming rate due to population growth and changing lifestyles.

Effective solid waste management is necessary for environmental sustainability and public health. We must categorize and manage waste effectively, adopt sustainable waste management strategies, and ensure widespread adoption of the three Rs (reduce, reuse, and recycle) to combat this issue. However, we must recognize the need for increased community involvement and public education to create a culture of responsible waste management practices. We must work together to create a cleaner, healthier, and more sustainable future. Failure to act now will lead to catastrophic environmental and public health consequences.

KEYWORDS

- **solid waste**
- **sustainable approach**
- **hazardous waste**
- **environment**
- **clean environment**

REFERENCES

1. Sk, M., Md, Qamar, S., & Sethy, T. (2023). *Solid Waste Management in Indian Perspectives: A Comprehensive Review*, *12*(1, No. 03), 35–45. https://www.researchgate.net/publication/373737929_SOLID_WASTE_MANAGEMENT_IN_INDIAN_PERSPECTIVES_A_COMPREHENSIVE_REVIEW
2. Soni, A., Das, P. K., & Kumar, P. (2023). A review on the municipal solid waste management status, challenges and potential for the future Indian cities. *Environment, Development and Sustainability*, *25*(12), 13755–13803.
3. Mohan, S. & Joseph, . (2021). Potential hazards due to municipal solid waste open dumping in India. *Journal of the Indian Institute of Science*, *101*, 1–14. 10.1007/s41745-021-00242-4.
4. Choudhury, M., Jyethi, D. S., Dutta, J., Purkayastha, S. P., Deb, D., Das, R., Roy, G., Sen, T., & Bhattacharyya, K. G. (2022). Investigation of groundwater and soil quality near to a municipal waste disposal site in Silchar, Assam, India. *International Journal of Energy and Water Resources*, *6*(1), 37–47. https://doi.org/10.1007/s42108-021-00117-5
5. Paul, S., Choudhury, M., Deb, U., Pegu, R., Das, S., & Bhattacharya, S. S. (2019). Assessing the ecological impacts of ageing on hazard potential of solid waste landfills: A green approach through vermitechnology. *Journal of Cleaner Production*, *236*, 117643. https://doi.org/10.1016/j.jclepro.2019.117643

6. Meena, M., Dotaniya, M., Meena, B., Rai, P., Antil, R., Meena, H., Meena, L., Dotaniya, C., Meena, V. S., Ghosh, A., Meena, K., Singh, A. K., Meena, V., Moharana, P., Meena, S. K., Srinivasarao, C., Meena, A., Chatterjee, S., Meena, D., & Meena, R. (2023). Municipal solid waste: Opportunities, challenges and management policies in India: A review. *Waste Management Bulletin*, *1*(1), 4–18. https://doi.org/10.1016/j.wmb.2023.04.001
7. Pandey, B., Reba, M., Joshi, P. K. et al. (2020). Urbanization and food consumption in India. *Scientific Reports*, *10*, 17241. https://doi.org/10.1038/s41598-020-73313-8
8. Jamal, H. (2020, January 16). *Types of Solid Wastes—Based on Origin and Composition.* About Civil.com, Retrieved from: https://www.aboutcivil.org/types-of-solid-wastes. (accessed: 16 January, 2020)
9. Adedayo, O. J., Tunde, T. T., Alex, O. T., & Olagbemiro, A. S. (2024). Quantification and characterization of solid waste at lead city University, Ibadan, Nigeria. *Asian Journal of Geological Research*, *7*(1), 41–57.
10. Joseph, A. M., Snellings, R., Van den Heede, P., Matthys, S., & De Belie, N. (2018). The use of municipal solid waste incineration ash in various building materials: A Belgian point of view. *Materials (Basel, Switzerland)*, *11*(1), 141. https://doi.org/10.3390/ma11010141
11. Karthikeyan, L., Suresh, V. M., Krishnan, V., Tudor, T., & Varshini, V. (2018). The management of hazardous solid waste in India: An overview. *Environments*, *5*(9), 103. https://doi.org/10.3390/environments5090103
12. Dugdhe, S., Shelar, P., Jire, S., & Apte, A. (2016). In *Efficient Waste Collection System International Conference on Internet of Things and Applications (IOTA),* Pune, India, 2016, pp. 143–147. https://doi.org/10.1109/IOTA.2016.7562711
13. Allesch, A., & Brunner, P. H. (2014). Assessment methods for solid waste management: A literature review. *Waste Management & Research*, *32*(6), 461–473. https://doi.org/10.1177/0734242X14535653
14. Das, S., Lee, S., Kumar, P., Kim, K., Lee, S. S., & Bhattacharya, S. S. (2019). Solid waste management: Scope and the challenge of sustainability. *Journal of Cleaner Production*, *228*, 658–678. https://doi.org/10.1016/j.jclepro.2019.04.323
15. Sahariah, B., Goswami, L., Farooqui, I. U., Raul, P., Bhattacharyya, P. K., & Bhattacharya, S. S. (2015). Solubility, hydrogeochemical impact, and health assessment of toxic metals in municipal wastes of two differently populated cities. *Journal of Geochemical Exploration, 157*, 100–109.
16. USEPA. (2016). Municipal Solid Waste Landfills (Regulations) [WWW Document]. URL Retrieved from: https://archive.epa.gov/epawaste/nonhaz/municipal/web/html/msw_regs.html (accessed: 2.27.18).
17. EPA History: Federal water pollution control act amendments of 1972 https://www.epa.gov/history/epa-history-federal-water-pollution-control-act-amendments-1972
18. Harborth, P., Fuß, R., Münnich, K., Flessa, H., & Fricke, K. (2013). Spatial variability of nitrous oxide and methane emissions from an MBT landfill in operation: Strong N_2O hotspots at the working face. *Waste Management*, *33*(10), 2099–2107. https://doi.org/10.1016/j.wasman.2013.01.028
19. Ghosh, S., & Hassan, S. (2011). Sanitary landfill. In: *Environmental and Engineering Geology*. Encyclopedia of Life Support Systems (EOLSS): Paris, France, pp. 22–22.

20. Arebey, M., Hannan, M. A., Basri, H., Begum, R. A., & Abdullah, H. (2011). Integrated technologies for solid waste bin monitoring system. *Environmental Monitoring and Assessment*, *177*(1–4), 399–408. https://doi.org/10.1007/s10661-010-1642-x
21. Ejaz, N., Akhtar, N., Nisar, H., & Naeem, U. A. (2010). Environmental impacts of improper solid waste management in developing countries: A case study of Rawalpindi City. *Social Work, 142*, 379–387.
22. Simatele, D., Dlamini, S., & Kubanza, N. S. (2017). From informality to formality: Perspectives on the challenges of integrating solid waste management into the urban development and planning policy in Johannesburg, South Africa. *Habitat International, 63*, 122–130.
23. Samiha, B. (2013). The importance of the 3R principle of municipal solid waste management for achieving sustainable development. *Mediterranean Journal of Social Sciences*, *4*, 129–135. 10.5901/mjss.2013.v4n3p129.
24. Yue, K. (2012). Comparative analysis of scrap car recycling management policies. *Procedia Environmental Sciences*, *16*, 44–50. https://doi.org/10.1016/j.proenv.2012.10.007
25. Mondal, T., Choudhury, M., Kundu, D., Dutta, D., & Samanta, P. (2023). Landfill: An eclectic review on structure, reactions and remediation approach. *Waste Management*, *164*, 127–142. https://doi.org/10.1016/j.wasman.2023.03.034
26. Yogini, C., Reddy, B. & Chitturi, C. (2020). Three 'R': An effective sustainable waste management approach. In Ghosh, S. (Ed.), *Solid Waste Policies and Strategies: Issues, Challenges and Case Studies*. Springer: Singapore. 10.1007/978-981-15-1543-9_1.
27. Fatimah, S., Deliana, Y., Rasmikayati, E., Trimo, L., & Hapsari, H. (2024, April). Towards ecovillage through rural waste management institution in cinanjung, sumedang regency, West Java. In *IOP Conference Series: Earth and Environmental Science*, Vol. 1318, No. 1, p. 012005. IOP Publishing.
28. Podlasek, A., Vaverková, M. D., Jakimiuk, A., & Koda, E. (2024). A comprehensive investigation of geoenvironmental pollution and health effects from municipal solid waste landfills. *Environmental Geochemistry and Health*, *46*(3), 97.
29. Sharma, B., Goswami, Y., Sharma, S., & Shekhar, S. (2021). Inherent roadmap of conversion of plastic waste into energy and its life cycle assessment: A frontrunner compendium. *Renewable and Sustainable Energy Reviews*, *146*, 111070.
30. Tang, Y. Y., Tang, K. H. D., Maharjan, A. K., Aziz, A. A., & Bunrith, S. (2021). Malaysia moving towards a sustainability municipal waste management. *Industrial and Domestic Waste Management*, *1*, 26–40. https://doi.org/10.53623/idwm.v1i1.51
31. Knox, A. (2005). *An Overview of Incineration and EFW Technology as Applied to the Management of Municipal Solid Waste (MSW)*. Energy Subcommittee of the Ontario Environment Industry Association.
32. Lu, J., Yang, X., Lai, Y., Wan, X., Gao, J., Wang, Y., Tan, L., & Deng, F. (2024). Utilization of municipal solid waste incinerator bottom ash (MSWIBA) in concrete as partial replacement of fine aggregate. *Construction and Building Materials*, *414*, 134918
33. Guo, F., Men, H., & Chen, W. (2024). Waste-to-energy incineration site selection framework based on heterogeneous fuzzy regret-PROMETHEE model considering life-cycle carbon emissions. *Environmental Science and Pollution Research*, *31*(3), 3722–3744.

34. Aasberg-Petersen, K., Christensen, T. S., Dybkjǣr, I. Sehested, J. Østberg, M., Coertzen, R. M., Keyser, M. J., & Steynberg, A. P. (2004). *Chapter 4 Synthesis Gas Production for FT Synthesis* 10.1016/S0167-2991(04)80461-0.
35. Boslaugh, S. E. (2024, May 2). *Pyrolysis. Encyclopedia Britannica*. https://www.britannica.com/science/pyrolysis
36. Czajczyńska, D., Anguilano, L., Ghazal, H., Krzyżyńska, R., Reynolds, A., Spencer, N., & Jouhara, H. (2017). Potential of pyrolysis processes in the waste management sector. *Thermal Science and Engineering Progress*, *3*, 171–197. https://doi.org/10.1016/j.tsep.2017.06.003
37. Nidoni, P. G. (2017). Incineration process for solid waste management and effective utilization of by-products. *International Research Journal of Engineering and Technology*, *4*(12), 378–382.
38. Ibrahim, H. A. (2020). Introductory chapter: Pyrolysis. In *Recent Advances in Pyrolysis*.
39. Belgiorno, V., De Feo, G., Della Rocca, C., & Napoli, R. M. (2003). Energy from gasification of solid wastes. *Waste Management*, *23*(1), 1–15. https://doi.org/10.1016/S0956-053X(02)00149-6. PMID: 12623097.
40. Elgazar, Y. G., Khalifeh, H. A., Alkhedher, M., & Ramadan, M. (2024). A review of hydrogen production from food waste through gasification process. *International Journal of Hydrogen Energy*, 67, 959-969.
41. Zhao, X., Li, K., E. Lamm, M., Celik, S., Wei, L., & Ozcan, S. (2021). *Solid Waste Gasification: Comparison of Single- and Multi-Staged Reactors*. Intech Open. https://doi.org/10.5772/intechopen.96157
42. Akindele, A. A., & Sartaj, M. (2017). The toxicity effects of ammonia on anaerobic digestion of organic fraction of municipal solid waste. *Waste Management*, *71*, 757–766. https://doi.org/10.1016/j.wasman.2017.07.026
43. Ganguly, P., & Das, P. (2022). Integral approach for second-generation bio-ethanol production and wastewater treatment using peanut shell waste: Yield, removal, and ANN studies. *Biomass Conversion and Biorefinery, 14*, 689–699.
44. Patra, J., Basu, A., Mishra, A., & Dhal, N. K. (2017). Bioconversion of Municipal Solid Wastes for Bioethanol Production. *Biosciences Biotechnology Research Asia*, *14*(3).

CHAPTER 2

Does Waste Matter? The Critical Importance of Waste Management: A Case Study of South Asia

QI HE[1], WANYING LIAO,[2]* JIAJUN XU,[3]* and HONGTAO WANG[4]

[1]*Institute of South Asian Studies, Sichuan University, Chengdu, Sichuan, China*

[2]*Faculty of Arts and Social Sciences, Universiti Malaya, Kuala Lumpur, Malaysia*

[3]*Institute for Advanced Studies, Universiti Malaya, Kuala Lumpur, Malaysia*

[4]*College of Urban and Environmental Sciences, Central China Normal University, Wuhan, Hubei, China*

*Corresponding author

ABSTRACT

Waste management plays a critical role in environmental protection and sustainable development. In South Asia, rapid demographic and economic growth have exacerbated systemic waste-related challenges. Therefore, it is necessary to examine the responses of South Asian countries to waste management challenges through historical cases, analyzing the current status of waste management, experiences with sustainable waste management, and policies for effective waste management in these countries. This chapter explores the recent situation, solutions, and possible impacts after case studies and reviews of waste management practices in eight South Asian countries; meanwhile, it identifies key barriers, including low

Data Analytics and Smart Technologies in Solid Waste Management: Advancing Sustainability.
Moharana Choudhury, Sushobhan Majumdar, and Avnish Chauhan (Eds.)

DOI: 10.1201/9781779640833-3

public awareness stemming from insufficient governmental outreach, poor working conditions, and the lack of formal recognition and protections for workers in the informal sector. Building on the regional discourse of sustainable waste management and resource recovery, it offers strategic recommendations from four interrelated perspectives—environmental awareness, technological adoption, government engagement, and regional and global cooperation—underscoring their collective importance in advancing sustainable development across South Asia. Ultimately, by offering context-specific insights into South Asia's waste governance landscape, this chapter contributes a regionally grounded yet globally relevant perspective to the evolving discourse on sustainable waste systems in the Global South.

2.1 INTRODUCTION

Rapid urban development and population growth in South Asia have exacerbated waste generation, making effective solid waste management strategies imperative [32]. However, inadequate waste collection and treatment infrastructure in South Asian cities has resulted in uncontrolled dumping and landfilling, resulting in serious health risks and environmental degradation [1, 30]. Over the past decade, South Asia has made impressive progress in maintaining strong economic growth, reducing poverty, and prioritizing health care and education. Research showed that with the deepening of global informatization and technology, South Asian countries have achieved rapid economic development, among which factors such as human capital, trade openness, economic freedom, and foreign debt have played positive roles in promoting economic growth in South Asia [2–5]. However, South Asia's economic resurgence has led to faster, unplanned urbanization and urban population expansion, further exacerbating waste. Therefore, waste management in South Asian countries is at a "crossroads," and addressing waste management issues requires integrated strategies based on evidence-based and sustainable management practices.

As the most significant country by area in South Asia, Indian cities generate 62 million tons of garbage (municipal solid waste [MSW]) annually, which is expected to reach 165 million tons by 2030 [6]. Of the 43 million tons of MSW collected annually, 31 million tons are dumped in landfills, and only 11.9 million tons are disposed of [6]. Indian cities and towns are full of garbage and MSW, with most towns and cities having only important places kept clean and other places full of uncollected garbage, which makes many parts of the cities/towns look unsightly [7]. Mountains of garbage tower on the outskirts of every major city in India (as a result of the continued accumulation of oversaturated garbage dumps) containing large amounts of viruses and bacteria that contaminate the rivers, air, and groundwater, creating a significant germ hazard and manufactured risk of disaster for the people living in the surrounding areas [8, 33]. In addition, the lack of technology for the collection, segregation, treatment, and disposal of MSW in some of the mountainous regions of India has resulted in a significant regional problem. For example, managing MSW in the highland and mountainous areas, with remote and complex topography and susceptibility to natural hazards and disasters, is more complex and challenging than in the plains [9].

Based on the above background information, it has been shown that with increasing urbanization and industrialization, effective waste management not only reduces environmental pollution and resource wastage but also creates healthier and more livable urban environments, which are essential for preserving the health of ecosystems as well as for safeguarding the well-being of human beings and other organisms [10, 34]. Therefore, this book chapter contributes to understanding waste management in South Asian countries through a case base of eight South Asian countries, covering the situation of waste management, the region's experience of sustainable waste management, and the promotion of effective policies.

2.2 DESCRIPTION OF THE STUDY AREA

The vast area of southern Asia that lies between the Indian Ocean and the central and western Himalayas is referred to as South Asia. This region includes India, Afghanistan, Nepal, Bhutan, Pakistan, Bangladesh, Sri Lanka, and the Maldives. The eight countries, with a combined population of approximately 1.9 billion (about 23% of the world's total), include India—a critically significant and one of the world's most populous countries, which generates substantial waste daily; its per capita waste generation stood at 0.57 kg/day as of 2016, the second highest in the region after the Maldives [11].

2.3 APPLICATION AND CASE STUDY

2.3.1 PROCESSES FOR WASTE MANAGEMENT IN SOUTH ASIAN COUNTRIES

The South Asia Co-operative Environment Programme, founded by the governments of South Asia in 1982 to promote regional environmental protection, has demonstrated its sustained commitment—particularly in addressing solid waste challenges—through key governance activities such as its 15th (November 2019, in Bangladesh) and 16th (June 2024, in Bhutan) Governing Council Meetings, which advanced regional waste management initiatives. Asia has faced major waste-related issues for years, such as plastic pollution and food waste. Plans such as the unified action plan (e.g. South Asian Regional Multisectoral Plastic Action Plan advocated by Bangladesh) aim to reduce plastic waste, improve collection, encourage recycling, and address the problem of marine plastic waste. Economic surges, rapid urbanization, and changing consumer preferences have increased the volume and variety of MSW [12].

2.3.2 SOUTH ASIAN COUNTRIES' EXPERIENCE IN SUSTAINABLE WASTE MANAGEMENT

Indore, India's cleanest city, encourages better hygiene and recycling habits through songs that promote sanitation, environmental protection, and targeted recycling measures. Meanwhile, residents help with waste sorting, truck-mounted bins are provided free of charge, sanitation violators are fined, and a sustainable waste collection system is established. A series of targeted waste recycling measures aims to educate Indian citizens about these environmental and hygiene initiatives to foster better citizen hygiene and recycling sorting habits [13]. In addition, the Indore Municipal Corporation has faced strong opposition from the media for giving up the rights of sanitation workers who were fired for not wanting to work overtime. However, in recent years, the achievements of urban sanitation governance in Indore have also led to less opposition in the Indian media [14].

In Nepal, the current system needs to fully recognize the critical role played by informal waste disposal workers, and establishing waste collection centers can help ensure that waste is correctly disposed of [15]. A study is supposed to use waste management and recommendations to foster a circular economy from the informal sector [16]. In addition, taxes related to the transportation of recycled materials need to be reassessed, and circular economy practices promoted through the informal sector emphasize the need for governments to assume their responsibilities [17, 18].

Pakistan has become a major destination for e-waste imports (with the United States, the European Union, Japan, the United Kingdom, and Australia among the primary sources), of which only about 2% of imported used electronics are reused, with the majority entering the informal recycling sector for the dismantling [19]. For example, waste in obsolete computers was dismantled to recycle their plastic and metal components, and all recycling tasks were completed manually without safety gear. Unexpectedly, in the academic community, there is less research on this issue in Pakistan, especially first-hand information on the quantities of e-waste imported into Pakistan.

The Maldives encounter distinct challenges in managing solid waste because of its island-setting location. In response to the waste management problems it faces, waste reduction at the source, including marine waste management, must be prioritized, and a balance between incineration and recycling methods must be struck. However, much waste is burned in the open due to space constraints, and harmful emissions contribute to excessive carbon emissions, posing health risks to approximately 500,000 Maldivians [20]. In addition, other South Asian countries such as Afghanistan, Bhutan, Bangladesh, and Sri Lanka are gradually adopting sustainable waste management practices, but challenges remain.

2.3.3 EFFECTIVE POLICIES TO PROMOTE WASTE MANAGEMENT IN SOUTH ASIAN COUNTRIES

The increasing amount of waste generated alongside rapid urbanization is contributing to hazardous environmental conditions. Growing accumulation of garbage might lead to dangers to the eco-environment and contribute to the spread of deadly illnesses. It is acknowledged that factors such as melting Himalayan glaciers, rising sea levels, and growing typhoon frequency might significantly influence South Asia. Based on the dire environmental situation, some of the steps that South Asian nations have already taken and developed some successful strategy recommendations for their national contexts are as follows [1].

In India, managing solid waste in cities is a big task for local government bodies and is a struggle due to limited money, skills, and land. The government started the "Swachh Bharat Mission" in 2014 to improve sanitation and waste disposal [21]. India banned the import of most plastic waste in 2019, but it still gets much waste from European and American countries. In India, waste management should start with source separation, and it will improve if people follow the rules [7, 22, 23]. In this case, if people effectively implement and adopt these solid waste management rules, they can revolutionize the waste management system in India and provide excellent opportunities for stakeholders to benefit from it [6]. In addition, the informal recycling sector in India, consisting of waste pickers, also plays a crucial role in waste segregation and recycling. Training and educating the waste collectors could make their informal recycling more organized [6].

In the meantime, this chapter has pinpointed several vital areas to achieve sustainable solid waste management in other South Asian countries, which include conducting environmental surveys, formulating effective policies, establishing municipal waste management systems, and developing waste treatment methods. South Asian countries could improve waste management to protect the environment by integrating these efforts with national policies on waste-to-energy, recycling, and other small-scale methods. In summary, it is imperative to further promote source reduction, resource utilization, and harmless disposal of bulk solid waste; strengthen full-chain governance; and promote the development of the comprehensive resource utilization industry. In addition, local government authorities need to work with international agencies, businesses, and community organizations to ensure the effectiveness and compatibility of waste management policies.

2.4 CHALLENGES AND OPPORTUNITIES

2.4.1 IDENTIFY THE CHALLENGES IN WASTE MANAGEMENT

South Asian countries actively promote sustainable waste management's 3R strategy (reduce, reuse, and recycle). However, in comparison, existing policies focus more on recycling and disposal than reducing waste generation. Therefore, an accountability framework for monitoring and reporting needs to be established to strengthen waste management regulations in South Asia. In such a manner, all countries and communities would place a premium on improving the technical, enforcement, and institutional capacity of regulators and operators.

1. Due to the lack of government guidance on waste management, the public lacks awareness of sustainable development and environmental protection and takes less action. Awareness about waste reduction, segregation, and proper disposal is weak in South Asian cities. In addition, with limited information on health and environmental impacts, citizens need more incentives to change

consumption habits. Promoting sustainable concepts in education, urban welfare programs, ecological activities, and so on, can help people regard waste as a valuable resource [24].

2. Integrating the work of the informal sector into formal systems is crucial while ensuring safety and social protection [25]. In South Asia, where material recovery and recycling occur daily, the informal sector collects waste and recyclables in various aspects of the waste management system, but often in a relatively poor working environment.
3. Lack of detailed recycling guidelines or inefficient waste recovery in waste management procedures [35]. The informal sector is actively involved in collecting recyclables and trash from various locations within the waste management system, such as disposal sites, transfer stations, and door-to-door pickup. It is important to note that many of these facilities also have manual labor and temporary storage centers, and their working environment is often risky. As a result, waste management development in South Asia relies heavily on improving the efficiency and safety conditions of material recycling operations.

2.4.2 OVERCOMING CHALLENGES AND OPPORTUNITIES TO PROMOTE SUSTAINABLE WASTE MANAGEMENT

Effective waste management is critical to protecting the environment, conserving resources, and safeguarding public health. A Roadmap for Sustainable Waste Management and Resource Circulation in South Asia, 2019–2030 provides detailed guidance to South Asian countries by integrating waste management with Sustainable Development Goals and National Development Goals, aiming to a waste management plan to reduce negative impacts. The roadmap highlights the need to use resources efficiently, promote cooperation between countries, engage local communities in waste reduction actions, and adopt modern technological tools [26]. In summary, ensuring the effectiveness and sustainability of waste management requires a range of measures, including policy changes, technological innovation, community engagement, and global collaboration, and South Asian countries need to collaborate with other countries to achieve cleaner and more resilient waste management approaches [27].

2.5 SUGGESTIONS FOR EFFECTIVE IMPLEMENTATION

2.5.1 ENVIRONMENTAL AWARENESS

Increasing environmental awareness, handling waste according to the classification system, preventing waste generation, emphasizing the concept of "polluter pays," and raising environmental awareness. Waste generators should be financially responsible for waste disposal to protect the environment and public health. Furthermore, the concept of extended producer responsibility, whereby manufacturers assume greater responsibility for the environmentally friendly disposal of their products at the end of their product life cycles, should be emphasized by collaborating with relevant parties such as local and state governments, businesses, community organizations, and the public sector and integrating efforts from informal and community groups. Specifically, it is ideal to use the life-cycle approach, including starting from the product's creation and following it through its various stages (from raw material extraction to manufacturing, packaging, transportation, distribution, sale, use, and disposal as it moves through the waste hierarchy and waste management system). Furthermore, it focuses on eco-innovation, eco-design, sustainable production, and resource efficiency and reevaluates based on the resources required at each stage. Based on working together to use resources efficiently, they will adopt a "comprehensive environmental strategy" for their products and services and minimize the use of hazardous substances in manufacturing and product development.

2.5.2 TECHNICAL ADVICE

One way to improve waste management is to ensure that waste collection and transport services are affordable for everyone. Waste should be disposed of in landfills that meet basic environmental and public health standards. Meanwhile, businesses and households should actively collect solid recyclables, such as paying attention to garbage classification in daily life and uniformly recycling recyclables to increase recycling rates [31]. Governments should advocate for communities to implement nonbinding performance indicators and minimum recycling service standards to help track the amount of recyclables collected. Property managers and residents in the community should also cooperate, such as by organizing environmental activities or competitions. Partnering with transportation hubs, medical facilities, and academic institutions can further advance recycling efforts, and increasing the number of homes that separate organic waste can also help boost recycling and composting rates.

Encourage waste reduction at the source by implementing mandatory waste separation. Governments could require households and businesses to sort waste at source to ensure the effect of compostable materials or recyclable solids. Based on the primary classification system, the government should add other types of recycling methods based on the public's environmental awareness and the accessibility of recycling facilities, such as the layout of trash cans, promoting innovative ways to reduce waste and increase recycling rates, such as offering incentives for various local, state, federal, and private initiatives.

By linking organic waste to renewable energy, South Asian governments have a significant opportunity to reduce carbon emissions in transport, heating, and electricity, including recycling waste. Promoting biogas production and composting processes to effectively recycle organic waste not only has enormous economic benefits but is also an effective means of managing the various risks posed by these activities. In addition, working with environmental protection agencies and industry stakeholders could ensure that waste management-related products comply with environmental protection standards. Not only that, but a special committee needs to be established to formulate and monitor the implementation of policies for each enterprise. Meeting requirements is supposed to be an integral part of waste management. However, the environment can be contaminated if management practices are inadequate. Optimal waste disposal strategies are needed to minimize environmental harm and maximize the recovery of our available resources. By using related technologies such as landfills, waste-to-energy, and energy utilization, the quality of these technologies needs to improve further. Investigating other financial return methods, improving how landfills are managed, and working with industry and local governments to develop specific standards could help achieve more efficient waste management and resource recovery.

2.5.3 GOVERNMENT INTERVENTION

2.5.3.1 STRENGTHENING COOPERATION AND COORDINATION

Practical cooperation based on consistent goals is critical to ensuring the robust operation and progressive development of a sustainable waste management system. At a national level, detailed waste management plans must be developed to help move toward resource management approaches that are more appropriate for society. For example, national policy documents specified at the macro level should consider the specific implementation of various aspects such as financial markets and social environment; waste management strategies should be developed within national frameworks and consider local characteristics at the local level, such as wastewater treatment and water resources treatment in riverine regions of India and Nepal, recycled plastic processing and reuse in Bangladesh, and wind energy resources in the Maldives. In addition, governments can encourage similar cooperation models through policy measures such as financial support for infrastructure development and land acquisition and by dialoguing with relevant stakeholders to promote collaboration and informed decision-making.

2.5.3.2 ENHANCE ORGANIZATIONAL AND PERSONNEL CAPABILITIES

Responsibility for waste management is shared between the Ministry of Environment and municipal and regional development authorities at the federal level. They are responsible for setting laws and national standards, providing guidance on municipal waste disposal, and clarifying each individual's roles and responsibilities. The local government plans to work with waste generators, service providers, and other community members to develop a waste management plan and promote efficient waste management methods by national regulations.

2.5.3.3 ESTABLISH A TECHNOLOGY SYSTEM FOR WASTE STREAM MANAGEMENT

Dealing with different types of waste, like food, plastic, e-waste, medical waste, and construction debris, should be a priority over managing regular household waste. However, many South Asian countries lack the funds and technology to handle these waste streams, and more information on waste streams is necessary. Creating inventories of waste streams, improving regulations, using the right technologies, and enhancing the capabilities of waste management agencies are essential steps.

2.5.4 GLOBAL AND REGIONAL COOPERATION

2.5.4.1 PROMOTE GLOBAL COOPERATION AND STRENGTHEN REGIONAL PARTNERSHIPS

International cooperation on a global scale is a significant trend as the world faces interrelated waste management challenges, including plastic pollution, e-waste, and infrastructure shortages [28]. South Asian countries can cooperate through regional bodies such as SAARC (South Asian Association for Regional Cooperation) and bring about concrete actions based on international cooperation to promote important innovations in waste management, including knowledge-sharing, capacity-building, and harmonization of waste management policies.

2.5.4.2 INNOVATIONS IN ARTIFICIAL INTELLIGENCE AND CYBER-GOVERNANCE

Use artificial intelligence (AI) and data analytics to improve recycling efficiency, forecast trash creation trends, and optimize waste collection routes. Protecting waste management systems from cyber threats and ensuring the security of waste-to-energy plants, intelligent bins, and digital platforms is critical [29]. Moreover, future academic summits should explore industry innovations such as global ethics for AI, cybersecurity for waste infrastructure, and data sharing for optimized waste management.

2.5.4.3 PEACE BUILDING INNOVATIONS FOR WASTE MANAGEMENT

Dealing with waste is vital in areas recovering from conflict, and it should be part of efforts to build peace and help communities become more muscular. Get local people involved in collecting waste, recycling, and teaching others about it, and these actions could help bring people together and boost the economy in places with conflicts. South Asian countries need to collaborate, develop new ideas, and set clear rules to manage waste well, which could become cleaner and more sustainable through effective waste management. In addition, it should develop plans that include specific targets and practical strategies for dealing with waste and resources, help share information and technology, and making regional cooperation aims to reduce waste generation and enhance recycling, thereby avoiding environmental, economic, or social harm.

2.6 CONCLUSION

This chapter examines practical cases and past waste management experiences in South Asian countries. Analyzing each country's policies and actual situation shows that the current waste management situation in South Asian countries is not optimistic, especially when dealing with waste; it faces considerable challenges, such as rapid economic growth. Meanwhile, population growth has also brought about environmental problems. Based on a series of case studies, this chapter finds that waste management in South Asia suffers from insufficient government help, insufficient awareness of waste problems, poor working conditions for workers in the waste industry, and inadequate rules for recycling waste. "A Roadmap for Sustainable Waste Management and Resource Circulation in South Asia, 2019–2030" points out the possibilities to optimize and improve the current situation. In addition, this chapter also provides valuable suggestions on how to increase awareness of waste management, provide technical support, and encourage government participation and cooperation with other countries. Strengthening waste management has become a meaningful way to solve the garbage problem in South Asia.

In general, waste management is critical to the South Asia region and may lead to public problems such as environmental pollution, resource reduction, and medical and health challenges. This chapter provides an innovative perspective on how South Asian countries can develop more effective waste management strategies and tools. Including targeted waste treatment can improve environmental conditions and the quality of life of people in South Asia, helping to maintain social stability and economic growth. Given that India and other South Asian countries face waste management issues similar to those faced by the global environment, solving such problems has a vital impact on the continued development of South Asia and nearby regions. It can also serve as a model for other developing countries. In conclusion, waste management in South Asia has significant environmental, economic, and social impacts and can provide valuable lessons for global environmental management. By paying attention to the research and practice of waste management, sustainable development can be promoted, and the expected progress of the environment, economy, and society can be achieved.

KEYWORDS

- **South Asian countries**
- **waste management**
- **sustainability**
- **environment**
- **case study**

REFERENCES

1. Khan, A. A., Ahmed, Z., & Siddiqui, M. A. (2012). Issues with solid waste management in South Asian countries: A situational analysis of Pakistan, *Journal of Environmental and Occupational Science*, *1*(2), 129–131. https://www.jenvoh.com/abstract/issues-with-solid-waste-management-in-south-asian-countries-a-situational-analysis-of-pakistan-47440.html
2. Mukherji, I. N. (1981). Economic growth and social justice in South Asia: An interpretation of empirical evidence. *India Quarterly*, *37*(3), 389–404. https://doi.org/10.1177/097492848103700304
3. Bhattacharjee, J., & Haldar, S. K. (2015). Economic growth in South Asia: Binding constraints for the future. *Journal of South Asian Development*, *10*(2), 230–249. https://doi.org/10.1177/0973174115588847
4. Mohsin, M., Ullah, H., Iqbal, N., Iqbal, W., & Taghizadeh-Hesary, F. (2021). How external debt led to economic growth in South Asia: A policy perspective analysis from quantile regression. *Economic Analysis and Policy*, *72*, 423–437. https://doi.org/10.1016/j.eap.2021.09.012
5. Ahmed, S., Mushtaq, M., Fahlevi, M., Aljuaid, M., & Saniuk, S. (2023). Decomposed and composed effects of economic freedom on economic growth in South Asia. *Heliyon*, *9*(2), E13478. https://doi.org/10.1016/j.heliyon.2023.e13478
6. Agarwal, R. (2020). Waste management crisis in India. *RECYCLING Magazine*. Retrieved from: https://www.recycling-magazine.com/2020/05/06/waste-management-crisis-in-india/ (accessed: 22 January, 2024)
7. Central Pollution Control Board. (2015). The national action plan for municipal solid waste management. https://cpcb.nic.in/uploads/MSW/Action_plan.pdf
8. Mumu. (2023). India faces a garbage mountain crisis. *Huxiu*. Retrieved from: https://m.huxiu.com/article/2205943.html (accessed: 22 January, 2024)
9. Thakur, A., Kumari, S., Sinai Borker, S., Prashant, S. P., Kumar, A., & Kumar, R. (2021). Solid waste management in Indian Himalayan Region: Current scenario, resource recovery, and way forward for sustainable development. *Frontiers in Energy Research*, *9*, 609229. https://doi.org/10.3389/fenrg.2021.609229

10. Mihai, F. C., Gündoğdu, S., Markley, L. A., Olivelli, A., Khan, F. R., Gwinnett, C., Gutberlet, J., Reyna-Bensusan, N., Llanquileo-Melgarejo, P., Meidiana, C., Elagroudy, S., Ishchenko, V., Penney, S., Lenkiewicz, Z., & Molinos-Senante, M. (2022). Plastic pollution, waste management issues, and circular economy opportunities in rural communities. *Sustainability*, *14*(1), 20. https://doi.org/10.3390/su14010020
11. Amin, S., Khandaker, M. K., Jannat, J., Khan, F., & Rahman, S. Z. (2023). Cooperative environmental governance in urban South Asia: Implications for municipal waste management and waste-to-energy. *Environmental Science and Pollution Research*, *30*, 69550–69563. https://doi.org/10.1007/s11356-023-27152-5
12. Kumarasiri, B., & Dissanayake, P. (2021). Barriers to implementing waste-to-energy projects in Sri Lanka: A PESTEL analysis. *Built Environment Project and Asset Management*, *11*(4), 544–558. https://doi.org/10.1108/BEPAM-03-2020-0051
13. Explained Desk. (2024). How Indore became ranked India's cleanest city, year after year. *The Indian Express*. Retrieved from: https://indianexpress.com/article/explained/indore-india-cleanest-city-reason-9105288/ (accessed: 21 January, 2024)
14. Xu, Q. (2018). Indore has been crowned India's "cleanest city" and citywide radio announcements will notify those who urinate and defecate in the streets. *Observers*. Retrieved from: https://www.guancha.cn/global-news/2018_05_18_457226.shtml (accessed: 21 January, 2024)
15. Cataldo, N. L., Oyinlola, M., Sigdel, S., Nguyen, D., & Gallego-Schmid, A. (2023). Waste management in Nepal: Characterization and challenges to promote a circular economy. *Circular Economy and Sustainability*, *2023*. https://doi.org/10.1007/s43615-023-00283-0
16. ADB. (2013). Solid waste management in Nepal: Current status and policy recommendations.https://www.adb.org/publications/solid-waste-management-nepal-current-status-and-policy-recommendations
17. Central Bureau of Statistics. (2021). Waste management baseline survey of Nepal 2020. https://unstats.un.org/unsd/envstats/Censuses%20and%20Surveys/Waste-Management-Baseline-Survey-of-Nepal-2020.pdf
18. Oh, Y. (2024). Insights into Nepal's waste management: My observations & reflections. *United Nations Development Programme (UNDP)*. Retrieved from: https://www.undp.org/nepal/blog/insights-nepals-waste-management-my-observations-reflections (accessed: 17 February, 2024)
19. Ilyas, S., Srivastava, R. R., Kim, H., & Abbas, Z. (2020). Electrical and electronic waste in Pakistan: the management practices and perspectives. In M. N. V. Prasad, M. Vithanage, & A. Borthakur (Eds.), Handbook of Electronic Waste Management: International Best Practices and Case Studies (pp. 263–281). *Butterworth-Heinemann*. https://doi.org/10.1016/B978-0-12-817030-4.00007-3
20. The World Bank. (2022). Maldives is turning waste to wealth, energizing youth, to safeguard its future. Retrieved from: https://www.worldbank.org/en/news/feature/2022/07/22/maldives-is-turning-waste-to-wealth-energizing-youth-to-safeguard-its-future (accessed: 17 February, 2024)
21. Ministry of Urban Development. (2016). Swachh Bharat mission municipal solid waste management manual part II: The manual. *Central Public Health and Environmental Engineering Organisation (CPHEEO)*. https://mohua.gov.in/upload/uploadfiles/files/Part2.pdf

22. Ha, K. O. (2022). Amazon packages burn in India, last stop in broken plastic recycling system. *Bloomberg*. Retrieved from: https://www.bloomberg.com/features/2022-india-plastic-recycling-pollution/ (accessed: 22 January, 2024)
23. Dayal, P. (2023). Paper trail. *CBC NEWS*. Retrieved from: https://www.cbc.ca/newsinteractives/features/paper-trail (accessed: 22 January, 2024)
24. Hondo, D., Arthur, L., & Gamaralalage, P. J. D. (2020). Solid waste management in developing Asia: Prioritizing waste separation. *ADBI Policy Brief, 2020*(7), 1–10. https://www.adb.org/publications/solid-waste-management-developing-asia
25. Dhokhikah, Y., & Trihadiningrum, Y. (2012). Solid waste management in Asian developing countries: Challenges and opportunities. *Journal of Applied Environmental and Biological Sciences*, *2*(7), 329–335. https://repository.unej.ac.id/xmlui/bitstream/handle/123456789/77179/Solid_waste_management_in_Asian_developi.pdf?sequence=1&isAllowed=y
26. Utilities One. (2024). Waste to energy: Opportunities and challenges. Retrieved from: https://utilitiesone.com/blog/waste-to-energy-opportunities-and-challenges (accessed: 17 February, 2024)
27. ISWA. (2021). The future of the waste management sector trends, opportunities and challenges for the decade. https://www.iswa.org/wp-content/uploads/2021/10/ISWA-2021f-Rev2-FK-1.pdf?v=06fa567b72d7
28. The Stimson Center. (2023). Future of International Cooperation Report 2023 - Building shared futures: Innovating governance for global and regional problem-solving. https://www.stimson.org/wp-content/uploads/2023/09/Future-of-International-Cooperation-Report.pdf
29. World Economic Forum. (2021). Principles for strengthening global cooperation. https://www3.weforum.org/docs/WEF_Global_Action_Group_Principles_2021.pdf
30. Mondal, T., Choudhury, M., Kundu, D., Dutta, D., & Samanta, P. (2023). Landfill: An eclectic review on structure, reactions and remediation approach. *Waste Management*, *164*, 127–142. https://doi.org/10.1016/j.wasman.2023.03.034
31. Paul, S., Choudhury, M., Deb, U., Pegu, R., Das, S., & Bhattacharya, S. S. (2019). Assessing the ecological impacts of ageing on hazard potential of solid waste landfills: A green approach through vermitechnology. *Journal of Cleaner Production*, *236*, 117643. https://doi.org/10.1016/j.jclepro.2019.117643
32. Choudhury, M., & Dutta, J. (2017). A comparative study of municipal solid waste management status for three major towns of Upper Assam-India. *International Journal of Waste Resources*, *7*(291), 2. https://doi.org/10.4172/2252-5211.1000291
33 Choudhury, M., Jyethi, D. S., Dutta, J., Purkayastha, S. P., Deb, D., Das, R., Roy, G., Sen, T. & Bhattacharyya, K. G. (2022). Investigation of groundwater and soil quality near to a municipal waste disposal site in Silchar, Assam, India. *International Journal of Energy and Water Resources*, *6*(1), 37–47. https://doi.org/10.1007/s42108-021-00117-5
34. Bhardwaj, L. K., Choudhury, M., Goswami, L., Rahman, F., & Xu, J. (2025). Micro- and Nanoplastics: Its Impact on Antarctica and Similar Ecosystems. In S. Goswami, M. Choudhury, & P. Samanta (Eds.), Micro-Nano Plastics Exposure, Environmental Degradation and Public Health Crisis: Perspectives and Concerns for Sustainable Development (pp. 127-135). Springer Nature. http://doi.org/10.1007/978-981-96-5047-7_11

35. Xu, J., Liao, W., Lim, N. H. H., Fong, C. S., Wang, H., & Xu, Y. (2024). Operational and Environmental Assessment for Night Markets under the Informal Economy Impacts: Insights from the Case Study of Selangor, Malaysia. 3C TIC, 13(1), 189-219. https://doi.org/10.17993/3ctic.2024.131.189-219

CHAPTER 3

A Case on Unseen Consequences of Waste: Environmental and Health Impacts of Inadequate Management

VIJAI,[1*] WORAKAMOL WISETSRI,[2] and MOHD ASIF SHAH[3]

[1]*School of Commerce, Vel Tech Rangarajan Dr. Sagunthala R&D Institute of Science and Technology, Chennai, Tamilnadu, India*

[2]*Department of Manufacturing and Service Industry Management, Faculty of Business and Industrial Development, King Mongkut's University of Technology North Bangkok, Bangkok, Thailand*

[3]*Department of Economics, Kebri Dehar University, Kebri Dehar, Ethiopia*

*Corresponding author

ABSTRACT

This chapter delves into the multifaceted and often underestimated consequences of inadequate waste management, shedding light on its profound environmental and health impacts. The surge in waste generation, fueled by rapid urbanization and evolving consumption habits, has given rise to unseen challenges. Beyond the visible pollution of landfills and water bodies, the study examines the insidious effects on soil, water, and air quality. Soil contamination from hazardous waste disrupts ecosystems and enters the food chain, while water pollution threatens aquatic life and compromises clean water sources. Air quality degradation results from open burning and emissions from poorly managed landfills, releasing harmful substances into the atmosphere. The health

Data Analytics and Smart Technologies in Solid Waste Management: Advancing Sustainability. Moharana Choudhury, Sushobhan Majumdar, and Avnish Chauhan (Eds.)

DOI: 10.1201/9781779640833-4

implications are equally significant, encompassing respiratory diseases, waterborne illnesses, and food chain contamination. Through case studies on electronic waste in developing countries and plastic pollution in oceans, the study underscores these challenges' global and interconnected nature. Mitigating the unseen consequences necessitates a comprehensive approach, integrating improved waste management practices, heightened public awareness, and targeted policy interventions. Addressing these hidden dimensions is crucial for safeguarding ecosystems, biodiversity, and human well-being in the face of escalating waste-related issues on a global scale.

3.1 INTRODUCTION

In the face of rapid urbanization, burgeoning industrialization, and shifting consumer behavior, the 21st century has ushered in an era of unprecedented waste generation, posing formidable challenges for effective waste management [14]. This section introduces the intricate and multifaceted landscape of waste management challenges that societies across the globe are grappling with [38]. The surge in population density in urban centers, coupled with the expansion of industries and the adoption of modern consumption patterns, has given rise to an exponential increase in the volume and complexity of waste. Beyond the conventional understanding of waste as a byproduct of daily life, the contemporary context demands a closer examination of the intricate interplay between waste management, environmental sustainability, and public health [12]. This introduction sets the stage for an in-depth exploration of the visible and unseen consequences of inadequate waste management practices, emphasizing the urgency of addressing this critical issue for the well-being of our planet and its inhabitants [17].

3.2 RAPID URBANIZATION AND INDUSTRIAL GROWTH

Rapid urbanization and industrial growth pose significant challenges to waste management systems worldwide [30]. As populations concentrate in urban areas and industries expand, the generation of waste escalates, straining existing infrastructure and exacerbating environmental degradation. Urbanization leads to increased consumption and disposal rates, while industrial growth intensifies the production of hazardous waste and emissions. In many developing regions, inadequate waste management infrastructure further compounds these challenges, resulting in pollution of air, water, and soil [39]. Moreover, rapid urbanization often leads to informal settlements with limited access to proper waste disposal services, exacerbating public health risks and environmental inequities. Addressing the consequences of rapid urbanization and industrial growth requires integrated strategies that prioritize sustainable urban planning, improved waste management infrastructure, and heightened environmental regulations [1]. Additionally, promoting resource efficiency, circular economy principles, and community engagement is essential for mitigating the environmental and health impacts of urbanization and industrialization [29].

3.3 CHANGING CONSUMPTION PATTERNS

Changing consumption patterns significantly influence waste generation and management practices globally. As societies transition toward more consumer-driven economies, the volume and composition of waste undergo substantial shifts. Increasing affluence and globalization contribute to a culture of disposable products and single-use packaging, leading to a surge in waste generation [3]. Moreover, evolving consumer preferences toward technology and fast fashion exacerbate the problem, resulting in the proliferation of electronic and textile waste. Addressing changing consumption patterns requires multifaceted approaches that promote sustainable lifestyles, product stewardship, and circular economy principles [11]. Encouraging responsible consumption habits, such as reducing, reusing, and recycling, can help minimize waste generation [36]. Additionally, industry initiatives to design products for longevity, repairability, and recyclability are essential for transitioning toward a more sustainable consumption paradigm. Public education campaigns and policy interventions that incentivize sustainable practices further support the shift toward more resource-efficient and environmentally conscious consumption patterns [18].

3.4 VISIBLE AND UNSEEN CONSEQUENCES

The ramifications of inadequate waste management extend far beyond the visible pollution that scars landscapes and waterways. While the unsightly presence of littered streets and polluted rivers is evident to the naked eye, the unseen consequences permeate ecosystems and pose insidious threats to both the environment and human health [3].

3.4.1 VISIBLE CONSEQUENCES

The visible consequences of poor waste management are stark and immediate. Littered landscapes not only mar the aesthetics of urban and natural environments but also contribute to the degradation of soil and water quality [34]. Improperly disposed waste, whether in the form of plastic bottles or discarded electronics, forms an unsightly mosaic that betrays the unsustainable practices of societies [19]. Water bodies bear the visible burden of waste mismanagement, with plastic debris floating on the surface and accumulating along coastlines. This not only harms marine life but also poses a direct threat to the livelihoods of communities dependent on fisheries [14]. The detrimental impact on flora and fauna in these visibly polluted ecosystems is a distressing testament to the consequences of neglectful waste disposal [4].

3.4.2 UNSEEN CONSEQUENCES

Beneath the surface, however, lie the unseen consequences that demand equal attention. Soil contamination, a less conspicuous but equally pressing issue, arises as hazardous substances leach into the earth from improperly disposed waste. These contaminants disrupt the delicate balance of microorganisms and nutrients essential for plant growth, affecting agricultural productivity and posing risks to ecosystems [16]. The loss of biodiversity stands as another unseen consequence, intricately linked to the encroachment of waste into natural habitats. As species face habitat degradation and fragmentation, the interconnected web of life weakens, and ecosystems become more vulnerable to collapse [35]. The decline of essential pollinators, disruptions in nutrient cycling, and the subtle unraveling of intricate ecological relationships represent the hidden toll of inadequate waste management. Moreover, the unseen threat extends to microscopic dimensions, with pollutants breaking down into microplastics and infiltrating food chains. This microscopic intrusion not only jeopardizes aquatic life but also poses potential risks to human health as these contaminants find their way into the food we consume. With the challenges of waste management, it is imperative to recognize and tackle both the visible and unseen consequences [20]. A holistic approach that combines visible pollution control with measures to prevent soil contamination, habitat degradation, and the loss of biodiversity is crucial. Only through such comprehensive strategies can societies hope to mitigate the far-reaching impacts of inadequate waste management on the environment and safeguard the well-being of current and future generations [2].

3.5 ENVIRONMENTAL IMPACTS SOIL CONTAMINATION

Soil contamination, a consequential byproduct of inadequate waste management practices, poses significant environmental challenges with far-reaching implications. As waste accumulates in landfills and open dumping sites, a cascade of contaminants infiltrates the soil, altering its composition and compromising the delicate balance that sustains ecosystems [5, 40]. The repercussions of soil contamination are multifaceted, affecting the environment and human health. Toxic substances from improperly disposed waste, such as heavy metals, pesticides, and industrial chemicals, seep into the soil, creating a dangerous cocktail that disrupts the intricate web of life beneath our feet [21]. One of the immediate consequences is the degradation of soil fertility. Vital nutrients essential for plant growth are replaced by harmful chemicals, impeding the ability of flora to thrive [15]. This degradation has a ripple effect, jeopardizing agricultural productivity and undermining the stability of natural ecosystems [6]. Furthermore, the infiltration of contaminants into groundwater exacerbates the problem, as polluted water tables become a source of toxic nourishment for plants and a potential threat to human populations reliant on groundwater for drinking and irrigation [33]. The insidious nature of soil contamination becomes evident as these contaminants make their way up the food chain, posing risks to terrestrial and aquatic life [23].

The unseen impacts of soil contamination extend to the microscopic realm, affecting soil-dwelling microorganisms crucial for nutrient cycling and maintaining soil health. As the microbial community faces disruptions from toxic substances, the entire ecosystem's resilience is compromised, amplifying the vulnerability of the environment to further degradation [6]. Mitigating the environmental impacts of soil contamination requires a concerted effort to address its root causes. Implementing sustainable waste disposal practices, including proper landfill management, waste treatment, and recycling, is essential to prevent the release of hazardous substances into the soil [41]. Additionally, soil remediation techniques, such as phytoremediation and bioremediation, can rehabilitate contaminated sites and restore the ecological balance of affected areas. Recognizing and addressing the environmental impacts of soil contamination is pivotal for preserving ecosystems, agricultural sustainability, and the planet's overall health. A holistic approach to waste management that considers the potential consequences on soil health is indispensable in fostering a resilient and balanced environment for present and future generations [1].

3.6 ENVIRONMENTAL IMPACTS: WATER POLLUTION

Water pollution, stemming from inadequate waste management practices, is a critical environmental challenge with profound and far-reaching consequences. As various forms of waste, including plastic, chemicals, and hazardous materials, find their way into water bodies, the intricate balance of aquatic ecosystems is disrupted, giving rise to a myriad of environmental impacts [24]. The most visible manifestation of water pollution is the degradation of water quality. Pollutants such as heavy metals, industrial effluents, and plastic debris introduce harmful substances into rivers, lakes, and oceans, jeopardizing the health of aquatic life and compromising the availability of clean water for human consumption. The contamination threatens marine organisms and has cascading effects on entire ecosystems. Aquatic biodiversity suffers because of water pollution, with sensitive species facing declines or extinctions due to the altered chemical composition of their habitats. The food chain disruption, as pollutants accumulate in organisms at different trophic levels, amplifies the impact, further destabilizing aquatic ecosystems [7].

Moreover, the unseen consequences of water pollution extend to the microscopic level. The presence of contaminants, including microplastics and chemical residues, has profound implications for plankton, the foundation of aquatic food webs. As these microscopic organisms are compromised, the entire structure of marine ecosystems is at risk, leading to imbalances that reverberate through the interconnected web of life [25]. The consequences of water pollution are not confined to aquatic environments.

Polluted water bodies serve as a source of drinking water for communities, and the contamination poses direct risks to human health. Waterborne diseases, ingestion of toxic substances, and long-term health impacts become prevalent in areas where clean water sources are compromised. Mitigating the environmental effects of water pollution necessitates a multifaceted approach. Improved waste management practices reducing single-use plastics, stringent regulations on industrial discharges, and developing effective wastewater treatment systems are essential steps. Additionally, conservation efforts to protect water ecosystems and promote sustainable practices that reduce the release of pollutants into water bodies are critical components of a comprehensive solution. Recognizing the environmental impacts of water pollution underscores the urgency of addressing the root causes through sustainable waste management and conservation measures [11]. Preserving the health of water ecosystems is vital not only for aquatic biodiversity, but also for still, it is also inseparable from the broader goal of ensuring a sustainable and healthy environment for current and future generations.

3.7 ENVIRONMENTAL IMPACTS—AIR QUALITY DEGRADATION

Air quality degradation resulting from inadequate waste management practices represents a silent yet pervasive threat to the environment and public health. As various pollutants are released into the atmosphere during waste disposal, combustion, and industrial activities, the delicate balance of the air we breathe is disrupted, leading to various environmental consequences [26]. Visible air pollution, characterized by smog, haze, and particulate matter, is a conspicuous outcome of air quality degradation. These pollutants, emitted from burning waste and fossil fuels, contribute to the formation of ground-level ozone and delicate particulate matter, which pose risks to respiratory health and contribute to the deterioration of air quality in urban and industrialized areas. Unseen consequences, however, extend beyond the visible manifestations of air pollution. Gaseous pollutants, including sulfur dioxide, nitrogen oxides, and volatile organic compounds, are released into the atmosphere during improper waste disposal and industrial processes. These pollutants undergo chemical reactions in the air, forming acid rain and contributing to the acidification of soils and water bodies. The far-reaching impact of acid rain extends to ecosystems, causing harm to vegetation, aquatic life, and soil health.

The environmental consequences of air-quality degradation are not confined to localized areas but transcend geographical boundaries. The dispersion of pollutants in the atmosphere contributes to regional and even global phenomena, such as smog episodes, transboundary haze pollution, and altering climate patterns. These consequences underscore the interconnected nature of air quality and the need for concerted international efforts to address the shared challenge of atmospheric pollution [8]. Mitigating the environmental impacts of air quality degradation necessitates a comprehensive approach. Implementing cleaner waste management practices, transitioning to renewable energy sources, and enforcing stringent emission standards for industries are crucial. Additionally, promoting public awareness about the importance of air quality and its interplay with waste management fosters a sense of responsibility. It encourages sustainable practices at both individual and community levels [27]. The environmental impacts of air quality degradation are complex and multifaceted, affecting ecosystems on local, regional, and global scales. Recognizing the interconnected nature of atmospheric pollution and adopting sustainable measures are imperative for safeguarding air quality and the planet's overall health and resilience [30].

3.8 ENVIRONMENTAL IMPACTS—LOSS OF BIODIVERSITY

In the intricate tapestry of ecosystems, the consequences of inadequate waste management extend beyond visible pollution to encompass a silent crisis—the loss of biodiversity [27]. As waste accumulates in natural habitats, disrupting the delicate balance of ecosystems, the intricate web of life faces unprecedented challenges. The consequences are profound, affecting species diversity, ecological functions, and the resilience of entire ecosystems. The accumulation of waste in natural environments transforms once-pristine habitats into landscapes tainted by discarded materials. Landfills, improperly disposed waste, and pollution alter the physical and chemical characteristics of soils, affecting the plants and organisms that form the foundation of ecosystems. The repercussions of this disruption are far-reaching, contributing to the decline of numerous species [9, 42].

Habitat degradation emerges as a critical factor in the loss of biodiversity. As waste encroaches upon natural spaces, it fragments habitats, restricts movement corridors for wildlife, and reduces the availability of essential resources [28]. This fragmentation isolates populations, making species more vulnerable to predation, disease, and environmental changes. Moreover, the toxic substances leaching from waste into soil and water directly threaten flora and fauna. Microorganisms, insects, and larger organisms struggle to adapt to the altered chemical composition of their surroundings, leading to population declines and, in some cases, local extinctions.

The interconnectedness of species within ecosystems becomes apparent as the loss of one species can trigger a cascading effect, impacting the entire food web. Pollinators vital for plant reproduction, decomposers crucial for nutrient cycling, and predators that maintain balance face threats from biodiversity loss, contributing to ecosystem instability [29]. The unseen consequences of waste mismanagement extend to microbial life, often overlooked but fundamental to ecological processes. Soil microorganisms are vital in nutrient cycling, decomposition, and soil health. As waste disrupts these microenvironments, the intricate relationships within the microbial community unravel; affecting the resilience of ecosystems [10]. Mitigating the loss of biodiversity necessitates a paradigm shift in waste management practices. This includes the adoption of sustainable waste disposal methods, the implementation of conservation measures to protect natural habitats, and the restoration of ecosystems impacted by waste. By recognizing the interconnectedness of biodiversity and waste management, societies can work toward preserving the richness and resilience of the natural world for future generations [31].

3.9 CASE STUDY—ELECTRONIC WASTE IN DEVELOPING COUNTRIES

Electronic waste (e-waste) poses a significant challenge in developing countries due to the rapid proliferation of electronic devices and the lack of adequate waste management infrastructure. With the accelerating pace of technological advancements and the increasing affordability of electronic gadgets, developing nations are experiencing a surge in e-waste accumulation. However, the disposal and management of this electronic refuse present complex issues, exacerbated by limited resources and inadequate recycling facilities. In many developing countries, informal recycling practices prevail, driven by the economic opportunities of extracting valuable materials from discarded electronics. However, these practices often involve manual dismantling without proper safety precautions, exposing workers to hazardous substances like lead, mercury, and cadmium. Consequently, informal e-waste recycling operations pose significant health risks to workers and nearby communities. Moreover, improper disposal methods contribute to environmental pollution, as toxic components from e-waste contaminate soil, water, and air. Burning electronic components to extract valuable metals releases harmful pollutants into the atmosphere, exacerbating air quality degradation. Addressing the challenges of e-waste in developing countries requires concerted efforts to establish formalized waste management systems, promote safe recycling practices, and raise awareness about the environmental and health impacts of improper e-waste disposal. International cooperation and support are essential to assist developing nations in managing the escalating e-waste crisis sustainably [32].

3.10 CONCLUSION AND COMPREHENSIVE SOLUTIONS

In conclusion, addressing the unseen consequences of inadequate waste management requires comprehensive solutions encompassing environmental stewardship, public health, and sustainable development. By recognizing the interconnected nature of waste management challenges, societies can adopt holistic approaches that mitigate environmental degradation and safeguard human well-being. Comprehensive solutions entail a combination of regulatory frameworks, technological innovations, public awareness campaigns, and international cooperation. Establishing stringent waste management regulations, including proper waste disposal protocols and recycling incentives, is essential for minimizing environmental pollution and mitigating the adverse impacts of waste mismanagement.

Technological innovations are crucial in enhancing waste management practices, such as advanced recycling technologies for e-waste and wastewater treatment systems for industrial effluents [43]. Embracing sustainable consumption patterns and promoting the circular economy can further reduce waste generation and resource depletion, fostering a more resilient and resource-efficient society. Public awareness campaigns foster a culture of environmental responsibility and encourage individuals to adopt sustainable behaviors. Education and outreach programs can empower communities to make informed choices about waste disposal and consumption habits, promoting environmental sustainability at the grassroots level. Furthermore, international collaboration and knowledge-sharing platforms are vital for effectively addressing global waste management challenges. By leveraging collective expertise and resources, countries can develop innovative solutions, implement best practices, and support capacity-building initiatives in developing nations. In essence, comprehensive solutions to waste management require a concerted effort from governments, industries, civil society, and individuals. By embracing sustainability as a guiding principle and implementing integrated strategies, societies can mitigate the unseen consequences of waste mismanagement and build a more resilient and equitable future for all.

KEYWORDS

- **unseen consequences**
- **waste**
- **global challenge**
- **rapid urbanization**
- **industrial growth**
- **biodiversity loss**
- **soil contamination**

REFERENCES

1. Ali, M., & Wang, W. (2017). *Chemosphere*, *164*, 96–105. https://doi.org/10.1016/j.chemosphere.2016.09.013.
2. Bhattacharya, P., et al. (2017). *Environmental Pollution, 220*(Pt A), 465–473. https://doi.org/10.1016/j.envpol.2016.09.070.
3. Christensen, T. H., et al. (2011). *Waste Management & Research, 29*(3), 191–220. https://doi.org/10.1177/0734242X10388402.
4. Eurostat. (2020). Municipal waste statistics. Retrieved from: https://ec.europa.eu/eurostat/statistics-explained/index.php/Municipal_waste_statistics

5. Eze, S., & Byrne, P. (2017). *Waste Management, 59*, 2–3. https://doi.org/10.1016/j.wasman.2016.09.031.
6. Hossain, M. S., et al. (2020). *Journal of Hazardous Materials*, *389*, 121814. https://doi.org/10.1016/j.jhazmat.2020.121814.
7. Kaza, S., et al. (2018). The World Bank. What a Waste 2.0: A Global Snapshot of Solid Waste Management to 2050. Retrieved from http://documents.worldbank.org/curated/en/288851518818494120/What-a-Waste-2.0-A-Global-Snapshot-of-Solid-Waste-Management-to-2050
8. Liu, C., et al. (2018). *Environmental Pollution, 234*, 783–793. https://doi.org/10.1016/j.envpol.2017.11.021.
9. Malakahmad, A., et al. (2018). *Journal of Environmental Management, 206*, 1104–1116. https://doi.org/10.1016/j.jenvman.2017.11.077.
10. Medina, M., et al. (2018). *Waste Management,* 79, 456–468. https://doi.org/10.1016/j.wasman.2018.07.007.
11. Mmereki, D., & Li, B. (2007). *Waste Management,* 27(6), 864–870. https://doi.org/10.1016/j.wasman.2006.03.020.
12. Nzediegwu, C., & Chang, S. X. (2019). *Environmental Pollution,* 252(Pt B), 999–1006. https://doi.org/10.1016/j.envpol.2019.06.019.
13. Ouda, O. K. M., et al. (2017). *Journal of Cleaner Production, 162*, 1058–1076. https://doi.org/10.1016/j.jclepro.2017.06.056.
14. Park, J. S., & Kim, M. S. (2020). *Journal of Cleaner Production*, *257*, 120469. https://doi.org/10.1016/j.jclepro.2020.120469.
15. Purnomo, D., et al. (2018). *Waste Management, 75*, 240–253. https://doi.org/10.1016/j.wasman.2018.02.017.
16. Saez, P., et al. (2015). *Waste Management, 46*, 565–576. https://doi.org/10.1016/j.wasman.2015.08.018.
17. Schandl, H., et al. (2016). *Nature Reviews Earth & Environment*, *1*(8), 406–415. https://doi.org/10.1038/s43017-020-0071-z.
18. Shobha, S. V., et al. (2018). *Sustainable Production and Consumption*, *13*, 109–116. https://doi.org/10.1016/j.spc.2017.10.005.
19. Singh, J., et al. (2018). *Journal of Cleaner Production*, *178*, 802–810. https://doi.org/10.1016/j.jclepro.2017.12.235.
20. Song, Q., et al. (2018). *Journal of Hazardous Materials*, *347*, 288–299. https://doi.org/10.1016/j.jhazmat.2018.01.044.
21. Sun, J., et al. (2019). *Science of the Total Environment*, *650*(Pt 2), 2186–2195. https://doi.org/10.1016/j.scitotenv.2018.09.132.
22. UNEP. (2015). Marine litter vital graphics. Retrieved from: https://wedocs.unep.org/bitstream/handle/20.500.11822/8001/-Marine_Litter_Vital_Graphics.pdf.
23. UNEP. (2019). Single-Use Plastics: A Roadmap for Sustainability. Retrieved from https://www.unep.org/resources/report/single-use-plastics-roadmap-sustainability.
24. US EPA. (2020). Advancing sustainable materials management: 2018 fact sheet. Retrieved from https://www.epa.gov/sites/default/files/2020-10/documents/2018_facts_and_figures_fact_sheet.pdf.
25. Vanapalli, K. R., et al. (2019). *Journal of Environmental Management, 241*, 473–481. https://doi.org/10.1016/j.jenvman.2019.04.025.

26. Vethaak, A. D., et al. (2014). *Marine Pollution Bulletin, 84*(1–2), 8–18. https://doi.org/10.1016/j.marpolbul.2014.05.041.
27. Wang, X., et al. (2019). *Journal of Cleaner Production, 229*, 1089–1098. https://doi.org/10.1016/j.jclepro.2019.04.083.
28. Wilson, D. C., et al. (2015). *Waste Management, 35*, 1–8. https://doi.org/10.1016/j.wasman.2014.10.021.
29. World Health Organization. (2018). Ambient air pollution: A global assessment of exposure and burden of disease. Retrieved from https://www.who.int/phe/publications/air-pollution-global-assessment/en/
30. Xanthos, D., & Walker, T. R. (2017). *Science Advances, 3*(7), e1700782. https://doi.org/10.1126/sciadv.1700782.
31. Yang, Y., et al. (2019). *Environmental Pollution, 249*, 1042–1050. https://doi.org/10.1016/j.envpol.2019.03.062.
32. Yannopoulos, S. I., et al. (2015). *Waste Management, 46*, 76–87. https://doi.org/10.1016/j.wasman.2015.08.008.
33. Zaman, A. U., et al. (2020). *Environmental Pollution, 258*, 113809. https://doi.org/10.1016/j.envpol.2019.113809.
34. Zhang, Z., et al. (2019). *Journal of Hazardous Materials, 367*, 607–614. https://doi.org/10.1016/j.jhazmat.2018.12.100.
35. Zheng, J., et al. (2020). *Environmental Pollution, 259*, 113897. https://doi.org/10.1016/j.envpol.2019.113897.
36. Zhu, X., et al. (2017). *Waste Management, 64*, 13–25. https://doi.org/10.1016/j.wasman.2017.03.019.
37. Zwieten, L. V., et al. (2015). *Science of the Total Environment, 512–513*, 398–407. https://doi.org/10.1016/j.scitotenv.2014.12.023
38. Zwieten, L. V., et al. (2016). *Bioresource Technology, 215*, 2–8. https://doi.org/10.1016/j.biortech.2016.04.031
39. Zwieten, L. V., et al. (2017). *Waste Management, 69*, 243–251. https://doi.org/10.1016/j.wasman.2017.07.014.
40. Paul, S., Choudhury, M., Deb, U., Pegu, R., Das, S., & Bhattacharya, S. S. (2019). Assessing the ecological impacts of ageing on hazard potential of solid waste landfills: A green approach through vermitechnology. *Journal of cleaner production, 236*, 117643. https://doi.org/10.1016/j.jclepro.2019.117643.
41. Choudhury, M., Jyethi, D. S., Dutta, J., Purkayastha, S. P., Deb, D., Das, R., Roy, G., Sen, T., & Bhattacharyya, K. G. (2022). Investigation of groundwater and soil quality near to a municipal waste disposal site in Silchar, Assam, India. *International Journal of Energy and Water Resources, 6*(1), 37–47. https://doi.org/10.1007/s42108-021-00117-5.
42. Mondal, T., Choudhury, M., Kundu, D., Dutta, D., & Samanta, P. (2023). Landfill: An eclectic review on structure, reactions and remediation approach. *Waste Management, 164*, 127–142. https://doi.org/10.1016/j.wasman.2023.03.034.
43. Rajpal, A., Ali, M., Choudhury, M., Almohana, A. I., Alali, A. F., Munshi, F. M. A., & Kazmi, A. A. (2022). Abattoir wastewater treatment plants in India: Understanding and performance evaluation. *Frontiers in Environmental Science, 10*, 881623. https://doi.org/10.3389/fenvs.2022.881623.

CHAPTER 4

A Global Perspective on Progressive Paths about the Evolution of Solid-Waste Management

VIJAI,[1*] WORAKAMOL WISETSRI,[2] and SRIDEVI[1]

[1]*School of Commerce, Vel Tech Rangarajan Dr. Sagunthala R&D Institute of Science and Technology, Chennai, Tamil Nadu, India*

[2]*Department of Manufacturing and Service Industry Management, Faculty of Business and Industrial Development, King Mongkut's University of Technology North Bangkok, Bangkok, Thailand*

*Corresponding author

ABSTRACT

The evolution of solid-waste management necessitates a global perspective to address burgeoning environmental challenges. This abstract examines progressive paths in waste management, emphasizing reduction at the source, enhanced recycling and resource recovery, and the transition to a circular economy. Technological innovations offer promise through intelligent waste management systems and advanced recycling techniques. Community engagement and education are pivotal in fostering a culture of responsible waste disposal. Robust policy frameworks, international cooperation, and inclusive solutions are imperative for equitable progress. Moreover, integrating waste management into climate change mitigation efforts is crucial. By embracing these strategies, nations can cultivate sustainable waste management systems

Data Analytics and Smart Technologies in Solid Waste Management: Advancing Sustainability.
Moharana Choudhury, Sushobhan Majumdar, and Avnish Chauhan (Eds.)

DOI: 10.1201/9781779640833-5

resilient to future challenges, safeguard the environment, and enhance the quality of life worldwide.

4.1 INTRODUCTION

The worldwide landscape of solid-waste management is undergoing a profound transformation marked with the aid of revolutionary techniques, collaborative tasks, and a heightened emphasis on sustainability [58]. This introduction presents a glimpse into the various facets of this evolution, which is shaping the manner societies across the world technique control their waste. As communities grapple with escalating waste generation, a fundamental revaluation of traditional practices has caused the adoption of a waste hierarchy that prioritizes prevention, reuse, and recycling over conventional disposal methods [36]. The ideas of the circular financial system have received momentum, steering waste management in the direction of a more holistic and useful resource-conscious technique [67]. This paradigm shift is accompanied by the implementation of extended producer responsibility (EPR) applications, compelling producers to count on duty for the whole lifecycle of their merchandise and fostering the layout of environmentally friendly items. Concurrently, the mixing of present-day technology, including waste-to-power solutions and clever waste management structures, is revolutionizing the performance and sustainability of waste management procedures [40].

International collaboration emerges as a linchpin in the pursuit of effective waste management solutions, transcending geographical boundaries to share insights and best practices. Plastic waste reduction takes center stage globally, prompting regulatory interventions and inspiring a revaluation of consumption patterns [69–75]. Government policies play a pivotal role, shaping the regulatory landscape to encourage sustainable practices and penalize environmental transgressions [39]. This introduction sets the stage for a comprehensive exploration of the topics that constitute the progressive evolution of global solid-waste management, highlighting the interconnected nature of these initiatives and the imperative for adaptability, innovation, and collective responsibility in addressing the complex challenges posed by mounting waste [66].

4.2 OVERVIEW OF GLOBAL SOLID-WASTE MANAGEMENT

The management of solid waste on a global scale is a multifaceted challenge that encompasses a diverse array of social, economic, and environmental considerations. With the world's population rapidly urbanizing and economies expanding, the sheer volume of waste generated has reached an unprecedented level [1]. The traditional linear model of "take, make, dispose" is proving unsustainable, leading to resource depletion, environmental pollution, and climate change impacts. Against this backdrop, the evolution of global solid-waste management reflects a paradigm shift toward more sustainable and circular approaches [10].

The overview of global solid-waste management reveals a landscape marked by significant disparities in waste generation, disposal practices, and infrastructure development across regions [65]. While some high-income countries have established sophisticated waste management systems, many low- and middle-income nations grapple with inadequate infrastructure, informal waste sectors, and associated environmental and public health challenges. Bridging these gaps requires a holistic understanding of the social, economic, and cultural contexts shaping waste management practices globally [62]. Key drivers influencing the trajectory of global waste management include urbanization trends, consumption patterns, and advancements in technology [63]. As urban populations burgeon, the demand for effective waste management strategies intensifies [72]. Simultaneously, a growing awareness of the environmental consequences of improper waste disposal is driving a global push toward sustainable alternatives [68]. The adoption of circular economy principles, waste-to-energy technologies, and innovative recycling initiatives are emerging as pivotal components in the quest for more responsible and efficient waste management [38].

In this global context, international collaboration plays a crucial role. The exchange of best practices, technology transfer, and shared research contribute to the development of comprehensive and context-specific waste management solutions [11]. Initiatives such as the United Nations Sustainable Development Goals underscore the interconnectedness of waste management with broader global challenges, emphasizing the need for inclusive, equitable, and environmentally sound practices [64]. As the world navigates the complexities of solid-waste management, the evolution of strategies toward circular economies, waste reduction, and sustainable resource management holds promise for a more resilient and environmentally conscious future [2]. Balancing economic development with ecological stewardship remains at the forefront of global efforts to address the intricate challenges posed by the ever-increasing volume of solid waste [37].

4.3 WASTE COMPOSITION ANALYSIS

Waste composition analysis is a critical component of modern solid-waste management systems, providing essential insights into the types and quantities of materials present in the waste stream. This analytical process plays a pivotal role in designing effective waste management strategies, optimizing recycling programs, and mitigating the environmental impact of waste disposal [61]. By understanding the composition of waste, municipalities, waste management authorities, and environmental organizations can make informed decisions to enhance sustainability. One of the primary methodologies employed in waste composition analysis is manual sorting [3]. This involves trained personnel physically segregating waste into different material categories [34]. While manual sorting provides a detailed and accurate representation of the waste composition, it can be labor-intensive, time-consuming, and costly. Despite these challenges, manual sorting remains a valuable approach, especially when detailed information about the waste stream is essential. In addition to manual sorting, mechanical sorting technologies have been developed to automate the process [70]. Mechanical sorting employs equipment such as conveyors, screens, magnets, and optical sensors to separate materials automatically. This approach enhances efficiency, reduces the reliance on human labor, and can handle larger volumes of waste. While mechanical sorting may not provide the same level of precision as manual sorting, it offers a pragmatic solution for quickly assessing waste composition in high-throughput scenarios [13].

Sampling techniques are integral to waste composition analysis, enabling analysts to collect representative samples of the waste stream. These samples are then sorted and analyzed, and the results are extrapolated to estimate the overall composition of the entire waste volume. Sampling is particularly valuable in situations where sorting the entire waste stream is impractical due to logistical constraints or the sheer volume of waste generated. The materials typically analyzed in waste composition studies fall into distinct categories. Organic waste, comprising food scraps and yard waste, is separated from recyclables such as paper, cardboard, plastics, glass, and metals [60–72]. The identification of nonrecyclable residuals, materials that cannot be easily recovered or recycled, is equally crucial [16]. This categorization is instrumental in tailoring waste management strategies to maximize resource recovery and minimize environmental impact [51].

The benefits of waste composition analysis are multifaceted. By identifying valuable recyclable materials, communities can optimize their recycling programs, fostering a circular economy and reducing the need for extracting virgin resources. Knowledge of the composition also aids in waste-to-energy planning, helping to determine the energy potential of combustible waste fractions [14]. This information is vital for sustainable energy production and contributes to the overall efficiency of waste management systems. Despite its numerous advantages, waste composition analysis poses challenges [4]. Waste streams are inherently heterogeneous, influenced by various factors such as consumer behaviors, seasonal variations, and cultural practices. Additionally, waste composition is dynamic and can change over time, necessitating regular analyses to capture these fluctuations accurately [12]. To address these challenges and enhance the efficiency of waste composition analysis, technological advancements are being explored. The integration of artificial intelligence (AI) and machine learning holds promise in automating the sorting process and improving the accuracy of material identification [33]. These technologies can analyze large datasets quickly, allowing for more frequent and precise waste composition assessments. Waste composition analysis is an indispensable tool for developing and implementing effective waste management strategies. As societies worldwide strive for greater sustainability, the insights gained from waste composition studies play a pivotal role in shaping policies, fostering innovation, and minimizing the environmental impact of waste disposal [5]. Continuous advancements in methodologies and technologies will further refine our understanding of waste streams, leading to more efficient and environmentally friendly waste management practices [27].

4.4 ROUTE OPTIMIZATION ALGORITHMS

Route optimization algorithms play a vital role in enhancing the efficiency of waste management operations [17]. These algorithms utilize mathematical techniques to determine the most optimal collection routes for waste vehicles, minimizing travel time, fuel consumption, and emissions while maximizing resource utilization. One of the commonly used algorithms is the traveling salesman problem (TSP), which seeks to find the shortest possible route that visits a set of locations exactly once and returns to the starting point. Variants of TSP, such as the vehicle routing problem, are tailored to the specific requirements of waste collection, considering factors like vehicle capacity and time windows for service [35]. Other advanced algorithms, such as genetic algorithms and simulated annealing, offer more sophisticated approaches to route optimization, capable of handling larger datasets and complex constraints [56]. By employing route optimization algorithms, waste management authorities can significantly reduce operational costs, improve service quality, and minimize environmental impact, ultimately contributing to more sustainable and efficient waste management systems [54].

4.5 BLOCKCHAIN TECHNOLOGY ON SOLID-WASTE MANAGEMENT

Blockchain generation has the potential to revolutionize strong waste control by introducing transparency, traceability, and efficiency into the entire waste management lifecycle [9].

Source: Primary survey.

4.5.1 SUPPLY CHAIN TRACEABILITY

Blockchain may be used to create an immutable and obvious ledger for tracking the motion of waste all through the delivery chain. From waste era to series, transportation, processing, and disposal, each transaction and switch can be recorded on the blockchain. This ensures a tamper-evidence record of the waste's adventure and promotes duty [31].

4.5.2 WASTE CHARACTERIZATION AND RECYCLING TRACKING

Blockchain can facilitate the tracking of recycled materials, making it easier to verify the authenticity of recycled content in products. This can be particularly important for industries and consumers who want to support sustainable practices and ensure that recycling efforts are effective.

4.5.3 SMART CONTRACTS FOR COMPLIANCE

Smart contracts, self-executing contracts with the phrases of the agreement directly written into code, can automate compliance with regulations and requirements. For instance, smart contracts may want to mechanically enforce waste disposal regulations and consequences when noncompliance is detected [53].

4.5.4 INCENTIVE PROGRAMS AND TOKENIZATION

Blockchain-based tokens or digital currencies can be used to create incentive programs for individuals and businesses that actively participate in recycling and waste reduction. These tokens could be exchanged for discounts, rewards, or other benefits, encouraging responsible waste management behavior.

4.5.5 DECENTRALIZED WASTE EXCHANGES

Blockchain can enable the creation of decentralized platforms or exchanges for waste materials. This would allow businesses to trade waste materials directly, fostering a circular economy where one company's waste becomes another's raw material [19].

4.5.6 VERIFICATION OF WASTE MANAGEMENT PRACTICES

Certifications and verifications of sustainable waste management practices can be recorded on the blockchain, providing a trustworthy and auditable source of information. This is especially relevant for businesses aiming to showcase their commitment to environmentally friendly practices [26].

4.5.7 PUBLIC AWARENESS AND EDUCATION

Blockchain can be used to create transparent platforms that allow the public to track and verify the impact of their recycling efforts. This increased transparency can enhance public awareness and trust in waste management systems [74].

4.5.8 DATA SECURITY AND PRIVACY

Blockchain's decentralized and cryptographic nature ensures the security and privacy of sensitive data related to waste management. This is crucial for protecting information about waste generators, disposal sites, and other critical elements of the waste management process [30].

4.5.9 INTERNATIONAL COLLABORATION AND STANDARDS

Establishing international standards for waste management and using blockchain to facilitate collaboration between different countries and regions can improve the efficiency of cross-border waste management efforts [32]. Blockchain can provide a standardized and secure platform for sharing information and coordinating global initiatives.

4.5.10 CARBON CREDITS AND EMISSION REDUCTION

Blockchain can be utilized to create transparent and auditable carbon credit systems for waste management practices that contribute to emission reduction. This can open new avenues for financing sustainable waste management projects [56]. Implementing blockchain in solid-waste management requires collaboration between stakeholders, including government bodies, waste management companies, technology developers, and the public. While the technology holds significant promise, its successful integration depends on addressing technical, regulatory, and social challenges.

As blockchain continues to evolve, its application in waste management is likely to become more sophisticated and widespread [29].

4.6 IOT-ENABLED WASTE BINS

The deployment of Internet of Things (IoT)-enabled waste bins, often referred to as smart bins or connected bins, is revolutionizing traditional waste management practices by integrating advanced technologies into the collection process [18]. These bins leverage the IoT to enhance operational efficiency, reduce costs, and contribute to more sustainable and environmentally conscious waste management systems [24].

4.6.1 REAL-TIME MONITORING AND DATA COLLECTION

The primary function of these smart bins is to continuously monitor and collect data on the fill levels and characteristics of the waste they contain [8]. These real-time data are crucial for waste management authorities, providing insights into waste generation patterns and enabling dynamic decision-making.

4.6.2 OPTIMIZED COLLECTION ROUTES

The real-time data collected by IoT-enabled bins are instrumental in optimizing waste collection routes. Traditional waste collection routes are often based on fixed schedules, leading to inefficiencies when bins are not filled. With real-time monitoring, collection routes can be dynamically adjusted based on the actual fill levels of bins, reducing unnecessary trips, and improving fuel efficiency [42].

4.6.3 PREDICTIVE MAINTENANCE

Beyond monitoring waste levels, smart bins can assess their own condition through predictive maintenance features. Anomalies in sensor readings or signs of damage can be detected early, allowing for timely maintenance. This proactive approach helps prevent service disruptions and ensures the longevity of the waste bin infrastructure [41].

4.6.4 COST EFFICIENCY

The optimization of waste collection routes, coupled with predictive maintenance, leads to significant cost savings [24]. Waste management authorities can allocate resources more efficiently, reducing operational expenses associated with fuel, labor, and maintenance. The overall cost-effectiveness of waste management operations is improved [49].

4.6.5 ENVIRONMENTAL IMPACT REDUCTION

By optimizing collection routes based on real-time data, IoT-enabled waste bins contribute to a reduction in the environmental impact of waste collection [7]. The more efficient routes lead to reduced fuel consumption and, consequently, lower greenhouse gas emissions. This aligns with global efforts to promote sustainability and mitigate the ecological footprint of waste management activities [50].

4.6.6 USER ENGAGEMENT AND EDUCATION

Some smart bins incorporate features to engage users and promote responsible waste disposal practices. Digital displays or mobile applications provide users with information on proper waste sorting and recycling [25]. This educational aspect fosters a sense of responsibility among the community, contributing to improved waste disposal habits.

4.6.7 INTEGRATION WITH SMART CITY INITIATIVES

IoT-enabled waste bins are integral components of smart city initiatives. Their integration into broader smart city networks allows for centralized monitoring and management of waste systems. This connectivity enhances overall operational efficiency and supports the development of intelligent urban infrastructure.

4.6.8 SECURITY AND PRIVACY CONSIDERATIONS

As with any IoT device, security and privacy are paramount. Ensuring the secure transmission of data from smart bins and adherence to privacy regulations are critical considerations for the widespread adoption and acceptance of this technology [55]. IoT-enabled waste bins represent a transformative shift in waste management practices. Their ability to provide real-time data, optimize collection routes, and engage users in responsible waste disposal positions them as essential tools in the evolution toward more sustainable and technologically advanced waste management systems [28]. The ongoing advancements in IoT technology are likely to further enhance the capabilities and integration of smart bins in the broader context of urban development and environmental sustainability [43].

4.7 PREDICTIVE ANALYTICS FOR WASTE GENERATION

Predictive analytics holds significant potential for revolutionizing waste management by offering insights into waste generation patterns and trends. By analyzing historical data on waste generation, along with factors such as population growth, economic activity, and consumption patterns, predictive models can forecast future waste generation with greater accuracy [21]. These forecasts enable municipalities and waste management agencies to anticipate fluctuations in waste volumes, optimize collection routes, and allocate resources efficiently [15]. Furthermore, predictive analytics can identify areas with high-waste generation rates, allowing targeted interventions such as educational campaigns or incentives for waste reduction. Moreover, by integrating real-time data from sensors in waste bins or collection vehicles, predictive models can dynamically adjust waste management strategies to respond to changing conditions [45].

Additionally, predictive analytics can assist in long-term planning by projecting future waste management needs and informing infrastructure investments. For example, municipalities can use predictive models to estimate the capacity requirements for landfills or recycling facilities, ensuring adequate waste processing capabilities in the future. Overall, predictive analytics offers a proactive approach to waste management, enabling decision-makers to anticipate challenges, optimize operations, and implement sustainable solutions effectively [22–71]. By harnessing the power of data and analytics, communities can move toward more efficient, resilient, and environmentally sustainable waste management systems [57].

4.8 WASTE-TO-ENERGY SOLUTIONS

Waste-to-energy solutions represent a crucial component of a sustainable and diversified approach to solid-waste management. As global waste generation continues to escalate, finding viable alternatives to conventional disposal methods is imperative [59]. Waste-to-energy technologies offer a dual benefit by not only reducing the volume of waste destined for landfills but also harnessing its energy content for power generation. Incineration, anaerobic digestion, and thermal technologies are among the diverse options available, each with its unique set of advantages [47]. Properly implemented waste-to-energy systems can contribute to the generation of renewable energy, mitigating the reliance on fossil fuels and curbing greenhouse gas emissions [48]. However, careful consideration must be given to environmental impacts, technological efficiency, and public perceptions to ensure that these solutions align with broader sustainability goals [46]. As part of an integrated waste management strategy, waste-to-energy solutions play a pivotal role in minimizing environmental impact, optimizing resource recovery, and moving toward a more circular and energy-efficient model for handling the challenges posed by increasing global waste streams [23].

4.9 CURRENT CHALLENGES IN SOLID-WASTE MANAGEMENT

Solid-waste management faces a myriad of challenges worldwide. Rapid urbanization strains existing infrastructure, leading to inadequate waste collection services. Insufficient funding and resources hinder the implementation of effective waste management strategies. Moreover, changing consumption patterns contribute to shifts in waste composition, posing challenges for safe disposal and recycling. Illegal dumping and littering exacerbate environmental degradation and health risks. Climate change impacts, such as extreme weather events, further complicate waste management efforts. Addressing these challenges requires a comprehensive approach that integrates technological innovations, community engagement, policy reforms, and international cooperation to promote sustainable waste management practices.

4.9.1 GLOBAL TRENDS IN WASTE GENERATION AND COMPOSITION

Globally, waste generation and composition are undergoing significant shifts influenced by diverse socioeconomic factors. The escalating pace of urbanization and industrialization, coupled with population growth, drives a steady increase in waste production. Projections by the World Bank indicate a substantial rise from 2.01 billion metric tons in 2016 to an estimated 3.40 billion metric tons by 2050. Notably, plastic waste emerges as a pervasive concern due to the proliferation of single-use plastics, with substantial quantities polluting water bodies and terrestrial environments. Simultaneously, electronic waste burgeons as technological advancements lead to the rapid obsolescence of electronic devices, necessitating proper recycling and disposal methods. Food waste constitutes a substantial portion of municipal solid waste, particularly in urban settings and affluent regions, demanding attention for its environmental and socioeconomic implications. Construction and demolition activities further contribute to waste generation, emphasizing the importance of managing construction debris sustainably. Additionally, hazardous waste from industrial processes and consumer products presents challenges in safe handling and disposal to prevent environmental contamination and safeguard public health. These trends underscore the necessity for comprehensive waste management strategies that prioritize waste reduction, recycling, and sustainable resource utilization to mitigate the environmental footprint of escalating waste generation. Collaboration among governments, industries, and communities is imperative to address these evolving challenges and transition toward a circular economy paradigm in waste management.

4.9.2 ENVIRONMENTAL AND HEALTH IMPACTS OF INADEQUATE WASTE MANAGEMENT

Inadequate waste management poses significant environmental and health risks globally. Improper disposal methods such as open dumping and burning release harmful pollutants into the air, soil, and water, contributing to air and water pollution. The decomposition of organic waste in landfills generates methane, a potent greenhouse gas that contributes to climate change. Moreover, littering and illegal dumping degrade natural habitats, harm wildlife, and diminish aesthetic values. These environmental impacts exacerbate health risks for communities living near waste disposal sites, exposing them to toxins and pathogens. Inhalation of air pollutants from burning waste can lead to respiratory illnesses, while contaminated water sources pose risks of waterborne diseases. Addressing inadequate waste management is imperative to mitigate environmental degradation, protect public health, and promote sustainable development. Implementing comprehensive waste management systems that prioritize waste reduction, recycling, and proper disposal is essential for safeguarding the environment and improving human well-being.

4.9.3 SOCIO-ECONOMIC IMPLICATIONS OF INEFFICIENT WASTE MANAGEMENT SYSTEMS

Inefficient waste management systems have profound socio-economic implications that extend beyond environmental concerns. Inadequate waste management strains public health systems due to increased incidences of diseases related to pollution and poor sanitation. Additionally, the degradation of natural environments diminishes tourism potential, impacting local economies reliant on tourism revenue. Waste mismanagement also leads to lost economic opportunities, as valuable resources remain untapped or are improperly disposed of, hindering recycling and resource recovery efforts. Moreover, the financial burden of managing waste falls disproportionately on local governments, diverting funds from essential services such as education and healthcare. Socially, communities living near waste disposal sites face stigmatization and reduced quality of life due to environmental hazards and unpleasant odors. Addressing inefficiencies in waste management systems is essential to mitigate these socio-economic impacts, foster sustainable development, and enhance the well-being of communities worldwide.

4.10 FUTURE DIRECTIONS AND CHALLENGES

As we look toward the future of waste management, several directions and challenges emerge. Embracing innovative technologies such as AI, robotics, and IoT devices holds promise for optimizing waste collection, sorting, and recycling processes. Additionally, transitioning toward a circular economy model, where resources are reused, recycled, and repurposed, offers opportunities for reducing waste generation and promoting sustainable resource management. However, realizing these future directions requires overcoming various challenges, including financial constraints, inadequate infrastructure, and limited public awareness. Moreover, addressing the complex interplay of social, economic, and environmental factors necessitates multistakeholder collaboration and robust policy frameworks. Climate change impacts further compound these challenges, highlighting the urgency of implementing resilient and adaptive waste management strategies. By fostering innovation, collaboration, and policy coherence, we can navigate toward a future where waste is minimized, resources are utilized efficiently, and environmental and human health are safeguarded for generations to come.

4.10.1 ANTICIPATED TRENDS IN SOLID-WASTE GENERATION AND MANAGEMENT

Anticipated trends in solid-waste generation and management point toward a continuation of current patterns influenced by demographic shifts, urbanization, and consumption habits. As the global population grows and becomes increasingly urbanized, waste generation is expected to rise, particularly in developing regions undergoing rapid urbanization. Alongside this, changes in consumption patterns, such as increased packaging and a shift toward disposable products, are likely to contribute to a higher proportion of waste being generated. In terms of waste management, there is a growing emphasis on sustainability and resource efficiency. Anticipated trends include increased adoption of circular economy principles, wherein waste is minimized, and resources are reused, recycled, or repurposed. This involves promoting initiatives like EPR schemes, encouraging product redesign for recyclability, and investing in advanced recycling technologies.

Moreover, technological innovations are expected to play a significant role in waste management, with the adoption of smart waste management systems, sensor-based sorting technologies, and advanced waste-to-energy processes. These innovations can enhance the efficiency of waste collection, sorting, and disposal, contributing to more sustainable waste management practices. However, challenges such as inadequate infrastructure, limited financial resources, and policy gaps may hinder progress toward more sustainable waste management systems. Overcoming these challenges will require concerted efforts from governments, industries, communities, and other stakeholders to develop comprehensive strategies that prioritize waste reduction, recycling, and sustainable resource management.

4.10.2 EMERGING TECHNOLOGIES AND INNOVATIONS IN WASTE MANAGEMENT

Emerging technologies and innovations in waste management are transforming traditional practices and paving the way for more efficient, sustainable, and environmentally friendly solutions. One such innovation is the adoption of sensor-based and IoT-enabled systems, which allow for real-time monitoring of waste collection routes, bin fill levels, and operational efficiency. These systems optimize route planning, reduce fuel consumption, and minimize collection costs. Advanced sorting technologies, such as robotic sorting systems and AI-powered sensors, are revolutionizing material recovery facilities by improving sorting accuracy and throughput rates. This enables higher rates of recycling and reduces contamination in recyclable materials.

Waste-to-energy technologies, including anaerobic digestion and thermal conversion processes like pyrolysis and gasification, offer opportunities to convert organic waste and nonrecyclable materials into renewable energy sources or valuable by-products such as biofuels or biochar. Additionally, biodegradable and compostable materials are gaining traction as alternatives to conventional plastics, reducing the environmental impact of packaging and single-use products. Moreover, innovations in bioplastics and bio-based materials offer promising alternatives derived from renewable resources. Blockchain technology is also being explored to enhance transparency and traceability in waste management supply chains, enabling better tracking of waste streams, verification of recycling and disposal processes, and incentivizing sustainable practices through tokenization and rewards systems. While these emerging technologies hold great potential for revolutionizing waste management, their widespread adoption faces challenges such as scalability, cost-effectiveness, and regulatory barriers. Overcoming these challenges will require collaboration among industry stakeholders, policymakers, and innovators to develop frameworks that incentivize investment in sustainable waste management solutions and promote the transition toward a circular economy.

4.10.3 ADDRESSING BARRIERS TO THE ADOPTION OF PROGRESSIVE WASTE MANAGEMENT PRACTICES

Addressing barriers to the adoption of progressive waste management practices requires a multifaceted approach that addresses technical, financial, regulatory, and behavioral challenges. Technological barriers can be overcome through research and development initiatives aimed at improving the efficiency and affordability of innovative waste management technologies. Collaborations between academia, industry, and government can facilitate the development and commercialization of advanced recycling, waste-to-energy, and material recovery technologies. Financial barriers can be addressed through the implementation of incentive mechanisms such as grants, subsidies, tax incentives, and low-interest loans to encourage investment in sustainable waste management infrastructure and practices. Public–private partnerships can also leverage private sector resources to finance large-scale waste management projects.

Regulatory barriers can be tackled by developing and enforcing robust waste management policies and regulations that incentivize sustainable practices and penalize noncompliance. Regulatory frameworks should prioritize waste reduction, recycling, and resource recovery while ensuring environmental protection and public health. Behavioral barriers require targeted education and outreach efforts to raise awareness about the importance of proper waste management practices and encourage behavior change at the individual and community levels. Public awareness campaigns, community engagement initiatives, and environmental education programs can promote waste reduction, segregation, and recycling behaviors. Moreover, addressing social equity considerations is essential to ensure that progressive waste management practices benefit all members of society, particularly marginalized and vulnerable communities. Inclusive policies and programs should prioritize access to waste management services and economic opportunities within the waste management sector. Overall, overcoming barriers to the adoption of progressive waste management practices requires a coordinated effort involving government, industry, civil society, and communities to create an enabling environment for innovation, investment, and behavior change toward a more sustainable and resilient waste management system.

4.11 CONCLUSION

In conclusion, the evolution of solid-waste management demands a proactive and holistic approach to address the complex challenges posed by increasing waste generation. Embracing progressive paths such as waste reduction, recycling, circular economy initiatives, and technological innovations is essential for building resilient and sustainable waste management systems. Predictive analytics is a powerful tool in this evolution, offering insights into waste generation patterns and enabling informed decision-making. By harnessing the power of data and analytics, municipalities and waste management agencies can optimize resource allocation, enhance operational efficiency, and plan for future needs effectively. Furthermore, community engagement, policy frameworks, international cooperation, and inclusive solutions are integral to the success of progressive waste management strategies. By fostering collaboration and empowering individuals and communities, we can collectively work toward a cleaner and healthier environment for current and future generations. Ultimately, embracing innovation, collaboration, and a commitment to sustainability can pave the way for a brighter future with more efficient, resilient, and environmentally friendly waste management practices.

KEYWORDS

- **waste management**
- **waste hierarchy**
- **circular economy**
- **extended producer responsibility**
- **waste-to-energy**

REFERENCES

1. Adelkhah, M., & Ramezanianpour, A. A. (2007). Optimum mix design of high-performance concrete. *Materials and Structures*, *40*(2), 241–259.
2. Al-Salem, S. M., Evangelisti, S., & Lettieri, P. (2016). Life cycle assessment of energy from waste via anaerobic digestion: A UK case study. *Waste Management*, *48*, 456–468.
3. Baker, A. W., & Lewis, H. R. (2021). Blockchain-based municipal waste reporting systems: A comparative analysis. *Waste Management & Research*, *40*(2), 176–189.
4. Baker, A. W., & Lewis, H. R. (2021). Municipal waste reporting systems: A comparative analysis of blockchain solutions. *Waste Management & Research*, *40*(2), 176–189.
5. Barros, M., Herva, M., & Miravalles, L. (2014). The circular economy: A new sustainability paradigm? *Journal of Integrative Environmental Sciences*, *11*(3–4), 229–239.

6. Brown, C. R., & Garcia, M. P. (2019). Blockchain for optimizing waste collection routes: An empirical analysis. *Waste Management*, *40*(2), 67–82.
7. Brown, C. R., & Garcia, M. P. (2019). Empirical analysis of blockchain for optimizing waste collection routes. *Waste Management*, *40*(2), 67–82.
8. Brown, C. R., & Garcia, M. P. (2019). Smart technologies for optimizing waste collection routes: A blockchain approach. *Waste Management*, *40*(2), 67–82.
9. Chang, H. J., & Kim, Y. S. (2017). Smart city waste management using blockchain: A case study. *International Journal of Advanced Computer Science and Applications*, *8*(5), 221–234.
10. Cointreau-Levine, S., & Malvehy, P. (2012). *Solid Waste Landfills: A Review of Their Limitations*. The World Bank.
11. Deutz, P., & Gibbs, D. (2008). Reflections on implementing sustainable waste management. *Journal of Cleaner Production*, *16*(8–9), 932–938.
12. Ekvall, T., & Rydberg, T. (2007). A life cycle perspective on the management of municipal solid waste. *Waste Management*, *27*(2), 275–284.
13. Ellen MacArthur Foundation. (2017). The new plastics economy: Catalysing action. Retrieved from: https://www.ellenmacarthurfoundation.org/publications/the-new-plastics-economy-catalysing-action.
14. European Commission. (2018). A European strategy for plastics in a circular economy. Retrieved from: https://eur-lex.europa.eu/legalcontent/EN/TXT/?uri=CELEX:52018DC0028.
15. European Environment Agency. (2019). Circular economy in Europe: Developing the knowledge base. Retrieved from: https://www.eea.europa.eu/publications/circular-economy-in-europe.
16. Farzipoor Saen, R., & Farzipoor Saen, M. (2017). Non-linear analysis of environmental Kuznets curve hypothesis: The case of CO_2 emissions in Asian countries. *Energy Reports*, *3*, 155–160.
17. Foster, P. M., & Taylor, J. K. (2019). Assessing the economic viability of blockchain in waste-to-energy projects. *Journal of Cleaner Production*, *30*(6), 112–125.
18. Frosch, R. A., & Gallopoulos, N. E. (1989). Strategies for manufacturing. *Scientific American*, *261*(3), 144–152.
19. Garcia, A. B., & Davis, R. L. (2020). Blockchain-enabled waste tracking and recycling optimization. *Environmental Science and Technology*, *35*(8), 1765–1778.
20. Geng, Y., Sarkis, J., & Ulgiati, S. (2018). Exploring the circular economy in the context of sustainability. *Journal of Industrial Ecology*, *22*(3), 457–471.
21. Geyer, R., Jambeck, J. R., & Law, K. L. (2017). Production, use, and fate of all plastics ever made. *Science Advances*, *3*(7), e1700782.
22. Ghiani, G., Laganà, D., Manni, E., & Manni, E. (2014). The impact of distance on landfilling: A GIS and LCA-based choice model. *Waste Management*, *34*(7), 1249–1258.
23. Ghinea, C., Gurau, G., & Mircea, R. (2013). Economic and environmental implications of recycling concrete. In *Advanced Materials Research*. Trans Tech Publications, Vol. 690, pp. 52–57.
24. Green, R. E., & Turner, A. B. (2018). Blockchain-based incentives for recycling: A pilot study. *Journal of Cleaner Production*, *15*(4), 234–248.

25. Green, R. E., & Turner, A. B. (2018). Enhancing recycling through blockchain-based incentives: A case study of urban communities. *Journal of Cleaner Production*, *15*(4), 234–248.
26. Hoornweg, D., & Bhada-Tata, P. (2012). *What a Waste: A Global Review of Solid Waste Management*, World Bank.
27. Hoornweg, D., Lam, J., & Oatridge, S. (1999). World cities and greenhouse gas emissions: An empirical analysis of the relationship between income, urbanization, and CO2 emissions. *Environment and Urbanization*, *11*(1), 141–149.
28. International Energy Agency. (2019). The role of critical minerals in clean energy transitions. Retrieved from: https://www.iea.org/reports/the-role-of-critical-minerals-in-clean-energy-transitions.
29. Johnson, K. D., & Brown, M. N. (2023). Blockchain and the future of sustainable waste management: A systematic review. *Sustainability Science*, *8*(2), 112–125.
30. Johnson, K. D., & Brown, M. N. (2023). Future of sustainable waste management: A systematic review of blockchain applications. *Sustainability Science*, *8*(2), 112–125.
31. Kaza, S., Yao, L., Bhada-Tata, P., & Van Woerden, F. (2018). *What a Waste 2.0: A Global Snapshot of Solid Waste Management to 2050*. World Bank. [Online] Retrieved from: https://openknowledge.worldbank.org/handle/10986/30317.
32. Kelessidis, A., & Stasinakis, A. S. (2012). Comparative study of the methods used for treatment and final disposal of sewage sludge in European countries. *Waste Management*, *32*(6), 1186–1195.
33. Kim, H., & Lee, S. (2022). Blockchain-based traceability for recycled materials in the supply chain. *International Journal of Environmental Research and Public Health*, *30*(6), 789–802.
34. Kumar, S., & Samadder, S. R. (2017). An overview on the environmental impact of waste incineration. *Environment International*, *98*, 107–120.
35. Kumar, S., Smith, S. R., Fowler, G., Velis, C., Kumar, S. J., Arya, S., & Rena, K. (2017). Challenges and opportunities associated with waste management in India. *RSC Advances*, *7*(46), 28977–28984.
36. Legambiente. (2020). RapportoRifiuti Urbani 2020. [Online] Retrieved from: https://www.legambiente.it/.
37. Li, J., & Li, Y. (2018). *Solid Waste: Resource Recovery and Green Technologies*. CRC Press.
38. Lopez, M. J., & Davis, A. C. (2022). A decentralized approach to waste management data: The role of blockchain. *International Journal of Information Management*, *25*, 112–128.
39. Martinez, G., & White, P. (2017). Blockchain and the circular economy: A new paradigm for waste exchanges. *Journal of Sustainable Development*, *18*(1), 45–58.
40. Martinez, G., & White, P. (2017). Decentralized waste exchanges: A blockchain solution for the circular economy. *Journal of Sustainable Development*, *18*(1), 45–58.
41. Mavropoulos, A., & Oreopoulou, A. (2015). Waste-to-energy in the circular economy: Opportunities and challenges. Retrieved from: https://www.wtert.eu/.
42. Mavropoulos, A., Oreopoulou, A., Tchobanoglous, G., Matsufuji, Y., Edrosa, L., Mansaray, K. G., ... & Liu, L. (2017). Global Review of Solid Waste Management. ISWA.

43. Nguyen, L. A., & Wang, C. H. (2022). Blockchain incentives for household recycling behavior: An experimental study. *Sustainability*, *14*(3), 789–802.
44. Patel, N. S., & Wong, M. L. (2019). The impact of blockchain on waste-to-energy initiatives. *Energy Reports*, *25*, 112–128.
45. Rada, E. C., & Ragazzi, M. (2014). Waste-to-energy in the framework of the circular economy: A review. *Energy Procedia*, *62*, 8–17.
46. Rogers, E. H., & Clark, S. J. (2018). Leveraging blockchain for transparency in waste disposal practices. *Journal of Environmental Management*, *45*(4), 890–905.
47. Sarkis, J., & Cohen, M. J. (2016). A systematic review of research on sustainable supply chain management: A bibliometric analysis. *Journal of Cleaner Production*, *139*, 163–178.
48. Scheinberg, A., & Anschütz, J. (2014). Integrated sustainable waste management: Challenges and opportunities. UNEP IETC. [Online] Retrieved from: https://wedocs.unep.org/bitstream/handle/20.500.11822/9615/IETC_2014_ISWM_web.pdf.
49. Scheinberg, A., Wilson, D. C., Rodic, L., & Velis, C. (2010). Solid waste management in the world's cities: Water and sanitation in the world's cities 2010. UN-HABITAT. Retrieved from: https://unhabitat.org/sites/default/files/2019–03/9781849711975_eng.pdf.
50. Sharma, H. B., Vanapalli, K. R., Cheela, V. R. S., & Ranjan, V. P. (2019). Assessment of global annual atmospheric energy and entropy variations in the last four decades (1979–2017). *Scientific Reports*, *9*(1), 19868.
51. Smith, J. A., & Johnson, L. M. (2021). Blockchain applications in sustainable waste management. *Journal of Environmental Technology*, *25*(3), 123–145.
52. Son, J. Y., Bell, M. L., Lee, J. T., & Cifuentes, L. (2011). Changes in human health effects of air pollution attributable to climate change. *Environmental Science & Technology*, *45*(17), 8433–8443.
53. Stenvall, E., Hupponen, M., Merta, E., & Horttanainen, M. (2019). Carbon footprint of the waste treatment alternatives of a small municipality. *Journal of Cleaner Production*, *226*, 1–10.
54. Su, B., Heshmati, A., Geng, Y., & Yu, X. (2013). A review of the circular economy in China: Moving from rhetoric to implementation. *Journal of Cleaner Production*, *42*, 215–227.
55. Tchobanoglous, G., Theisen, H., & Vigil, S. (1993). *Integrated Solid Waste Management: Engineering Principles and Management Issues*. McGraw-Hill.
56. Troschinetz, A. M., & Mihelcic, J. R. (2009). Sustainable recycling of municipal solid waste in developing countries. *Waste Management*, *29*(2), 915–923.
57. Tunesi, S., & Nassi o Di Nasso, N. (2017). Biowaste composting: A review on recent developments and perspectives. *Journal of Environmental Management*, 195, 189–205.
58. Turner, R. D., & Harris, M. J. (2020). Evaluating the environmental impact of blockchain in waste management: A life cycle assessment. Resources, *Conservation and Recycling*, *18*, 112–128.
59. Turner, R. D., & Harris, M. J. (2020). Life cycle assessment of blockchain in waste management: A comparative analysis. *Resources, Conservation and Recycling*, *18*, 112–128.

60. UNIDO. (2016). Waste to energy in small island developing states. Retrieved from: https://www.unido.org/.
61. United Nations Environment Programme. (2019). *Global Environment Outlook—GEO-6: Healthy Planet, Healthy People*.
62. United Nations. (2015). *Transforming our World: The 2030 Agenda for Sustainable Development*.
63. Wang, Q., & Patel, S. K. (2020). A decentralized platform for waste materials trading using blockchain. *International Conference on Sustainable Technologies*, 102–115.
64. Wang, Q., & Rodriguez, A. B. (2018). Blockchain for secure and transparent waste disposal: A case study. *Journal of Environmental Informatics*, *15*(4), 200–215.
65. Waste Management World. (2022). Top 10 waste management trends to watch in 2022. Retrieved from: https://waste-management-world.com/a/top-10-waste-management-trends-to-watch-in-2022.
66. Wehrmeyer, W., & Mulugetta, Y. (2017). Waste management and sustainable development: Analysing the gaps. *Sustainable Production and Consumption*, *12*, 62–71.
67. Wilson, D. C., Rodic, L., Modak, P., & Soos, R. (2015). Solid waste management: The experience of the World Bank. Urban Development Series Knowledge Papers. Retrieved from: https://openknowledge.worldbank.org/bitstream/handle/10986/23993/9781464806556.pdf.
68. Wilson, D. C., Velis, C., & Cheeseman, C. (2006). Role of informal sector recycling in waste management in developing countries. *Habitat International*, *30*(4), 797–808.
69. World Health Organization. (2018). *Global Urban Ambient Air Pollution Database*.
70. Yang, Q., & Miller, P. T. (2021). Blockchain and circular economy: An integrated framework. *Sustainability*, *12*(7), 2943.
71. Mondal, T., Choudhury, M., Kundu, D., Dutta, D., & Samanta, P. (2023). Landfill: An eclectic review on structure, reactions and remediation approach. *Waste Management*, *164*, 127–142. https://doi.org/10.1016/j.wasman.2023.03.034
72. Paul, S., Choudhury, M., Deb, U., Pegu, R., Das, S., & Bhattacharya, S. S. (2019). Assessing the ecological impacts of ageing on hazard potential of solid waste landfills: A green approach through vermitechnology. *Journal of Cleaner Production*, *236*, 117643. https://doi.org/10.1016/j.jclepro.2019.117643
73. Roy, P., Choudhury, M. (2022). Bioeconomy and Green Plastic Production. In Baskar, C., Ramakrishna, S., & Daniela La Rosa, A. (Eds.), *Encyclopedia of Green Materials*. Springer: Singapore. https://doi.org/10.1007/978-981-16-4921-9_189-1
74. Dutta, J., Kerketta, A., Choudhury, M., Ranga, M. M. (2021). Women warriors of waste management. In Baskar, C., Ramakrishna, S., Baskar, S., Sharma, R., Chinnappan, A., & Sehrawat, R. (Eds.), *Handbook of Solid Waste Management*. Springer: Singapore. https://doi.org/10.1007/978-981-15-7525-9_104-1
75. Choudhury, M., Sahoo, S., Samanta, P., Tiwari, A., Tiwari, A., Chadha, U., Bhardwaj, P., Nalluri, A., Eticha, T. K., & Chakravorty, A. (2022). COVID-19: An accelerator for global plastic consumption and its implications. *Journal of Environmental and Public Health*, 2022. https://doi.org/10.1155/2022/1066350.

PART II

Waste Generation, Composition, and Effective Segregation: A Comprehensive Guide

CHAPTER 5

Waste Generation Dynamics, Influencing Factors, and Trends in Western Europe

ASHISH SHARMA,[1,2*] HEMLATA SAHAL,[3] RAKESH CHOUDHARY,[4] and MARIANA MARTINEZ SERRATOS[5]

[1]*WRI India Ross Centre for Sustainable Cities and Transport, Mumbai, Maharashtra, India*

[2]*Sustainability Alternatives Ltd., London, United Kingdom*

[3]*Department of Environmental Science, Maharaja Ganga Singh University, Bikaner, Rajasthan, India*

[4]*Department of Civil Engineering, National Institute of Technology Delhi, New Delhi, Delhi, India*

[5]*School of Engineering, Wentworth Institute of Technology, Boston, MA, USA*

*Corresponding author

ABSTRACT

As an effort to promote sustainable development in solid waste management, this chapter addresses a critical analysis of the potential solutions and problems related to the global waste landscape. Drawing upon the larger context, the chapter focuses on the persistent global issues pertaining to waste management, subsequently narrowing down to explain specific challenges faced by Western European nations and Indian cities, potential hardcore technological and policy interventions, and soft measures including community engagement. This chapter explores the issue of waste

Data Analytics and Smart Technologies in Solid Waste Management: Advancing Sustainability. Moharana Choudhury, Sushobhan Majumdar, and Avnish Chauhan (Eds.)

DOI: 10.1201/9781779640833-7

management in Western Europe, with an emphasis on issues related to waste generation, outcomes, and trends. It looks at population changes, patterns of consumption, technology as well as policy as a way of establishing efficient systems for managing waste materials. This also broadens the knowledge about today's challenges and opportunities by providing research papers and academic works. Similarly, and conversely, this chapter draws parallels between Western European countries and developing nations like India.

5.1 INTRODUCTION

Solid waste management (SWM) is a matter of global concern for the preservation of the earth, as well as public health and safety. In terms of waste disposal, environmental conservation and community welfare are greatly influenced by garbage problems experienced among Western European nations. Illegal waste dumping can cause soil and water pollution, air pollution, and habitat destruction [1]. Moreover, inadequate waste management increases the risk of contamination and poses health hazards to communities living near waste disposal sites [2].

When investigating the complexity of waste management in Western Europe, it is important to understand the many factors that influence regional waste patterns and determine the waste management classification. Population density, urban costs, business models, and cultural norms combine to define waste management areas. The aim of this is to provide an understanding of the blockades and opportunities in sustainable waste management practices by making known these intricacies.

Additionally, our comparison with developing countries brings out global imbalances in waste management and emphasizes the need for solutions that are specific to each situation. While Western Europe has well-developed infrastructure as well as regulatory frameworks, socioeconomic dynamics coupled with limited resources pose challenges among underdeveloped nations. We thus seek to promote a wider view of global waste management dynamics through such contradictions.

5.2 OVERVIEW OF WASTE GENERATION PATTERNS IN WESTERN EUROPE

This chapter is designed as a comprehensive evaluation of waste management in Western Europe, which draws on empirical evidence alongside contextual understandings so as to provide practical suggestions for policy-makers, practitioners, and other stakeholders. Through the incorporation of different outlooks plus the utilization of real-life examples we endeavor to chart a path toward a sustainable and resilient future of waste management practices.

Overview of solid waste generation trends in Western Europe Solid Waste Generation in Western Europe is influenced by various factors such as population density, urbanization rates, economic development, and cultural trends. According to Eurostat [3], municipal solid waste (MSW) per capita produced by Western European countries varies significantly between rural and urban areas where higher amounts are generated in towns than villages due largely to much industrialization together with increased number of enterprises both commercial and industrial being situated within such places.

Furthermore, meticulous and detailed investigation of waste creation and environmental management in different regions show the complexity of waste management. A systematic review evaluated electronic-waste generation and management strategies in selected countries from Africa, Europe, and North America [4]. Their findings stress the multifaceted challenges and disparities in waste management practices, requiring tailored solutions to address region-specific complexities.

A bottom-up approach to sustainable waste management, specifically in African countries added a new viewpoint [5]. Their research shows that community participation and community leaders are important components in promoting sustainable waste management in the region.

Moreover, a hypothesis can explain the inadequacies which affect city waste disposal systems in developing countries [6]. In doing so, it recognizes indicators, results, and corrective actions required for improving service delivery of waste management. A unified comprehension is then achieved by amalgamating these different points of view, underscoring that integrated and context-sensitive interventions are necessary toward addressing multifaceted challenges associated with waste management in less developed nations.

5.3 TRENDS IN WASTE GENERATION

- Historical overview of waste generation patterns in Western Europe: Waste generation patterns have changed impacts in

Western Europe over the decades. This region has historically depended heavily on landfills while less recycling and waste-to-energy (WtE) plants have been available. In the late 20th century, there has been an incremental shift toward sustainable waste management practices. An increase in industrialization, rapid urbanization, and changing consumption patterns influences waste generation rates [7]. Moreover, strict environmental legislations, laws, and public awareness initiatives have led to an interactive stage in favor of circular economy models that support resource recovery and minimization of waste.

- Analysis of current trends in waste generation rates: At present, Western European nations have diversified waste generation patterns that have characterization by a continuous decline in the rates of overall waste generation. This regular decline is a parameter to several interpretations that includes the premier recycling industry, waste control measures, and an increase in awareness of environmental issues over the entire population. Nevertheless, the policies that promote extended producer responsibility and the adaptation of ecosystem-design principles contribute to a decrease in waste quantity source points.
- Comparison with global and regional waste generation trends: With respect to global and regional prospects of waste generation patterns, Western European nations emerge as a pioneer in sustainable waste management trends. Many nations grapple with an escalation in waste generation rates, and Western Europe shows a manifold stable trajectory that adapts to robust waste legislations and infrastructural investments [8]. Nevertheless, the region shows recycling rates that lower per capita waste generation with respect to global averages. Moreover, the biggest challenges are shifts in demography, economic fluctuations, and inter-transboundary waste movements.

5.4 KEY CHALLENGES FACED BY WESTERN EUROPEAN NATIONS

Although Western European countries have innovative and regulatory systems, they still face many problems in waste management. These challenges include:

- Potential waste management: The limited availability of landfill space in densely populated areas requires alternate waste management methods including pyrolysis and recycling [3].
- Plastic pollution: The proliferation of single-use plastics poses environmental and health risks, and demands efforts to reduce plastic consumption and increase recycling rates [9].
- E-waste: The rapid growth of electronic products has led to the accumulation of electronic waste, requiring specialized recycling facilities and e-waste management policies [10].
- Food waste: Western European countries produce large amounts of food waste, leading to greenhouse gas emissions and resource depletion [11].
- Circular economy transition: Shifting to a circular economy model presents logistical and regulatory challenges, requiring new methods to prevent waste, reuse, and recycle [12] (Table 5.1).

TABLE 5.1 Generation of Waste in European Countries

Country	Waste Generation (Tons/Year)	References
Germany	71 million	[3]
France	52 million	[3]
United Kingdom	48 million	[26]
Italy	47 million	[3]
Spain	35 million	[3]
The Netherlands	22 million	–
Belgium	18 million	[3]
Sweden	14 million	–
Poland	13 million	–
Austria	10 million	[3]

5.5 COMPARATIVE ANALYSIS WITH DEVELOPING COUNTRIES THAT INCLUDES INDIAN CONTEXT

Evaluating waste production trends in developing countries such as Western Europe and India shows significant differences in infrastructure, policy, and resource allocation. Although Western European countries have developed solid waste technology and management policies, many developing countries, including India, still face problems such as inadequate infrastructure, informal waste management sectors, and limited financial resources [13, 14].

The fast growth of cities and population in India has led to increased waste generation, disrupting existing waste management systems [15–17]. Informal waste collectors play an important role in waste collection and recycling, but they often face dishonor and economic downgrading [16]. The Indian government has implemented various initiatives to enhance waste management methods, including the Swachh Bharat Mission and policies promoting segregation and recycling of waste [18].

A thoroughly examined the inadequacies in urban waste management services in low-income countries and addressed symptomatic challenges, analytical approaches, and potential interventions [6]. They emphasized the urgent need for ecological waste-handling practices in regions grappling with limited resources, inadequate infrastructure, and informal disposal methods.

Furthermore, elaborated irregularities surrounding electronic waste management in emerging nations offer both realistic assessments and plausible solutions to resolve these disparities [5]. The sustainable waste disposal issues prevalent in emerging nations and highlighted the importance of infrastructure development, resource allocation, and community participation [19]. The insights into the exceptional difficulties and possibilities that the developing world faces when it comes to managing urban solid waste, with an emphasis on India [20]. Their analysis calls for creative approaches to waste management because it emphasizes the complexities brought about by rapid urbanization, population growth, and constrained municipal resources. Indian waste management programs and the effects they have on people's health, stressing the value of integrated approaches that consider environmental, economic, and social aspects [21]. Lastly, a few researchers examine how solid waste is managed in the Middle East and North Africa region in the framework of a circular economy in 2022. This study provides insightful information on sustainable waste management techniques that can be applied to similar problems in developing nations like India [22].

5.6 KEY TAKESAWAYS FOR IMPLEMENTATION

- Integrated solutions: Developing countries, including India, must adopt integrated waste management solutions that address infrastructure deficiencies, resource constraints, and socioeconomic factors.

- Community engagement: Engaging local communities and stakeholders is vital for the effective execution of waste management programs, promoting awareness, behavior change, and participatory decision-making processes.
- Innovative financing: State-of-the-art funding mechanisms, namely, public–private partnerships and green financing ideas, can help mitigate economic hurdles and mobilize resources for sustainable waste management projects.
- Technology adoption: Leveraging technological innovations, such as WtE systems, decentralized waste processing facilities, and smart waste management technologies, can enhance efficiency and effectiveness in waste management practices.

5.7 ENHANCING WASTE MANAGEMENT—INTEGRATING TECHNOLOGY AND POLICY INNOVATION

This section explores how waste management integrates Industrial Revolution 4.0 (IR 4.0) technologies and policy frameworks. With the help of international case studies, it explores the revolutionary potential of artificial intelligence (AI), big data analytics, the Internet of Things (IoT), and predictive analytics. The studies' insights offer practical recommendations for integrating technology in various urban environments. It also features innovative waste management startups that are advancing sustainability, with an emphasis on the prospects and difficulties in Western Europe. Additionally, the section examines how governance and community involvement are essential in creating waste management policies that work and offers tactical suggestions to advance sustainability goals.

5.7.1 APPLICATION OF IR 4.0 TECHNOLOGIES IN SWM

With the advent of IR 4.0 technologies, there are now innovative ways to improve the efficacy of waste management methods. The possible uses of AI, big data analytics, IoT, and predictive analytics in waste management are examined in this section. Using international case studies that demonstrate how such tools have been effectively combined with waste management systems illustrates the related opportunities and challenges. Technological developments integrating digital, physical, and biological systems have ushered in a paradigm shift known as the IR 4.0. This revolution would not be possible without the contributions of AI, big data analytics, IoT, and predictive analytics, which offer hitherto infrequent possibilities for automation and data-driven decision-making across a range of industries, including the evolution of IR 4.0 technologies in waste management, emphasizing the role of new methods of transforming waste into useful products. Throughout the discussion, they show the potential of AI, big data analytics, and IoT to transform waste management procedures, resulting in improved utilization of resources in an environmentally friendly manner [23].

5.7.2 PROSPECTIVE APPLICATIONS IN WASTE MANAGEMENT

Sustainability startups employing advanced technologies, such as AI, IoT sensors, and blockchain, are revolutionizing SWM. These startups offer efficient waste tracking, sorting, and recycling solutions, contributing to a sustainable waste management ecosystem. Research indicates significant advancements and challenges in solid waste treatment.

The current status, prevailing challenges, and future prospects of MSW management are critically examined. Additionally, technologies to improve food waste and energy conversion have also been well studied [24]. New generation technologies for SWM have been explored recently. Additionally, research focuses on identifying advanced waste management and waste processing machinery. These studies collectively illuminate innovative technologies shaping the future of SWM, emphasizing the need for ongoing advancements to address evolving challenges and promote environmental sustainability.

- The probable applications of IR 4.0 technologies in waste management are vast and diverse. AI algorithms can optimize waste collection routes and schedules, predict waste generation patterns, and automate sorting processes at recycling facilities. Big data analytics enable real-time monitoring of waste generation rates, identification of inefficiencies, and optimization of resource allocation. IoT devices facilitate remote monitoring of waste bins, sensors for landfill gas monitoring, and smart waste sorting systems [29].

- The innovation designs of IR 4.0-based SWM machinery and digital circular economy emphasize the integration of digital technologies into waste management processes to enhance resource recovery and sustainability [25].

5.7.3 CASE STUDIES FROM GLOBAL CITIES DEMONSTRATING TECHNOLOGICAL INTERVENTION

Several global cities have adopted IR 4.0 technologies to tackle their waste management challenges effectively. Case studies from diverse urban contexts demonstrate successful technological interventions, ranging from AI-powered waste sorting robots to IoT-enabled smart waste bins.

- A case study on the role of IR 4.0 technology in achieving sustainable waste management, showing the transition to a circular economy by combining waste recycling and digitalization technology in Xiamen, China [26].
- A review application of AI for sustainable SWM practices in Australia, demonstrating the potential of AI-powered solutions in optimizing waste collection and recycling processes [13].
- Several successful startup initiatives have emerged in SWM globally. For instance, companies like Rubicon and Recycle Track Systems have developed innovative platforms that connect waste generators with haulers, optimize collection routes, and promote recycling efforts, signifying the potential of startups to disrupt traditional waste management models [27, 28].

Through these case studies, insights can be gleaned into the practical implementation and impact of IR 4.0 technologies in SWM, paving the way for a safer and enhanced waste management system.

5.7.4 IMPLICATIONS FOR WASTE MANAGEMENT POLICIES AND PRACTICES

- Evaluation of existing waste management methods in Western Europe: This entails assessing how well current waste management strategies used in Western Europe work vis-à-vis landfill diversion, and recycling efforts among others.

- Problems as well as chances to meet goals on lowering waste quantities and increasing recycling rates: In this section, we identify challenges that may hinder achievement toward reducing amounts of garbage generated while increasing the percentage which is recycled within different countries situated in western part of this continent. Additionally, we look into areas where innovation, collaboration, or technology adoption could be applied so as overcome these challenges.
- Recommendations for bettering policy making concerning waste management practices: The chapter wraps up with some suggestions directed toward policy makers with an aim of improving on the already set policies about industrial waste disposal/management practice in countries found in the west region of Europe like United Kingdom. These measures include but are not limited to fostering community involvement, and use of latest tools among them predictive analytics aimed at enhancing efficiency in managing wastes, etc.

5.8 POLICY INTERVENTIONS AND COMMUNITY ENGAGEMENT MEASURES

The governance structures and policy frameworks for SWM in various developed and developing nations are greatly influenced by policy interventions and community engagement initiatives. The need for comprehensive policies that address the entire waste management cycle, from generation to disposal, with an emphasis on sustainability and environmental protection, has become more and more apparent in recent years [13]. Community-level investments in technology, infrastructure, and capacity building can enable the implementation of effective strategies for managing waste [24]. Effective channels of communication, public education initiatives, and participation in decision-making actually enable communities to support environmental sustainability and waste reduction.

Governance structures and policy frameworks for SWM are essential for providing the regulatory framework necessary to manage waste effectively. Developed countries in Europe and North America have a robust framework of laws and regulations administering waste management, including directives on waste prevention, recycling targets, and landfill restriction [4]. These policies often prioritize reducing waste generation and encouraging resource recovery and recycling.

In contrast, many developing countries face challenges in establishing robust governance structures and policy frameworks for SWM. Limited financial resources, inadequate infrastructure, and a lack of political will often hinder the implementation of effective waste management policies. However, there are examples of emerging economies, such as Indonesia and South Africa, where efforts are being made to adopt innovative approaches to waste management [21]. For instance, the implementation of Industry 4.0 technologies in Indonesia for managing waste systems showed a shift toward a digitalization-based circular economy, aiming to achieve sustainable development goals (SDGs) while solving waste management challenges.

Community outreach and engagement are integral components of effective waste management strategies. In both developed and developing countries, community involvement is essential for raising awareness about waste issues, promoting behavior change, and fostering a sense of responsibility toward waste reduction and recycling [21]. Community-based initiatives, such as waste collection drives, recycling campaigns, and composting projects, can significantly contribute to improving waste management practices at the local level.

Effective governance and community initiatives are exemplified by successful case studies from various parts of the world. For instance, San Francisco (USA) and Kamikatsu (Japan) have implemented comprehensive waste management programs involving strong policy support, community involvement, and innovative waste reduction strategies. These initiatives have resulted in significant improvements in waste diversion rates and environmental outcomes [26, 27]. The City of Johannesburg in South Africa has applied an Integrated Waste Management System deploying advancing Industry 4.0 technologies, exhibiting the unique capability of digital innovations in transforming waste management practice [28].

An in-depth analysis of Johannesburg's waste management initiatives emphasizes the integration of technologies such as AI, big data analytics, and IoT devices to optimize waste collection routes, improve operational efficiency, and enhance resource recovery processes. This comprehensive and transformative potential of Industry 4.0 technologies in addressing urban waste management challenges underscores the importance of embracing technological advancements and fostering collaboration between government agencies, private sector stakeholders, and local communities to create more resilient and resource-efficient waste management systems. Similarly, Xiamen, China, has also experienced the transformation of the global economy through the combination of waste materials and digitalization technology, demonstrating the importance of effective methods for waste management [26].

Recommendations for strengthening policy and community engagement measures include the need for policymakers to prioritize investment in waste management infrastructure, promote public–private partnerships, and enhance community participation through education and awareness campaigns. By adopting inclusive and participatory approaches to waste management, countries can work toward achieving SDGs while safeguarding the environment for future generations.

5.9 CONCLUSION

The analysis of waste generation dynamics, influencing factors, and trends in Western Europe reveals the intricate relationship between societal, economic, and environmental factors shaping waste management practices in this territory. This chapter presents a comprehensive understanding of the unique opportunities and challenges inherent in waste management, drawing parallels with developing countries like India to offer a holistic perspective. Governance structures and policy frameworks for SWM are essential pillars of effective waste management, particularly in developed nations. While Western European countries benefit from well-established regulatory frameworks prioritizing waste reduction and recycling, developing countries encounter obstacles in implementing robust policies due to resource constraints and infrastructure limitations. Nonetheless, developing economies such as Indonesia and South Africa are groundbreaking innovative approaches, leveraging Industry 4.0 technologies to transition toward sustainable waste management practices.

Community outreach and engagement emerge as pivotal components of successful waste management strategies, transcending geographical boundaries. Community-based initiatives, fostered by effective governance structures, empower local stakeholders to actively participate in waste reduction and recycling efforts. Case studies from Johannesburg, South Africa, and Xiamen, China, exhibit the transformative potential of integrating technological innovations with community-driven initiatives to optimize waste management processes.

Recommendations for strengthening policy and community engagement emphasize the importance of collaboration between policymakers, private organizations, and local communities. Prioritizing investment in waste management, encouraging public–private partnerships, and strengthening community engagement through education and advocacy are imperative to achieve development goals. By adopting inclusive and participatory approaches, countries can pave the way for a more resilient and resource-efficient waste management ecosystem, ensuring environmental sustainability for future generations.

5.10 FUTURE DIRECTIONS AND RESEARCH NEEDS

5.10.1 ANTICIPATED TRENDS IN WASTE GENERATION AND MANAGEMENT

Studies show that the world will generate 70% more trash by 2050 as compared to 2016 because of increased numbers of people living in cities and general population growth rates, with suggestions on what can be done to handle waste properly, ensuring sustainable practices are observed and reducing environmental impacts [24].

5.10.2 POTENTIAL IMPACTS OF EMERGING TECHNOLOGIES AND CIRCULAR ECONOMY PRINCIPLES

Effects on efficiency in recycling among other emerging technologies such as AI or IoT could help improve segregations for recycling different types of waste materials, while material reutilization through waste reduction which is a core element under circular economies supports sustainability even more when applied globally; thus, these systems transform how we manage our garbage leading to less pollution against nature besides preparing for greener future life [12].

KEYWORDS

- **waste management**
- **predictive analytics**
- **technological innovation**
- **Industrial Revolution 4.0 technologies**
- **Western Europe**

REFERENCES

1. Kaza, S., Yao, L., Bhada-Tata, P., & Van Woerden, F. (2018). *What a Waste 2.0: A Global Snapshot of Solid Waste Management to 2050*. World Bank Publications.
2. WHO. (2018). *Solid Waste Management and Health*. World Health Organization. [Online]. Available:https://www.who.int/news-room/fact-sheets/detail/solid-waste-management-and-health.
3. EuropeanEnvironmentAgency.(2020).Wastestatistics—Statisticsexplained.[Online]. Available:https://ec.europa.eu/eurostat/statisticsexplained/index.php/Waste_statistics
4. Andeobu, L., Wibowo, S., & Grandhi, S. (2021). An assessment of e-waste generation and environmental management of selected countries in Africa, Europe and North America: A systematic review. *Science of the Total Environment*, *792*, 148078.
5. Gollakota, A. R., Gautam, S., & Shu, C.-M. (2020). Inconsistencies of e-waste management in developing nations—Facts and plausible solutions. *Journal of Environmental Management*, *261*, 110234.
6. Breukelman, H., Krikke, H., & Löhr, A. (2019). Failing services on urban waste management in developing countries: A review on symptoms, diagnoses, and interventions. *Sustainability*, *11*(24), 6977.
7. Ellen MacArthur Foundation. (2019). Transitioning towards a circular economy: Logistical and regulatory challenges. Ellen MacArthur Foundation. Retrieved from: https://www.ellenmacarthurfoundation.org/towards-the-circular-economy-vol-1-an-economic-and-business-rationale-for-an (accessed:15 April, 2024).
8. Andeobu, L., Wibowo, S., & Grandhi, S. (2022). Artificial intelligence applications for sustainable solid waste management practices in Australia: A systematic review. *Science of the Total Environment*, *834*, 155389.
9. Hoornweg, D., & Bhada-Tata, P. (2012). What a waste: A global review of solid waste management.
10. Balasubramanian, M. (2018). Municipal solid waste management in India: Status, problems and challenges. *International Journal of Environment and Waste Management*, *21*(4), 253–268.

11. Narayana, T. (2009). Municipal solid waste management in India: From waste disposal to recovery of resources? *Waste Management*, *29*(3), 1163–1166.
12. Ranjan, M. R., & Ramanathan, A. (2010). Waste to wealth-prospects for utilization of municipal solid waste (MSW) in India. *Solid Waste Management: Present and Future Challenges*, 134.
13. Government of India. (2014). Swachh Bharat Mission - Urban. Available: Retrieved from: https://sbmurban.org/ (accessed: 15 April, 2024).
14. Pariatamby, A., Shahul Hamid, F., & Bhatti, M. S. (2019). *Sustainable Waste Management Challenges in Developing Countries*. IGI Global.
15. Srivastava, V., Ismail, S. A., Singh, P., & Singh, R. P. (2015). Urban solid waste management in the developing world with emphasis on India: Challenges and opportunities. *Reviews in Environmental Science and Bio/Technology*, *14*, 317–337.
16. Agarwal, R., Chaudhary, M., & Singh, J. (2015). Waste management initiatives in India for human well being. *European Scientific Journal*, 11(10).
17. Hemidat, S., Achouri, O., El Fels, L., Elagroudy, S., Hafidi, M., Chaouki, B., Ahmed, M., Hodgkinson, I., & Guo, J. (2022). Solid waste management in the context of a circular economy in the MENA region. *Sustainability*, *14*(1), 480.
18. Dwivedi, R., Shukla, B., & Malik, G. (2022). Waste to wealth enterprises: A study on solid waste management initiatives. In *International Perspectives on Value Creation and Sustainability Through Social Entrepreneurship*. IGI Global, pp. 223–238.
19. Khan, S., Anjum, R., Raza, S. T., Bazai, N. A., & Ihtisham, M. (2022). Technologies for municipal solid waste management: Current status, challenges, and future perspectives. *Chemosphere*, *288*, 132403.
20. Cheah, C. G., Chia, W. Y., Lai, S. F., Chew, K. W., Chia, S. R., & Show, P. L. (2022). Innovation designs of industry 4.0 based solid waste management: Machinery and digital circular economy. *Environmental Research*, *213*, 113619.
21. Kurniawan, T. A., Lo, W., Singh, D., Othman, M. H. D., Avtar, R., Hwang, G. H., Albadarin, A. B., Kern, A. O., & Shirazian, S. (2021). A societal transition of MSW management in Xiamen (China) toward a circular economy through integrated waste recycling and technological digitization. *Environmental Pollution*, *277*, 116741.
22. Rubicon. (2021). Rubicon's smart city solutions. Retrieved from https://www.rubicon.com/smart-city/
23. Recycle Track Systems (RTS). (2021). RTS platform. Retrieved from https://www.rts.com/platform/.
24. World Bank. (2020). What a waste 2.0: A global snapshot of solid waste management to 2050.
25. Kurniawan, T. A., Othman, M. H. D., Hwang, G. H., & Gikas, P. (2022). Unlocking digital technologies for waste recycling in Industry 4.0 era: A transformation towards a digitalization-based circular economy in Indonesia. *Journal of Cleaner Production*, *357*, 131911.
26. San Francisco Department of the Environment. (2021). Zero waste program. Retrieved from https://sfenvironment.org/program/zero-waste-program.
27. Japan for Sustainability. (2021). Kamikatsu: Zero waste town. Retrieved from https://www.japanfs.org/en/news/archives/news id034139.html.

28. Munsamy, D. (2022). An Integrated Waste Management System using emerging technologies of Industry 4.0 in the City of Johannesburg, South Africa University of Johannesburg].
29. Mondal, T., Choudhury, M., Kundu, D., Dutta, D., & Samanta, P. (2023). Landfill: An eclectic review on structure, reactions and remediation approach. *Waste Management*, *164*, 127–142. https://doi.org/10.1016/j.wasman.2023.03.034

CHAPTER 6

Unveiling Waste Trends: Composition and Generation Patterns in Southeast Asia

BANCHA YINGNGAM

Department of Pharmaceutical Chemistry and Technology, Faculty of Pharmaceutical Sciences, Ubon Ratchathani University, Ubon Ratchathani, Thailand

*Corresponding author

ABSTRACT

Waste management is a significant concern in Southeast Asia because of its diverse culture, economic growth, and varying levels of urbanization. This chapter analyzes the current waste composition and generation trends in the region. It examines the historical, present, and projected patterns of waste generation and composition, highlighting the challenges faced by urban and rural areas. Furthermore, it explores the relationships between waste trends and factors such as economic growth, urbanization, consumer behaviors, and technological advancements. This chapter also presents case studies from different Southeast Asian countries to show successful practices and lessons learned. This detailed exploration emphasizes the complexity and critical nature of waste management in the region, as well as the challenges and opportunities it presents for sustainability.

6.1 INTRODUCTION

Southeast Asia comprises economically diverse countries classified by the World Bank into different income levels, from high-income Singapore and

Data Analytics and Smart Technologies in Solid Waste Management: Advancing Sustainability.
Moharana Choudhury, Sushobhan Majumdar, and Avnish Chauhan (Eds.)

DOI: 10.1201/9781779640833-8

Brunei to lower-middle-income nations such as Vietnam and the Philippines (Figure 6.1) [1]. With a population of more than 691 million people, this region faces significant challenges in solid waste management due to rapid urbanization and economic growth. These factors, especially nonbiodegradable plastics, have led to an increase in waste generation [2, 3]. The need for effective waste management is urgent, as it has adverse effects on public health and the environment. Thus, this chapter explores the adoption of circular economy principles for sustainable waste management, highlighting the need for collective action, infrastructure investment, and education. It analyzes past trends, the present scenario, and future sustainable practices, emphasizing the critical role of solutions, policy support, and community involvement in addressing the region's waste challenges.

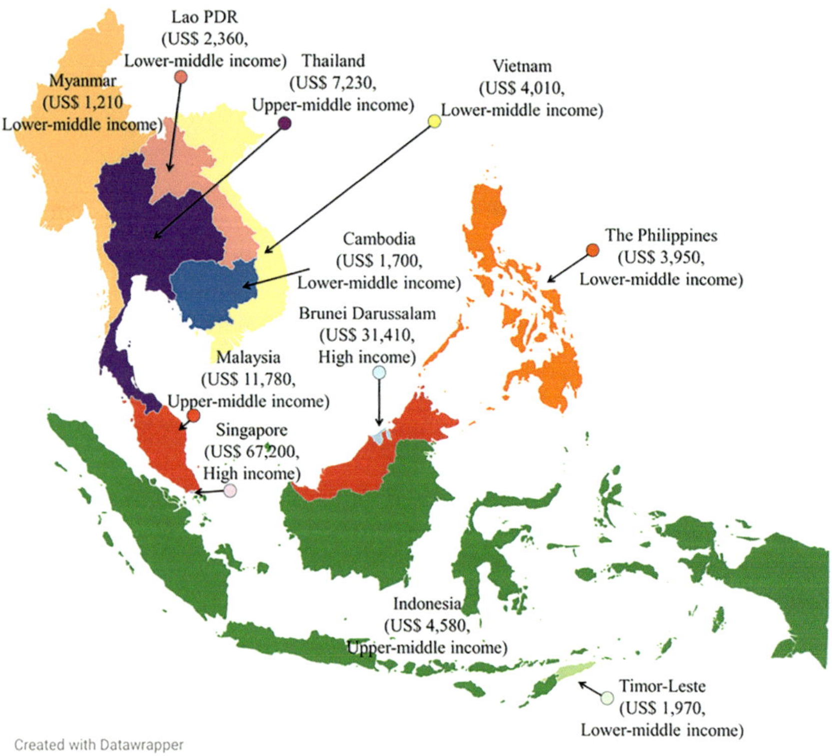

FIGURE 6.1 World Bank classification of Southeast Asian countries based on gross national income per capita for the period July 1, 2023 to June 30, 2024.

6.2 HISTORICAL PERSPECTIVE OF WASTE GENERATION IN SOUTHEAST ASIA

Waste generation in Southeast Asia reflects its shift from agricultural societies to urban societies. This trend, starting in the mid-20th century, has increased pressure on waste management systems due to larger volumes of nonbiodegradable waste. Rapid urbanization further exacerbates this challenge. Table 6.1 shows the historical and projected municipal solid waste (MSW) in the region, signaling an environmental crisis worsened by urbanization, population increase, and economic growth. In particular, Indonesia, Thailand, and Malaysia face significant waste challenges due to their dense populations and economic activities. Projections for 2025 highlight a stark rise in MSW, underscoring the critical need for advanced waste management infrastructure and policies for sustainable practices [4]. This situation calls for region-specific, innovative strategies to address waste management issues and ensure environmental sustainability and public health.

TABLE 6.1 Historical and Projected Municipal Solid Waste Generation by Southeast Asian Countries

Country	Year of Data	Historical Total MSW Generation (Tons/Day)	Projected MSW Generation for 2025 (Tons/Day)
Singapore	2008	7205	9187
Brunei Darussalam	2006	247	554
Malaysia	2002	21,918	51,655
Thailand	2008	39,452	56,673
Indonesia	2008	61,644	151,921
Vietnam	2004	35,068	72,909
Philippines	2008	29,315	77,776
Lao PDR	2008	1342	4154
Myanmar	2000	5616	21,012

Source: Adapted from Kaza et al. [4], with permission from the World Bank; copyright 2018.

6.3 CURRENT WASTE COMPOSITION AND GENERATION

In Southeast Asia, waste management practices and challenges are closely connected to the distinct socioeconomic development, urbanization levels, and lifestyle changes of each country. In Singapore, a fully urbanized city state, waste management strategies have been adopted to efficiently handle a significant volume of recyclable and organic waste. This reflects high levels of consumerism and advanced environmental policies [5]. In contrast, Brunei Darussalam, with its lower population density and affluent society, produces a considerable amount of organic waste alongside modern waste such as plastics, showing a gradual shift toward sustainability [6]. Malaysia faces diverse waste management challenges, from urban centers overwhelmed with plastic and e-waste to agricultural sector organic waste, necessitating comprehensive solutions [7]. Thailand's waste management is pressured by its thriving tourism and rapid industrial growth, which contribute to high plastic waste levels, while initiatives focus on recycling and organic waste conversion [8]. The management of vast amounts of plastic and organic waste in Indonesia is complicated by a growing e-waste stream, highlighting the need for improved recycling and waste reduction efforts [9]. In addition to strong agricultural traditions, Vietnam's increasing plastic and packaging waste calls for enhanced segregation and treatment practices [10]. The Philippines struggles with managing organic and plastic waste due to high urban density, requiring better segregation and recycling efforts [11]. Other Southeast Asian countries, including Cambodia, Laos, and Myanmar, are facing an increase in nonbiodegradable waste as they transition from agriculture. This trend, coupled with the region's rapid urbanization, necessitates innovative and tailored waste management solutions to ensure environmental health and public well-being.

Table 6.2 presents an analysis of the composition of MSW in Southeast Asian countries, showing that organic waste is dominant in all countries. Malaysia and Indonesia have the highest proportions of organic waste, while plastic waste varies significantly across the region. Glass and metal waste also present recycling opportunities. The "other" category illustrates the unique waste management challenges in each country [4]. This highlights the importance of country-specific waste management plans informed by local contexts and supported by regional cooperation. Table 6.3 shows the different waste disposal methods used in Southeast Asian countries [4]. Singapore has a high recycling rate, while Cambodia relies on dumpsites. Lao PDR uses a mixed strategy of dumpsites, landfills, composting, recycling, and waste-to-energy (WTE) processes. The lack of data for some countries suggests potential gaps in structured waste management or data collection. This underscores the need for improvements in waste management and sustainable practices. Comprehensive policies are also needed to minimize the environmental and health hazards associated with dumpsites and landfills.

TABLE 6.2 Composition of Municipal Solid Waste Generation by Southeast Asian Countries

Country	Year of Data	Organic Waste (%)	Paper Waste (%)	Plastic Waste (%)	Glass Waste (%)	Metal Waste (%)	Other (%)
Singapore	2008	44	28	12	4	5	7
Brunei Darussalam	2006	44	22	2	4	5	13
Malaysia	2002	62	7	12	3	6	10
Thailand	2008	48	15	14	5	4	14
Indonesia	2008	62	6	10	9	8	4
Vietnam	2004	60	2	16	7	6	9
Philippines	2008	41	19	14	3	5	18
Lao PDR	2008	46	6	10	8	12	21
Cambodia	2000	55	3	10	8	7	17
Myanmar	2000	54	8	16	7	8	7

Source: Adapted from Kaza et al. [4], with permission from the World Bank; Copyright 2018.

TABLE 6.3 MSW Disposal Methods by Southeast Asian Countries

Country	Year of Data	Dumps (%)	Landfills (%)	Compost (%)	Recycled (%)	WTE (%)	Other (%)
Singapore	2008	–	15	–	47	–	49
Thailand	2008	–	–	–	14	–	85
Lao PDR	2008	46	6	10	8	12	21
Cambodia	2000	100	–	–	–	–	–

Source: Adapted from Kaza et al. [4], with permission from the World Bank; Copyright 2018.

6.4 FACTORS AFFECTING WASTE COMPOSITION AND GENERATION

Economic growth and urbanization significantly impact waste composition and generation. Urban expansion leads to increased consumption and waste, particularly from nonorganic materials such as plastics and e-waste. This strains existing systems and demands innovative recycling.

Consumerism adds to the problem with disposable goods. Effective policies, strong enforcement, and public–private partnerships are crucial for tackling these challenges. When analyzing waste generation and management dynamics, it is clear that several key factors have a significant impact on waste trends. First, urbanization and population growth have dramatic effects on the amount of waste produced. As more people migrate to urban areas in search of better opportunities, the concentration of waste in these regions increases significantly. This poses challenges to existing waste management systems. Second, economic development and consumption patterns are closely linked to waste generation. With economic growth, there is an increase in the consumption of goods and services, resulting in higher volumes of waste, especially in the form of packaging and disposable products. Finally, cultural practices and societal attitudes toward waste have a profound influence on waste trends. In societies where recycling and reuse are deeply ingrained, less waste is sent to landfills. On the other hand, in cultures that prioritize convenience and disposable items, higher levels of waste are generated. Together, these factors—urbanization and population growth, economic development and consumption patterns, and cultural practices and societal attitudes—form a complex network that affects waste generation and management practices worldwide.

6.5 WASTE MANAGEMENT STRATEGIES

In Southeast Asia, waste management strategies are influenced by economic development, infrastructure, and regulatory differences across regions. Along with community-led recycling and composting initiatives, urban areas have shifted toward advanced collection and treatment methods, including sanitary landfills, recycling programs, and WTE plants. Rural waste management often relies on open dumping and burning due to limited infrastructure, with the informal recycling sector playing a crucial but underintegrated role in waste reduction [12]. Table 6.4 provides an overview of waste management policies in Southeast Asian countries, showing collective regional efforts to address the environmental impacts of waste. Policy efforts in Southeast Asian countries aim to reduce waste generation and tackle plastic pollution [13]. Initiatives range from Malaysia's plan to eliminate single-use plastics to Vietnam's focus on marine plastic pollution and Singapore's recycling targets [14, 15].

TABLE 6.4 Key Waste Management Policies in Southeast Asian Countries

Country	Policy Name	Year of Implementation	Impact	Reference
Indonesia	National Strategy on Solid Waste Management (https://ptsmi.co.id/cfind/source/files/smi-insight-2018-waste-management.pdf)	2018	Aims to reduce the generation of solid waste and enhance the management and recycling efforts	[13]
Thailand	Roadmap Toward Zero Waste (https://www.pcd.go.th/wp-content/uploads/2021/10/pcdnew-2021-10-19_08-59-54_995414.pdf)	2020	Targets significant waste reduction by 2027, with an emphasis on recycling and waste separation	[16]
Malaysia	Roadmap Toward Zero Single-Use Plastics (https://www.pmo.gov.my/wp-content/uploads/2019/07/Malaysia-Roadmap-Towards-Zero-Single-Use-Plastics-2018-2030-m-min.pdf)	2018–2030	Plans to phase out single-use plastics and encourage the adoption of sustainable alternatives	[14, 15]
Philippines	Solid Waste Management Act (RA 9003) Amendments (https://www.officialgazette.gov.ph/2001/01/26/republic-act-no-9003-s-2001/)	2020 (Ongoing updates)	Enhances local government units' waste management capabilities and public participation	[17]
Vietnam	National Action Plan on Ocean Plastic Waste Management (https://www.undp.org/vietnam/publications/national-action-plan-management-marine-plastic-litter-2030)	2019–2030	Focuses on reducing ocean plastic waste through comprehensive waste management policies	[18]
Singapore	Zero Waste Masterplan (https://www.mse.gov.sg/resources/zero-waste-masterplan.pdf)	2019	Sets ambitious targets for waste reduction, aiming for a 70% overall recycling rate by 2030	[19]
Cambodia	National Environment Strategy and Action Plan (https://www.adb.org/publications/cambodia-national-environment-strategy-action-plan)	2016–2030	Includes strategies for improving solid waste management practices across the country	[20]

TABLE 6.4 *(Continued)*

Country	Policy Name	Year of Implementation	Impact	Reference
Myanmar	Myanmar Sustainable Development Plan (https://themimu.info/sites/themimu.info/files/documents/Core_Doc_Myanmar_Sustainable_Development_Plan_2018_-_2030_Aug2018.pdf)	2018–2030	Outlines broad environmental goals, including improving waste management to reduce pollution	[21]
Laos PDR	National Strategy on Integrated Solid Waste Management (https://laoepf.org.la/wp-content/uploads/2022/11/20221114_EWMP_P175996_SIA-SMP-Vol-I_-Main-Report_Draft-Rev4_Clean.pdf)	2017	Targets the improvement of waste collection and treatment, promoting recycling and environmental sustainability	[22]
Brunei	Brunei Darussalam National Climate Change Policy (https://www.preventionweb.net/publication/brunei-darussalam-national-climate-change-policy)	2020	Includes waste management strategies to reduce greenhouse gas emissions and enhance recycling efforts	[23]

These policies, along with efforts in countries such as the Philippines, Cambodia, Myanmar, Laos, and Brunei, signify a regional shift toward sustainable practices for conserving the environment and promoting public health.

The 2018 report by Kaza et al. [4] on Southeast Asia's MSW emphasizes the disparities in waste generation and disposal. The region has an MSW generation rate of 0.27 tons per capita per year, with only 59% of MSW reaching designated disposal sites. Composting contributes 5% to MSW treatment, highlighting its environmental benefits. However, 27% of waste management practices are unspecified, suggesting a reliance on informal or less regulated methods. This situation highlights the urgent need for waste management strategies that are tailored to address specific challenges and opportunities in urban and rural areas. Using Thailand as a case study, the author presents a nuanced approach to waste management. As shown in Figure 6.2, in 2021, Thailand experienced a reduction in MSW generation to 24.98 t due to the impact of the COVID-19 pandemic on tourism and daily life, leading to a decrease in waste production. Only 16% of MSW underwent source segregation, and 6% was managed within households, particularly in areas lacking waste services. This process left 78% of the sample to be collected for disposal. Despite efforts, 25% of MSW was improperly disposed of, with plastic waste comprising 28% of MSW and predominantly ending up in landfills. This highlights the challenges of waste management and the environmental impact of plastic [24]. Furthermore, Ubon Ratchathani University implemented an effective waste segregation system using color-coded containers for different waste types—organic, recyclable, and general waste—facilitating efficient waste processing and recycling (Figure 6.3). This initiative, supported by educational campaigns, promotes environmental responsibility and serves as a sustainable waste management model, demonstrating the importance and feasibility of targeted waste management practices in enhancing environmental sustainability.

6.6 CASE STUDIES AND BEST PRACTICES

Across Southeast Asia, various successful waste management programs have illustrated innovative, sustainable approaches. Singapore's integrated system, which is renowned for its WTE facilities, recycling programs,

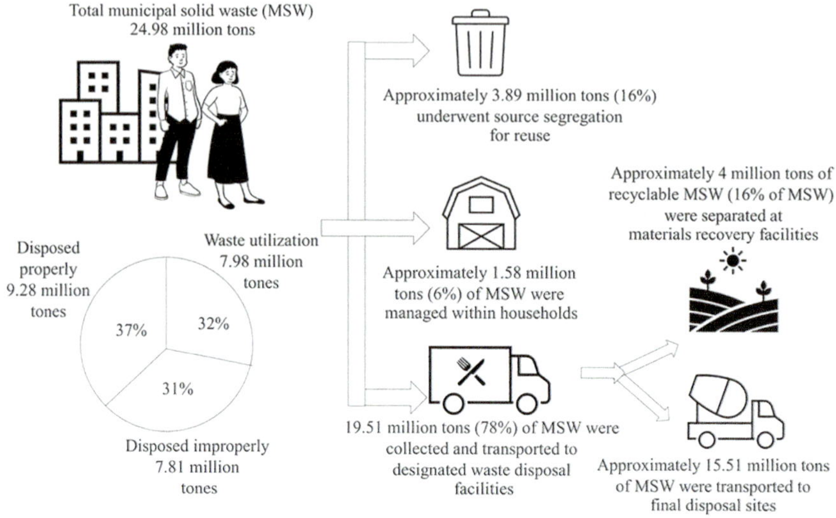

FIGURE 6.2 Overview of MSW management in Thailand in 2021.

Source: Created with Microsoft PowerPoint, 31 March 2024, 18:00.

FIGURE 6.3 Separate waste containers at Ubon Ratchathani University, Thailand.

Source: Ubon Ratchathani University, Thailand, 31 March 2024, 16:00.

and public campaigns, has significantly minimized landfill use [25]. Thailand's community-based initiative in Phuket and the Philippines' Ecological Solid Waste Management Act demonstrate the efficacy of local involvement and regulatory mandates in enhancing recycling and reducing waste [26]. Conversely, failures reveal crucial lessons, such as the need for greater community education and infrastructure to handle rapid waste generation increases [27]. A comparative analysis underscores the diversity in regional waste management, with urbanized nations such as Singapore, Malaysia, and Thailand leveraging technology and infrastructure, while countries with fewer resources face challenges with landfill reliance. The success stories from Indonesia and the Philippines emphasize community participation's value, and initiatives in Thailand and Vietnam showcase circular economy models' potential [28]. These insights highlight the importance of context-specific, adaptable strategies that integrate technological, policy, and community dimensions for effective waste management in Southeast Asia.

In the field of waste management in Southeast Asia, there are challenges and opportunities. One challenge is the region's diverse and rapidly changing waste composition and generation patterns. Urbanization and economic development strain existing waste management systems, leading to environmental degradation and public health concerns. Differing regulations and limited public awareness further complicate the identification of effective waste management solutions. However, these challenges also present opportunities for innovation and collaboration. The need for improved waste management is driving technological advancements, such as advanced recycling methods and energy recovery solutions. Moreover, Southeast Asian countries are increasingly cooperating and sharing best practices in waste management, leading to cost-effective solutions. It is crucial to involve communities and grassroots movements in advocating for sustainable living and waste reduction to achieve long-term behavioral change. Therefore, addressing waste trends in Southeast Asia requires a comprehensive approach that includes technological advancements, regulatory measures, and societal strategies.

6.7 FUTURE TRENDS AND PREDICTIONS

The future of solid waste management in Southeast Asia is changing due to economic growth, urbanization, and population increases, leading to an increase in waste generation. This is particularly evident in urban areas, where consumer lifestyles are shifting toward producing more nonorganic waste, such as plastics and electronics. Rural and agricultural regions also face significant challenges related to organic waste. To address this growing issue, waste management strategies are needed, with a focus on a shift to a circular economy and resource recovery. This approach emphasizes waste reduction, recycling, and the utilization of advanced technologies, including WTE solutions and smart waste management systems. Community involvement and the use of sustainable materials are also becoming increasingly important. Policy changes, including stricter waste disposal regulations and support for a circular economy, are crucial for driving improvements in waste management. International cooperation is essential for developing a more sustainable and efficient waste management sector [29–31].

However, the successful implementation and adaptability of these policies and initiatives at the national level will be crucial. It is essential to prioritize educational initiatives on waste reduction and support innovation in waste management technologies. Strengthening international collaborations will also enhance sustainable practices across the region. Adopting a multistakeholder approach involving governments, businesses, and communities is vital for effective waste management. Investment in research for new recycling methods and biodegradable materials will further support these efforts. Emphasizing local solutions tailored to specific regional challenges will ensure the effectiveness of these strategies. Engaging the youth through education and innovation challenges can inspire future solutions, making sustainable waste management a shared responsibility and priority.

6.8 CONCLUSION

Exploring solid waste management in Southeast Asia highlights the challenges of rapid urbanization, economic growth, and diverse cultural practices. These challenges have led to increased quantities and varieties of waste. Urban regions need help managing plastic, electronic, and hazardous waste, while rural areas face challenges with organic waste due to poor infrastructure. Successful strategies emphasize integrating technological innovations, community participation, and robust policy frameworks. On the other hand, failures indicate the dangers of neglecting these aspects.

Comprehensive waste management policies tailored to local conditions are recommended to address these issues. This should prioritize waste reduction, recycling, and the adoption of circular economy principles. Essential measures include investing in infrastructure, fostering public–private partnerships, implementing sustainable waste regulations, and enhancing community involvement and awareness to encourage sustainable waste practices. Future research should investigate new waste management technologies, evaluate waste policy effectiveness, and explore community-led models. This will help ensure that waste management practices remain responsive to changing needs and trends.

KEYWORDS

- **municipal solid waste**
- **sustainable solutions**
- **urban waste**
- **waste management**
- **waste technology**

REFERENCES

1. World Bank Group country classifications by income level. Retrieved from: https://blogs.worldbank.org/en/opendata/new-world-bank-group-country-classifications-income-level-fy24 (accessed: 25 March, 2024).
2. Brindhadevi, K., Barceló, D., Lan Chi, N. T., & Rene, E. R. (2023). E-waste management, treatment options and the impact of heavy metal extraction from e-waste on human health: Scenario in Vietnam and other countries. *Environmental Research*, *217*, 114926.
3. Chen, C.-C., Sujanto, R. Y., Tseng, M.-L., Fujii, M., & Lim, M. K. (2021). Sustainable consumption transition model: Social concerns and waste minimization under willingness-to-pay in Indonesian food industry. *Resources, Conservation and Recycling*, *170*, 105590.
4. Kaza, S., Yao, L., Bhada-Tata, P., & Van Woerden, F. (2018). *What a Waste 2.0: A Global Snapshot of Solid Waste Management to 2050*. World Bank: Washington, DC.
5. Lee, K. J. L., & Wong, S. F. (2023). Comparative environmental and socioeconomic assessment on mixed plastic waste management: A Singapore case study. *Science of The Total Environment*, *893*, 164884.

6. Pg Aliuddin, D. A. N. B., Reza, M. S., Abdalla, A. M., & Azad, A. K. (2023). In *Future Perspective of House Hold Waste Management in Brunei: A Comparison with Successful Swedish System, AIP Conference Proceedings*. AIP Publishing.
7. Shakil, N. S. M., Azhar, N. A. Z. M., & Othman, N. (2023). Solid waste management in Malaysia: An overview. *Information Management and Business Review*, *15*(1 (I) SI), 86–93.
8. Permana, A. S., As' ad, S., & Potipituk, C. (2023). The multipronged approach of solid waste management toward zero waste to landfill site: An Indonesia and Thailand experience. In *Pragmatic Engineering and Lifestyle: Responsible Engineering for a Sustainable Future*, Emerald Publishing Limited, pp. 113–129.
9. Suryawan, I. W. K., & Lee, C.-H. (2023). Citizens' willingness to pay for adaptive municipal solid waste management services in Jakarta, Indonesia. *Sustainable Cities and Society*, *97*, 104765.
10. Huynh, T. L., Le, T. K. O., Wong, Y. J., Phan, C. T., & Trinh, T. L. (2023). Toward sustainable composting of source-separated biodegradable municipal solid Waste—Insights from long an province, Vietnam. *Sustainability*, *15*(17), 13243.
11. Olalo, K. F., Nakatani, J., & Fujita, T. (2022). Optimal process network for integrated solid waste management in Davao City, Philippines. *Sustainability*, *14*(4), 2419.
12. Chandrappa, R., & Das, D. B. (2024). Mathematical modeling for solid waste management. In Chandrappa, R., & Das, D. B. (Eds.), *Solid Waste Management: Principles and Practice*. Springer International Publishing: Cham, pp. 771–805.
13. Kurniawan, T. A., Avtar, R., Singh, D., Xue, W., Dzarfan Othman, M. H., Hwang, G. H., Iswanto, I., Albadarin, A. B., & Kern, A. O. (2021). Reforming MSWM in Sukunan (Yogjakarta, Indonesia): A case-study of applying a zero-waste approach based on circular economy paradigm. *Journal of Cleaner Production*, *284*, 124775.
14.` Ma, Z. F., Ibrahim, Y. S., & Lee, Y. Y. (2020). Microplastic pollution and health and relevance to the Malaysia's roadmap to zero single-use plastics 2018–2030. *Malaysian Journal of Medical Sciences*, *27*(3), 1–6.
15. Loy, A. C. M., Lim, J. Y., How, B. S., Yiin, C. L., Lock, S. S. M., Lim, L. G., Alhamzi, H., & Yoo, C. (2023). Rethinking of the future sustainable paradigm roadmap for plastic waste management: A multination scale outlook compendium. *Science of The Total Environment*, *881*, 163458.
16. Phupadtong, A., Chavalparit, O., Suwanteep, K., & Murayama, T. (2023). Municipal emission pathways and economic performance toward net-zero emissions: A case study of Nakhon Ratchasima municipality, Thailand. *Journal of Environmental Management*, *347*, 119098.
17. Pintor, L. L., & Ancheta, A. A. (2022). Meta-analysis of studies on solid waste management in the Philippines. In *Effective Waste Management and Circular Economy*, CRC Press, pp. 243–253.
18. Nguyen, T. T., Duong, T. P. A., & Nguyen, T. N. A. (2021). Policy and legislations on microplastics pollution management in Vietnam. *VNU Journal of Science: Policy and Management Studies*, *37*(2), 36–43.
19. Rezvani Ghomi, E., Khosravi, F., Tahavori, M. A., & Ramakrishna, S. (2021). Circular economy: A comparison between the case of Singapore and France. *Materials Circular Economy*, *3*(1), 2.

20. Sokharavuth, P., Thiv, S., Nara, C., Him, C., Sokyimeng, S., Henze, D. K., Holmes, R., Kuylenstierna, J. C. I., Malley, C. S., Michalopoulou, E., & Slater, J. (2023). Air pollution mitigation assessment to inform Cambodia's first clean air plan. *Environmental Research*, *220*, 115230.
21. Palmer, E. (2021). Regulating infrastructure: Human rights and the sustainable development goals in Myanmar. *Human Rights Law Review*, *21*(3), 588–619.
22. Popli, K., Park, C., Han, S.-M., & Kim, S. (2021). Prediction of solid waste generation rates in urban region of Laos using sociodemographic and economic parameters with a multi linear regression approach. *Sustainability*, *13*(6), 3038.
23. Sukri, R. S., & Wah, D. (2022). Climate change adaptation in brunei darussalam. In Pereira, J. J., Zain, M. K., & Shaw, R. (Eds.), *Climate Change Adaptation in Southeast Asia*. Springer Singapore: Singapore, pp. 25–41.
24. The Pollution Control Department, Ministry of Natural Resources and Environment, Thailand, Action Plan on Plastic Waste Management Phase II (2023–2027), Thailand (accessed: 6 April, 2024).
25. Kua, H. W., He, X., Tian, H., Goel, A., Xu, T., Liu, W., Yao, D., Ramachandran, S., Liu, X., Tong, Y. W., Dai, Y., Peng, Y., & Wang, C.-H. (2022). Life cycle climate change mitigation through next-generation urban waste recovery systems in high-density Asian cities: A Singapore Case Study. *Resources, Conservation and Recycling*, *181*, 106265.
26. Babel, S., Ta, A. T., & Habarakada Liyanage, T. U. (2020). Current situation and challenges of waste management in Thailand. In Pariatamby, A., Shahul Hamid, F., & Bhatti, M. S. (Eds.), *Sustainable Waste Management Challenges in Developing Countries*. IGI Global: Hershey, PA, USA, pp. 409–440.
27. Dickella Gamaralalage, P. J., Ghosh, S. K., & Onogawa, K. (2022). Source separation in municipal solid waste management: Practical means to its success in Asian cities. *Waste Management & Research*, *40*(3), 360–370.
28. Thongplew, N., Onwong, J., Kotlakome, R., & Suttipanta, N. (2022). Approaching circular economy in an emerging economy: A solid-waste reutilization initiative in a small fresh market in Thailand. *Sustainability: Science, Practice and Policy*, *18*(1), 665–678.
29. Mondal, T., Choudhury, M., Kundu, D., Dutta, D., & Samanta, P. (2023). Landfill: An eclectic review on structure, reactions and remediation approach. *Waste Management*, *164*, 127–142.
30. Paul, S., Choudhury, M., Deb, U., Pegu, R., Das, S., & Bhattacharya, S. S. (2019). Assessing the ecological impacts of aging on hazard potential of solid waste landfills: A green approach through vermitechnology. *Journal of Cleaner Production*, *236*, 117643.
31. Choudhury, M., Jyethi, D. S., Dutta, J., Purkayastha, S. P., Deb, D., Das, R., Roy, G., Sen, T., & Bhattacharyya, K. G. (2022). Investigation of groundwater and soil quality near to a municipal waste disposal site in Silchar, Assam, India. *International Journal of Energy and Water Resources*, *6*(1), 37–47.

CHAPTER 7

Unblocking Resources: Techniques and Technologies for Material Recovery in Solid Waste Management

KONDLE RAVI,[1*] KUSHAGRA SHARMA,[2] BOLAGAM RAVIKUMAR,[1] VIMAL CHOUDHARY,[1] MOHD ASIF SHAH,[3] and ADNAN LAKDAWALA[4]

[1]*Department of Horticulture, School of Agriculture, Lovely Professional University, Phagwara, Punjab, India*

[2]*Department of Horticulture, Jawaharlal Nehru Krishi Vishwa Vidyalaya, Jabalpur, Madhya Pradesh, India*

[3]*Department of Economics, Kebri Dehar University, Kebri Dehar, Ethiopia*

[4]*George Washington University, Washington, DC, USA*

*Corresponding author

ABSTRACT

Rapid population growth, urbanization, industrialization, technological development, andinformatics lead to the generation of wastes from residential, commercial, manufacturing,mining, highly radioactive, agricultural, healthcare, and electrical sources. Globally, recyclingthese hazardous materials has become a significant challenge. Increasing amounts of solid waste and their environmental impacts make it necessary to adopt advanced methods for efficientdisposal. Solid waste management includes gathering, transporting, sorting, destroying, processing, recycling, and regulating waste. The main goal is to protect the ecosystem fromnegative effects, keep the environment clean, and safeguard humans from harm. Various modern techniques are used for waste management, including

Data Analytics and Smart Technologies in Solid Waste Management: Advancing Sustainability. Moharana Choudhury, Sushobhan Majumdar, and Avnish Chauhan (Eds.)

DOI: 10.1201/9781779640833-9

microbiological reprocessing, recycling recovery, and safe disposal in landfills, composting, waste-to-energy processes, destruction, pyrolysis, ionized gas extraction, and marine disposal. The modern approach to solid waste management is considered sustainable and supports future generations living in harmony with nature. This chapter aims to present multiple waste management techniques implemented worldwide. It also seeks to identify the most effective strategies by critically analyzing research findings from various studies.

7.1 INTRODUCTION TO MATERIAL RECOVERY IN SOLID WASTE MANAGEMENT

The prudent management of solid waste is an indispensable component of environmental stewardship, with the retrieval of materials from waste streams emerging as a cornerstone in sustainable waste practices [1]. The concept of "waste" extends beyond mere refuse, potentially yielding valuable products like bioplastics [2–33]. Over the past five decades, the global population has doubled, predominantly transitioning to urban living, with more than half of the world's population currently residing in urban areas. This demographic shift, increased urbanization, and extended life expectancy have more than doubled the requirements for water and crude oil. In India, rapid development has resulted in a substantial increase in municipal solid waste (MSW) generation, presenting a significant environmental challenge. The escalating waste production poses potential environmental, economic, and social issues, but recognizing it as a resource can address various demands [3]. The surge in solid waste generation is a global concern, with projections indicating a potential doubling of volume in the next decade, especially in low- and middle-income countries with high consumption rates and inadequate waste management practices. We anticipate reaching 2.4 billion tons by 2025, so urgent attention and effective waste management strategies are crucial [4]. Looking ahead, the global generation of solid waste is projected to reach 27 billion tons by 2050, with an estimated 2.2 billion tons being MSW from urban areas in 2025.

Developing countries face escalating challenges in managing MSW, with improper disposal in sanitary landfills posing environmental and health risks, including flies, noise, odor, groundwater pollution, and greenhouse gas emissions [5, 6]. This intricate process encompasses the systematic gathering, processing, and subsequent reintroduction of discarded materials into circulation to mitigate environmental repercussions and preserve finite resources [7, 8]. Exemplary case studies drawn from successful material recovery initiatives globally underscore the efficacy of integrated waste management systems, resilient public–private partnerships, and the integration of cutting-edge technologies in achieving commendable material recovery rates [9]. The surge in consumer product production, coupled with widespread global distribution and affordability, has presented challenges in effectively managing MSW. This is particularly evident in the escalating volumes of discarded electronic consumer products, adding complexity to waste management systems [10].

7.2 THE IMPORTANCE OF UNBLOCKING RESOURCES IN WASTE MANAGEMENT

Waste management is a pivotal facet of environmental sustainability, with an escalating recognition of unlocking resources within waste streams as a central strategy in optimizing waste management practices [11]. The global MSW generation reached 2.01 billion tons in 2019 and is projected to increase to 3.40 billion by 2050 [5]. MSW is identified as a notable contributor to greenhouse gas emissions, exerting a considerable influence on climate change. This knowledge forms the basis for developing effective strategy plans to improve waste separation, recycling initiatives, and overall waste management practices. This expansive chapter delves into the crucial aspects of unblocking resources, fundamental principles, cutting-edge technologies, attendant challenges, and the prospective economic and environmental advantages intricately linked with this strategic approach. Resource unblocking entails identifying, extracting, and recovering valuable materials are from waste streams, diverting them away from conventional disposal pathways like landfills or incineration [12]. Fundamental principles governing this process encompass applying advanced sorting techniques and material recovery methodologies and incorporating state-of-the-art technologies, including robotics and artificial intelligence. Challenges inherent in resource unblocking, such as contamination issues and the requisite for specialized infrastructure, underscore the intricate nature of this endeavor. Addressing these challenges are calls for coordinated efforts in public education, policy interventions, and strategic investments in research and development.

7.3 TRADITIONAL TECHNIQUES FOR MATERIAL RECOVERY

Material recovery, a fundamental aspect of waste management, has undergone evolutionary changes over time, marked by the development of traditional techniques geared toward extracting valuable resources from waste streams. Traditional material recovery techniques have a rich historical context, tracing to ancient civilizations like Rome, which demonstrated resource-conscious approaches to reclaiming metals from discarded items [13]. Presently, anaerobic digestion is a sophisticated waste management process that employs anaerobic bacteria to decompose organic matter in MSW. Manual sorting and handpicking are enduring methods of material recovery involving workers' manual separation of recyclable materials from waste through visual inspection. Although straightforward, this method remains relevant, particularly in regions with limited access to advanced sorting technologies, often employed for bulky items or uncomplicated waste streams. Source separation, another traditional technique, entails sorting waste at its origin, typically at households or businesses, fostering a cleaner and easily manageable waste stream [14]. Composting, effective for organic waste recovery, faces challenges related to the contamination of nonbiodegradable materials [15].

7.4 EMERGING TECHNOLOGIES IN SOLID WASTE MANAGEMENT

The traditional "take-make-dispose" model inherent in the linear economy is gradually giving way to an approach that emphasizes the comprehensive utilization of materials available in urban areas, including solid waste [16]. In recent times, advanced waste management approaches are the following.

7.4.1 DRONES AND REMOTE SENSING

Drones, equipped with cameras and sensors, can offer real-time data on landfill conditions, detect illegal dumping sites, and assess the overall health of waste management infrastructure [17].

7.4.2 BLOCKCHAIN FOR WASTE TRACEABILITY

Blockchain technology is emerging as a valuable tool for enhancing traceability and transparency in waste management processes. Through the utilization of blockchain, every stage of the waste management lifecycle can be meticulously recorded [18].

7.4.3 ROBOTIC WASTE SORTING

Robotic waste sorting represents a promising technology that involves the use of robots equipped with advanced sensors. These robots can identify and separate recyclables from mixed waste streams [19].

7.4.4 PRECISION WASTE COLLECTION AND SATELLITE IMAGING FOR ILLEGAL DUMPING DETECTION

High-resolution satellite images are utilized to detect and monitor illegal dumping sites, providing valuable information for enforcement and remediation efforts [20].

7.4.5 AEROBIC DIGESTION FOR ORGANIC WASTE

Aerobic digestion, as an innovative technology for managing organic waste, engages in the controlled breakdown of organic materials in the presence of oxygen. This process yields compost and biogas, offering a dual benefit of waste management and energy generation.

7.5 INTEGRATION OF DIGITAL SOLUTIONS IN RESOURCE UNBLOCKING

Integrating digital solutions in waste management practices marks a transformative shift toward optimizing resource recovery from waste streams. Leveraging technologies such as smart sensors, artificial intelligence, robotics, blockchain, digital twins, augmented reality, data analytics, remote monitoring systems, and public engagement platforms enhance the efficiency and effectiveness of waste sorting and recovery processes. These advancements address the challenges posed by improper waste handling practices like open dumping, burning, and landfilling, which threaten public health and environmental integrity. Using digital solutions introduces a comprehensive approach to waste management, employing technologies such as smart sensors and artificial intelligence to enhance sorting accuracy and optimize resource recovery processes [21]. Robotics automation minimizes reliance on manual labor and improves workplace safety, while blockchain ensures traceability and transparency in waste management operations [22]. Digital twin technology enables real-time monitoring and analysis of waste sorting facilities, facilitating operational optimization and resource recovery maximization. Augmented reality aids operators in material identification and sorting tasks, contributing to increased precision and efficiency. Ensuring interoperability among diverse technologies and establishing industry standards are crucial for a seamless waste management ecosystem.

7.6 CASE STUDIES: SUCCESSFUL APPROACHES TO MATERIAL RECOVERY

Material recovery is a pivotal facet of sustainable waste management, and its effective pursuit often entails the adoption of innovative strategies and technologies. This chapter explores case studies that illuminate successful material recovery initiatives, spotlighting varied approaches that have yielded positive outcomes in terms of elevated recycling rates, diminished environmental impact, and economic benefits. These case studies exemplify the pragmatic application of efficient material recovery practices in diverse contexts, offering insights into the multifaceted nature of successful waste management strategies.

7.6.1 SAN FRANCISCO, USA: PIONEERING ZERO-WASTE GOALS

San Francisco has emerged as a leader in material recovery endeavors, pioneering ambitious zero-waste objectives. The city has instituted a comprehensive waste management strategy encompassing source separation, rigorous recycling mandates, and inventive programs to foster waste reduction. By combining public awareness campaigns, resilient recycling infrastructure, and collaborations with local businesses, San Francisco accomplished a notable 80% waste diversion rate by 2020 [23]. The success of San Francisco's approach emphasizes the significance of adopting a holistic and community-centered strategy in attaining elevated material recovery rates.

7.6.2 OSLO, NORWAY: ADVANCED WASTE-TO-ENERGY FACILITIES

Oslo's waste management strategy places a significant emphasis on advanced waste-to-energy (WtE) facilities as a central element of material recovery. The city has committed investments in state-of-the-art incineration plants designed not only to diminish the volume of waste destined for landfills but also to produce heat and electricity. The generated energy serves the community through district heating, establishing a sustainable energy source [24]. Oslo's case serves as an illustration of how WtE technologies can play a pivotal role in material recovery by concurrently addressing waste volume reduction and contributing to energy generation. Grass sourced from urban roadside verges represents an underutilized potential resource for bioenergy recovery. There are two plausible bioenergy recovery techniques: first direct anaerobic digestion of the entire parent material and, second, the integrated generation of solid fuel and biogas from biomass procedure.

7.6.3 TAIPEI, TAIWAN: EFFECTIVE SOURCE SEPARATION

Taipei, Taiwan, distinguishes itself through its highly successful source separation program. The city has instituted a comprehensive waste separation system, mandating residents to meticulously sort their waste into various categories, including recyclables, food waste, and general waste. Various segregated solid digestate materials obtained from dairy manure and food waste feedstocks present valuable components for the commercial cultivation of *Pleurotus ostreatus* These sorted materials undergo processing in advanced recycling facilities. Taipei accomplished an impressive 55% recycling rate in 2019, underscoring the efficacy of source separation in augmenting material recovery [25].

7.6.4 SEOUL, SOUTH KOREA: INNOVATIVE FOOD WASTE RECYCLING

Seoul has effectively tackled the challenge of organic waste through the implementation of innovative food waste recycling programs. The city introduced smart bins equipped with weighing capabilities, charging residents based on the amount of food waste disposed of. The collected food waste undergoes processing into compost and bioenergy, thereby reducing the environmental impact associated with organic waste [26].

7.7 CHALLENGES AND CONSIDERATIONS

While the economic and environmental impacts of resource unblocking are significant, challenges and considerations must be addressed to maximize these benefits.

7.7.1 TECHNOLOGICAL BARRIERS

The successful implementation of resource unblocking relies on the integration of sophisticated technologies for precise sorting and processing. The upfront costs associated with acquiring and maintaining these advanced technologies may pose a significant barrier, particularly for smaller-scale waste management entities [27].

7.7.2 CONTAMINATION ISSUES

Contamination within waste streams poses a challenge as it can compromise the quality of recovered materials [28]. Mitigating contamination requires scientific approaches coupled with public awareness and education initiatives aimed at enhancing waste sorting practices at the source.

7.7.3 POLICY AND REGULATORY FRAMEWORKS

The efficacy of resource unblocking is intricately linked to the presence of supportive policy frameworks and regulations. Governments, from a scientific governance perspective, play a pivotal role in incentivizing businesses and individuals to embrace sustainable waste management practices through measures such as extended producer responsibility (EPR) and recycling targets.

7.8 COMMUNITY ENGAGEMENT AND PUBLIC AWARENESS

The success of waste management initiatives hinges on the active involvement of communities and heightened public awareness. Collaborative efforts among communities, local authorities, and stakeholders are essential for effective waste management strategies. This chapter delves into the significance of community engagement and public awareness in waste management, elucidating their pivotal roles in advocating responsible waste practices, promoting recycling, and cultivating a culture of sustainability such as

1. Community-centric decision-making: Community engagement entails the active involvement of residents, businesses, and organizations in the decision-making processes and actions related to waste management [29].
2. Engaging schools and educational institutions: Integrating waste management education into school curricula emerges as an effective strategy to instill sustainable values in future generations.
3. Leveraging technology for community engagement: Technology emerges as a potent tool for enhancing community engagement and public awareness in waste management [30]. Mobile apps, social media platforms, and online resources can offer accessible information, interactive tools, and real-time updates on waste management practices.

7.9 FUTURE TRENDS IN SOLID WASTE MANAGEMENT

The future trajectory of solid waste management reflects a shift toward innovative technologies and sustainable practices. WtE technologies, including advanced incineration and gasification processes, are expected to gain prominence for their ability to extract energy [31]. Intelligent waste management systems, equipped with sensors and Internet of Things devices, are poised to optimize resource allocation and enhance operational efficiency by enabling real-time monitoring of waste levels and collection routes. Additionally, decentralized and community-based solutions are projected to empower local communities to take ownership of their waste management practices. The proliferation of EPR programs will incentivize sustainable product design and accountability throughout the product lifecycle. Moreover, innovations in bio-based and biodegradable materials and advancements in plastic waste management technologies aim to address the environmental challenges associated with plastic pollution while fostering a more sustainable paradigm for plastic use and disposal.

7.10 CURRENT CHALLENGES IN MATERIAL RECOVERY

Material recovery in solid waste management is an indispensable facet of sustainable resource utilization, yet it contends with a spectrum of challenges that hamper its efficiency and effectiveness. Foremost among these challenges is contamination within waste streams, wherein nonrecyclable materials intermingle with recyclables, degrading the quality and market value of the recovered materials [14]. Furthermore, the persistent presence of policy gaps, marked by inconsistent recycling policies and inadequacies in EPR frameworks, impedes establishing a conducive regulatory environment for material recovery [32].

7.11 CONCLUSION: A SUSTAINABLE PATH FORWARD

In conclusion, the trajectory of solid waste management points toward a transformative paradigm characterized by technological advancements, the incorporation of circular economy principles, sustainable practices, and heightened global collaboration. These emerging trends collectively delineate a sustainable trajectory, where innovations in WtE technologies, intelligent waste management systems, and community-based solutions underscore dedication to resource efficiency and environmental stewardship. Concurrently, implementing EPR programs, using bio-based and biodegradable materials, and integrating robotics and automation in waste sorting signify an evolving landscape prioritizing responsible product lifecycle management. As stakeholders navigate these complexities, strategic interventions informed by the amalgamation of these trends promise to propel the field of solid waste management toward a more resilient, efficient, and sustainable future, contributing significantly to academic discourse and policy formulation in esteemed publications.

KEYWORDS

- **solid waste**
- **recovery material**
- **renewable energy**
- **unblocking resources**
- **waste management**

REFERENCES

1. Abdel Kader, S. S., & Qutb, S. (2023). A sustainable solid waste management framework for Egyptian cities based on synthesis the integrated solid waste management approach and urban environmental governance tool: Applying on New Damietta city. *Aswan University Journal of Environmental Studies*, *4*(5), 344–378.
2. Rosenboom, J. G., Langer, R., & Traverso, G. (2022). Bioplastics for a circular economy. *Nature Reviews Materials*, *7*(2), 117–137.
3. Tayebi-Khorami, M., Edraki, M., Corder, G., & Golev, A. (2019). Re-thinking mining waste through an integrative approach led by circular economy aspirations. *Minerals*, *9*(5), 286.
4. Khan, S., Anjum, R., Raza, S. T., Bazai, N. A., & Ihtisham, M. (2022). Technologies for municipal solid waste management: Current status, challenges, and future perspectives. *Chemosphere*, *288*, 132403.
5. Sharma, K. D., & Jain, S. (2020). Municipal solid waste generation, composition, and management: The global scenario. *Social Responsibility Journal*, *16*(6), 917–948.
6. Choudhury, M., Jyethi, D. S., Dutta, J., Purkayastha, S. P., Deb, D., Das, R., Roy, G., Sen, T., & Bhattacharyya, K. G. (2022). Investigation of groundwater and soil quality near to a municipal waste disposal site in Silchar, Assam, India. *International Journal of Energy and Water Resources*, *6*(1), 37–47. https://doi.org/10.1007/s42108-021-00117-5
7. Mondal, T., Choudhury, M., Kundu, D., Dutta, D., & Samanta, P. (2023). Landfill: An eclectic review on structure, reactions and remediation approach. *Waste Management*, *164*, 127–142. https://doi.org/10.1016/j.wasman.2023.03.034
8. Xavier, L. H., Ottoni, M., & Abreu, L. P. P. (2023). A comprehensive review of urban mining and the value recovery from e-waste materials. *Resources, Conservation and Recycling*, *190*, 106840.
9. Rosenzweig, C., Solecki, W. D., Romero-Lankao, P., Mehrotra, S., Dhakal, S., & Ibrahim, S. A. (Eds.). (2018). *Climate Change and Cities: Second Assessment Report of the Urban Climate Change Research Network*. Cambridge University Press.
10. Ravichandran, C., & Venkatesan, G. (2021). Toward sustainable solid waste management–challenges and opportunities. *Handbook of Advanced Approaches Towards Pollution Prevention and Control*, 67–103.
11. Broadhurst, J., Amaral Filho, J., Moyo, A., Nwaila, P., N'Gandu, H. S., Shongwe, B., Sibanda, L., Stander, H.-M., & Harrison, S. T. L. (2019). Resource efficient and socially responsible approaches for the integrated management of mine waste: Understanding the risks, opportunities, enablers and barriers.
12. Maiurova, A., Kurniawan, T. A., Kustikova, M., Bykovskaia, E., Othman, M. H. D., Singh, D., & Goh, H. H. (2022). Promoting digital transformation in waste collection service and waste recycling in Moscow (Russia): Applying a circular economy paradigm to mitigate climate change impacts on the environment. *Journal of Cleaner Production*, *354*, 131604.
13. Kadhila, T. (2022). Transitioning to sustainable waste management models: The case of Langebaan and Swakopmund Municipalities.

14. Bibri, S. E., Krogstie, J., Kaboli, A., & Alahi, A. (2024). Smarter eco-cities and their leading-edge artificial intelligence of things solutions for environmental sustainability: A comprehensive systematic review. *Environmental Science and Ecotechnology*, *19*, 100330.
15. Mohanty, A. K., Wu, F., Mincheva, R., Hakkarainen, M., Raquez, J. M., Mielewski, D. F., & Misra, M. (2022). Sustainable polymers. *Nature Reviews Methods Primers*, *2*(1), 46.
16. Rashid, S., & Malik, S. H. (2023). Transition from a linear to a circular economy. In *Renewable Energy in Circular Economy*. Springer International Publishing: Cham, pp. 1–20.
17. Dobrilovic, D., Jotanovic, G., Stjepanovic, A., Jausevac, G., & Perakovic, D. (2022). A model of UAV-based waste monitoring system for Urban areas. In *Cyber Security, Privacy and Networking: Proceedings of ICSPN 2021*. Singapore: Springer Nature Singapore, pp. 309–319.
18. Ahmad, R. W., Salah, K., Jayaraman, R., Yaqoob, I., & Omar, M. (2021). Blockchain for waste management in smart cities: A survey. *IEEE Access*, *9*, 131520–131541.
19. Koskinopoulou, M., Raptopoulos, F., Papadopoulos, G., Mavrakis, N., & Maniadakis, M. (2021). Robotic waste sorting technology: Toward a vision-based categorization system for the industrial robotic separation of recyclable waste. *IEEE Robotics & Automation Magazine*, *28*(2), 50–60.
20. Devesa, M. R., & Brust, A. V. (2021). Mapping illegal waste dumping sites with neural-network classification of satellite imagery. *arXiv preprint arXiv:2110.08599*.
21. Paul, S., Choudhury, M., Deb, U., Pegu, R., Das, S., & Bhattacharya, S. S. (2019). Assessing the ecological impacts of ageing on hazard potential of solid waste landfills: A green approach through vermitechnology. *Journal of Cleaner Production*, *236*, 117643. https://doi.org/10.1016/j.jclepro.2019.117643
22. Kurniawan, T. A., Othman, M. H. D., Hwang, G. H., & Gikas, P. (2022). Unlocking digital technologies for waste recycling in Industry 4.0 era: A transformation towards a digitalization-based circular economy in Indonesia. *Journal of Cleaner Production*, *357*, 131911.
23. Silva, A., Rosano, M., Stocker, L., & Gorissen, L. (2017). From waste to sustainable materials management: Three case studies of the transition journey. *Waste Management*, *61*, 547–557.
24. Lausselet, C., Cherubini, F., Oreggioni, G. D., del Alamo Serrano, G., Becidan, M., Hu, X., Rørstad, P. K., Strømman, A. H., & Strømman, A. H. (2017). Norwegian Waste-to-Energy: Climate change, circular economy and carbon capture and storage. *Resources, Conservation and Recycling*, *126*, 50–61.
25. Knickmeyer, D. (2020). Social factors influencing household waste separation: A literature review on good practices to improve the recycling performance of urban areas. *Journal of Cleaner Production*, *245*, 118605.
26. Jang, Y. C., Lee, G., Kwon, Y., Lim, J. H., & Jeong, J. H. (2020). Recycling and management practices of plastic packaging waste towards a circular economy in South Korea. *Resources, Conservation and Recycling*, *158*, 104798.
27. Batista, M., Caiado, R. G. G., Quelhas, O. L. G., Lima, G. B. A., Leal Filho, W., & Yparraguirre, I. T. R. (2021). A framework for sustainable and integrated municipal

solid waste management: Barriers and critical factors to developing countries. *Journal of Cleaner Production*, *312*, 127516.
28. Patil, Y. B., KT, S., & Prakash, A. (2018). Conceptual frameworks for the drivers and barriers of integrated sustainable solid waste management: A TISM approach. *Management of Environmental Quality: An International Journal*, *29*(3), 516–546.
29. McAllister, J. (2015). Factors influencing solid-waste management in the developing world.
30. Masoabi, T. G. (2020). Understanding public knowledge and awareness of e-waste management practices in Maseru, Lesotho (Doctoral dissertation, North-West University (South Africa)).
31. Kumar, A., Singh, E., Mishra, R., Lo, S. L., & Kumar, S. (2023). Global trends in municipal solid waste treatment technologies through the lens of sustainable energy development opportunity. *Energy*, *275*, 127471.
32. Kirchherr, J., Yang, N. H. N., Schulze-Spüntrup, F., Heerink, M. J., & Hartley, K. (2023). Conceptualizing the circular economy (Revisited): An analysis of 221 Definitions. *Resources, Conservation and Recycling*, *194*, 107001.
33. Roy, P., Choudhury, M. (2022). Bioeconomy and green plastic production. In Baskar, C., Ramakrishna, S., & Daniela La Rosa, A. (Eds.), *Encyclopedia of Green Materials*. Springer: Singapore. https://doi.org/10.1007/978-981-16-4921-9_189-1

CHAPTER 8

A Comprehensive Study on Assessing the Significance and Impact of Waste Segregation at the Source

AVNISH CHAUHAN[1*], CHANDAN MAHATA[2,3], SAMUELE BARRILI[4], and DEEP GUPTA[5]

[1]*Department of Environmental Science, Graphic Era Hill University, Dehradun, Uttarakhand, India*

[2]*Department of Agriculture and Biological Engineering, University of Illinois Urbana-Champaign, Urbana, IL, USA*

[3]*Department of Biological Systems Engineering, Virginia Tech, Blacksburg, VA, USA*

[4]*M4W—Marketing for Waste, Italy*

[5]*School of Smart Agriculture, COER University Roorkee, Haridwar, Uttarakhand, India*

*Corresponding author

ABSTRACT

The review offers a systematic assessment of the relevance and effects of waste separation at the source and its impact on the environment and society. Considering the global waste problem, this study examines the theoretical underpinnings of source segregation's efficacy and trash management's historical evolution. By using a mixed-methods approach, this study considers and assesses waste segregation practices by looking at community attitudes and barriers with both quantitative recycling rate

Data Analytics and Smart Technologies in Solid Waste Management: Advancing Sustainability.
Moharana Choudhury, Sushobhan Majumdar, and Avnish Chauhan (Eds.)

DOI: 10.1201/9781779640833-10

analysis and qualitative investigation. This chapter greatly assists legislators, waste management experts, and communities by providing evidence-based approaches for optimizing waste segregation. Waste segregation at the source is a significant requirement of waste management solutions to address today's environmental problems. This study is vital for advancing sustainable waste management practices and, ultimately, a healthier and cleaner environment for current and future generations.

8.1 INTRODUCTION

According to study, worldwide roughly 2010 million tons of solid waste are produced annually and of which 33% are not governed in a method that is eco-friendly in nature which poses threat to environment. Worldwide, the average daily trash production per person ranges widely from 0.11 to 4.54 kg. Approximately, 49.8 million tons of solid trash are produced in India each year [1]. Global estimates of the amount of garbage generated annually are close to 2 billion tons, and by 2050, that number is predicted to rise to 3400 million tons, a 70% increase. By 2050, it is predicted that trash creation would have increased thrice in countries like India, which are located in the fastest-growing regions of the world—South Asia and sub-Saharan Africa [2].

Globally, solid waste management is a chief concern for everyone in this world. Worldwide, each person produces 0.74 kg of waste each day on average. Global trash generation is predicted to increase by 70% by 2025, from 2011 million tons in 2016 to 3400 million tons annually [2]. For the preservation of the environment and human welfare, effective solid waste management is essential.

However, inappropriate waste management can cause contaminants in the biosphere, which can have harmful effects on human health [3]. The waste sorting at the source may facilitate processing and handling, improve when planned and executed well, waste reuse, recycling, and recovery can reduce the quantity of waste that needs to be disposed of by as much as 65% of the entire volume of garbage produced [4, 5]. One the most crucial methods for managing urban garbage is waste sorting at the source, which is also the top environmental priority [6].

In the 21st century, the most important issues confronting society all over is waste management. The volume of waste produced has risen to unprecedented levels due to expansion of urbanization, industrialization, and population growth. This has put serious hazards to the environment and public health as well as straining the capacity of current disposal techniques [7]. The idea of trash segregation at the source has drawn more attention in response to this worsening situation as a long-term way to lessen the negative impacts of careless garbage disposal [8].

8.2 WORLD SCENARIO OF WASTE GENERATION AND SEGREGATION

This study examines the volume of garbage produced, recycling rates, and alternative disposal techniques including landfilling and incineration, as well as the effects of poor management such as unlawful waste disposal (all numbers are expressed in kg/person). The Global Waste Index 2022 is the outcome and all details are provided in Table 8.1. Figures 8.1–8.3 present the projected waste generation in the world, showing global waste composition in percentage in the world and global treatment as well as disposal of waste in percentage, respectively [41].

8.3 TYPES OF WASTE AND THEIR CHARACTERISTICS

Solid waste comprises diverse materials that are given below.

8.3.1 MUNICIPAL SOLID WASTE

Household and commercial waste such as paper, plastics, food scraps, and glass [9].

8.3.2 INDUSTRIAL WASTE

Waste generated from manufacturing processes, including chemicals, metals, and solvents [10].

8.3.3 CONSTRUCTION AND DEMOLITION WASTE

Waste resulting from construction projects such as concrete, wood, and asphalt [11].

TABLE 8.1 Waste Generation in Top Ten Countries in the World (Global Waste Index 2022)

Position in the World	Country	Total Waste Generated (kg)	Recycling (kg)	Incineration (kg)	Landfills (kg)	Open Dump (kg)	Unaccounted Waste (kg)	Recycled (%)	Final Score (%)
1	South Korea	400.00	243.00	88.00	46.00	0.0	0.0	60.80	100
2	Denmark	845.00	300.00	382.00	07.00	0.0	0.0	35.60	94.90
3	Germany	632.00	302.00	204.00	05.00	0.0	13.00	47.80	90.40
4	Switzerland	706.00	210.00	333.00	00.00	0.0	0.0	29.80	89.30
5	Finland	596.00	168.00	345.00	03.00	0.0	0.10	28.20	89.30
6	Norway	726.00	256.00	337.00	17.00	0.0	6.0	35.30	88.50
7	Japan	336.00	66.00	268.00	02.00	0.0	0.0	19.60	86.90
8	The Netherlands	535.00	148.00	224.00	07.00	0.0	0.0	27.70	86.50
9	Sweden	431.00	87.00	259.00	03.00	0.0	0.0	20.20	84.80
10	Luxembourg	790.00	232.00	257.00	31.00	0.0	2.0	29.40	93.50

Source: https://sensoneo.com/global-waste-index/#:~:text=Rank-,Rank%202019,0%20kg

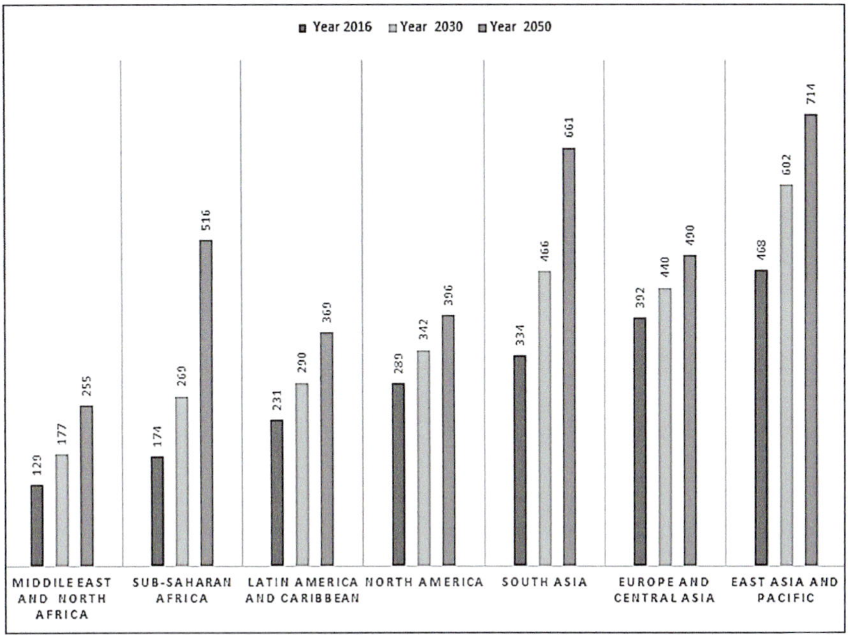

FIGURE 8.1 Projected waste generation in the world.

Source: Compiled by the authors.

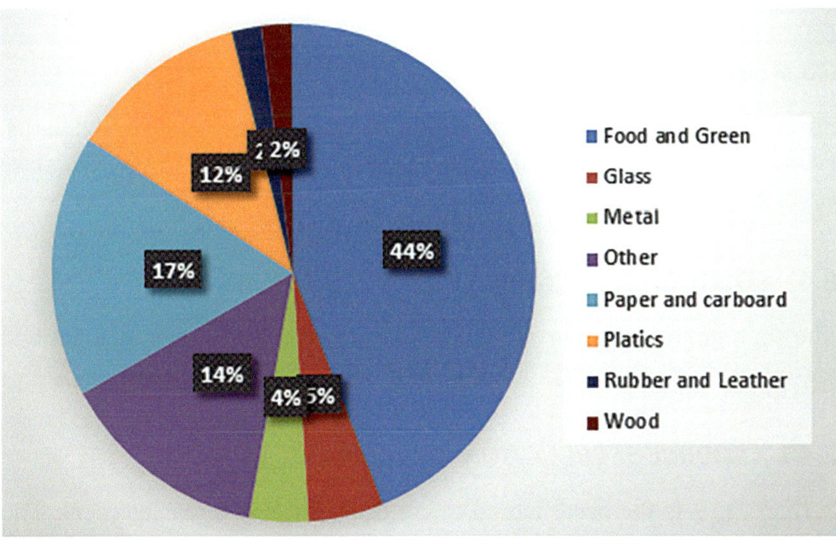

FIGURE 8.2 Global waste composition.

Source: Compiled by the authors.

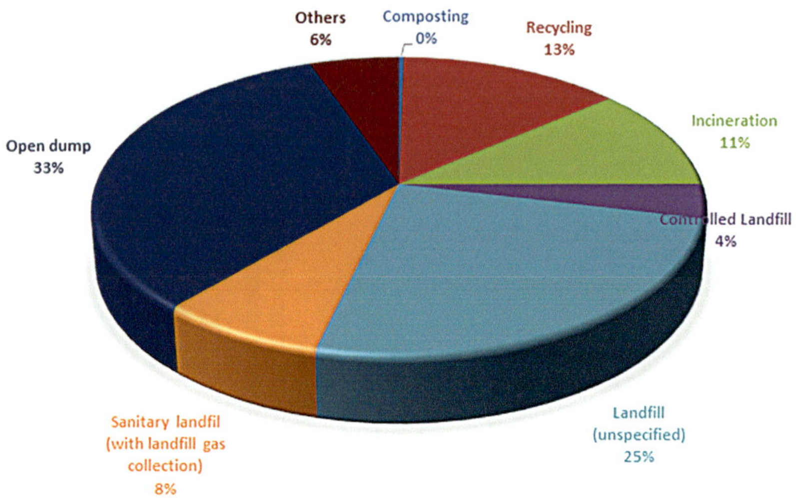

FIGURE 8.3 Global treatment and disposal of waste.

Source: Compiled by the authors.

8.3.4 ELECTRONIC WASTE (E-WASTE)

Discarded electronics containing valuable metals and hazardous substances [12].

8.3.5 BIOMEDICAL WASTE

From healthcare facilities, including infectious materials and sharps. Proper management is vital to mitigate environmental and health risks associated with each waste type.

8.4 SOLID WASTE SEGREGATION TECHNIQUES

The segregation plays an important role in solid waste management and various sources of segregation are as follows.

8.4.1 SOURCE SEGREGATION

Perhaps, this is the best method to control solid waste management. The separation of waste is at its point of origin. The sources can be classified into recyclables and nonrecyclables [36].

8.4.2 MANUAL SORTING

This is a labor-intensive method in which waste is sorted manually along with visual inspection.

8.4.3 MECHANICAL SORTING

This technique is an automatic process in nature which uses machinery such as conveyors and optical sorters to separate waste based on size, material type, and other characteristics.

8.4.4 GRAVITY SEPARATION

The utilizing differences in weight and density to separate materials, commonly used for separating recyclables [37].

8.4.5 BIOLOGICAL TREATMENT

This involves the use of biological processes, such as composting and anaerobic digestion, to treat organic waste and produce compost or biogas. Moreover, these methods promote efficient waste management and resource recovery [38].

This is methodical separation of diverse wastes at the site of generation, usually within homes, businesses, or industrial facilities, known as waste segregation at the source. The primary goal is to facilitate the recycling, repurposing, and responsible disposal of materials, thereby minimizing the volume of waste that is disposed of in landfills or incinerated [13].

8.5 BENEFITS OF SOURCE-BASED WASTE SEGREGATION

Separation or sorting of different types of garbage at the site or source-based waste segregation has several advantages such as social, economic, and public health benefits [14].

The proactive approach to waste management encourages resource conservation, pollution avoidance, and community empowerment [18].

The determination of the importance and impact of waste segregation at the source requires analyzing the outcomes in terms of environmental, economic, and social issues [39]. The extensive solid waste segregation programs have been found to significantly reduce landfill waste, enhance recycling rates, and improve resource efficiency [15]. The waste segregation may help in the reduction of poverty and local economic development by generating jobs in the recycling and waste management industries [16].

The sustainable use of resources as well as sorting of solid waste can reduce solid waste generation. The use of resources in a sustainable manner can assist a sense of environmental awareness and civic involvement by giving people the power to take accountability for their trash generation and consumption patterns. The sorting of waste also helps in preserving ecosystems, biodiversity, and public health [17].

The benefits of source-based waste segregation are numerous and include the following.

8.5.1 ENVIRONMENTAL BENEFITS

8.5.1.1 WASTE REDUCTION

The segregation of waste at source can reduce the volume of solid waste and which further decreases the load of waste that is sent to landfills or incinerators by separating recyclable, organic, and hazardous materials at the source [19].

8.5.1.2 RESOURCE CONSERVATION

The sorting of recyclable materials (such as metals, paper, plastics, and glass) at the source waste segregation conserves valuable resources and also reduces the demand on raw materials. Recycling of these waste materials further reduces energy consumption, greenhouse gas emissions, and environmental pollution [20].

8.5.1.3 POLLUTION PREVENTION

The effective waste segregation helps in prevention of the contamination of air, water, and soil with hazardous substances by minimizing the release of pollutants from improperly disposed waste. The separation of hazardous materials includes batteries, electronic waste, and household chemicals, prevents toxic leachate from entering groundwater, and protects ecosystems from pollution [21].

8.5.1.4 ENERGY RECOVERY

The source-based waste segregation facilitates the recovery of energy from organic waste through composting or anaerobic digestion. Organic waste such as food scraps and yard trimmings can be converted into biogas or compost, which can be used as renewable energy sources or soil amendments [22]. Hence, this reduces the dependence on fossil fuels and contributes to climate change mitigation efforts.

8.5.2 ECONOMIC BENEFITS

8.5.2.1 COST SAVINGS

Waste segregation at the source streamlines waste management procedures and cuts down on waste volumes to lower the costs of garbage collection, transportation, and disposal. Reducing landfill tipping fees, using less fuel, and making more money from the sale of recyclables are all ways that municipalities and waste management bodies can cut costs [23].

8.5.2.2 REVENUE GENERATION

Source-based waste segregation creates economic opportunities for recycling industries and waste management enterprises by providing a consistent supply of high-quality recyclable materials. Revenue can be generated through the sale of recovered materials, such as paper, plastics, glass, and metals, to manufacturers and processors for recycling into new products [24].

8.5.2.3 JOB CREATION

The expansion of recycling and resource recovery infrastructure associated with source-based waste segregation generates employment opportunities in collection, sorting, processing, and recycling activities. This creates jobs in local communities and stimulates economic growth, particularly in sectors such as recycling, waste management, and green technology [25].

8.5.3 SOCIAL BENEFITS

8.5.3.1 COMMUNITY ENGAGEMENT

The source-based waste segregation raises the community engagement and participation in waste management initiatives by empowering individuals to take responsibility for their waste generation habits. The public awareness, education programs, and outreach activities can muster residents, businesses, schools, and institutions to adopt sustainable waste segregation practices and reduce waste generation [26].

8.5.3.2 BEHAVIORAL CHANGE

The source-based waste segregation also promotes behavioral change by raising awareness of the environmental, social impacts of waste mismanagement and encouraging individuals to adopt more sustainable consumption. By adopting a culture of waste reduction, reuse, and recycling, waste segregation contributes to the development of environmentally responsible attitudes and behaviors [27].

8.5.3.3 PUBLIC HEALTH IMPROVEMENT

The proper waste segregation decreases public health risks associated with improper waste disposal such as disease transmission, contamination of food and water supplies, and exposure to hazardous substances. By preventing the accumulation of waste in residential areas and public spaces, waste segregation contributes to cleaner and healthier communities [28].

8.6 SIGNIFICANCE AND CHALLENGES OF WASTE SEGREGATION

In spite of the various initiatives and efforts in waste management in India, numerous challenges and barriers persist, hindering the effective implementation and sustainability of waste management practices. This type of challenge arises from a combination of social, economic, institutional, and infrastructural factors, which need to be addressed systematically to achieve meaningful progress in waste management. Some of the key challenges and barriers are as follows.

8.6.1 LACK OF INFRASTRUCTURE

The chief challenges in waste management, which associates with the inadequate infrastructure, are collection, segregation, treatment, and disposal of solid waste. Many cities as well as many towns in India suffer from lack of proper waste collection systems that lead to haphazard dumping and littering [29]. Existing treatment facilities such as sanitary landfilling and composting plants often suffer from insufficient capacity, outdated technology, and poor maintenance, exacerbating environmental pollution and health risks [40].

8.6.2 INADEQUATE WASTE SEGREGATION

The procedure of waste segregation at the source is key as well as crucial for waste management. Many households and establishments lack awareness and incentives to segregate waste properly, leading to mixed waste streams that are difficult and costly to process. In addition, cultural norms and behavioral practices often discourage waste segregation, making it challenging to implement segregation programs in effective manner [30].

8.6.3 INFORMAL WASTE SECTOR

The informal waste sector, comprising waste pickers, recyclers, and informal waste collectors, presents both opportunities and challenges in waste management. Informal waste workers play an important role in the collection and recycling of waste. To improve overall solid waste management, it is necessary to integrate the informal sector into solid waste management systems and offer assistance for skill development and livelihood enhancement [31].

8.6.4 POLICY AND REGULATORY GAPS

The presence of solid waste management rules, regulations, enforcement, and compliance remains major challenges. Irregularities in policy implementation, weak regulatory enforcement, and limited accountability mechanisms decline the effectiveness of waste management. The lack of harmonization between central- and state-level regulations, as well as overlapping mandates among different government agencies, further complicates the governance of waste management [32].

8.6.4 FINANCIAL CONSTRAINTS

Availability of limited financial resources and funding constraints pose significant barriers to the development and maintenance of solid waste management infrastructure as well as in services. Various municipalities and local authorities across the world struggle to allocate sufficient budgets for waste management, leading to inadequate investments in waste collection, treatment, and disposal facilities [42]. High capital costs associated with establishing modern waste management infrastructure prevent investment from the private sector, exacerbating the funding gap [33].

8.6.5 SOCIOCULTURAL ATTITUDES AND BEHAVIORS

Deep-rooted sociocultural attitudes and behaviors, such as littering, open defecation, and informal disposal of waste, pose significant challenges to effective waste management. Change in these ingrained behaviors requires targeted education, awareness campaigns, and community engagement initiatives. Addressing social stigma and cultural taboos associated with waste segregation and sanitation practices is essential for fostering sustainable waste management habits [34].

8.6.6 RAPID URBANIZATION AND POPULATION GROWTH

Rapid urbanization and population growth in India can exacerbate the challenges of waste management, placing immense pressure on existing infrastructure and services. Urban areas face rising volumes of waste due to uncontrolled population growth, migration, and changing consumption patterns. Moreover, unplanned urbanization and inadequate urban planning exacerbate issues such as land scarcity, encroachment, and land-use conflicts, which may hinder the development of sustainable waste management solutions [35].

8.7 CONCLUSION

Different nations behave differently when it comes to waste separation. While social and individual responsibility for this matter exists in most countries, this is not the case in others. The literature research attributes this to a lack of facilities required for waste separation and a need for more education in this field. As a result, education in these nations should target various age groups. This will raise their level of awareness and foster a favorable attitude in them. In addition, environmental equipment and facilities ought to be supplied. Other approaches, such as taxing more trash generators and exempting fewer waste generators from taxes, should be used in nations with more excellent knowledge of recycling.

The systematic review's findings have demonstrated that waste segregation behavior must be maintained and continued by applying ecological techniques at various scales and sporadic interventions over time. Strategies including feedback, theory-based treatments, and engagement successfully encourage waste separation behavior. Specific interventions have been successful over time by considering organizational, social, and environmental elements in addition to individual factors. As a result, interventions utilizing various tactics, temporal fluctuations, and increased resource accessibility through cooperation among various organizations are required.

KEYWORDS

- **waste segregation**
- **waste management**
- **environmental impact assessments**
- **policy implications**

REFERENCES

1. Swachh Bharat Mission [Internet]. [accessed: 11 November, 2021]. Retrieved from: http://swachhbharaturban.gov.in/dashboard/
2. Kaza, S., Yao, L., Bhada-Tata, P., & Van Woerden, F. (2018). *What a Waste 2.0: A Global Snapshot of Solid Waste Management to 2050*. World Bank Publications.
3. Kihila, J. M., Wernsted, K., & Kaseva, M. (2021). Waste segregation and potential for recycling—A case study in Dar es Salaam City, Tanzania. *Sustainable Environment*, *7*(1), 1935532.
4. Gyimah, P., Mariwah, S., Antwi, K. B., & Ansah-Mensah, K. (2021). Households' solid waste separation practices in the Cape Coast Metropolitan area, Ghana. *GeoJournal*, *86*(2), 567–583.
5. Bui, T. D., Tseng, J. W., Tseng, M. L., & Lim, M. K. (2022). Opportunities and challenges for solid waste reuse and recycling in emerging economies: A hybrid analysis. *Resources, Conservation and Recycling*, *177*, 105968.
6. Armijo De Vega, C., Ojeda Benítez, S., & Ramírez Barreto, M. E. (2008). Solid waste characterization and recycling potential for a university campus. *Waste Management*, *28*(1), S21–S26. https://doi.org/10.1016/j.was man.2008.03.022
7. Malinauskaite, J., Jouhara, H., Czajczyńska, D., Stanchev, P., Katsou, E., Rostkowski, P., Thorne, R. J., Colón, J., Ponsá, S., Al-Mansour, F., Anguilano, L., Krzyżyńska, R., López, I. C., Vlasopoulos, A. & Spencer, N. (2017). Municipal solid waste management and waste-to-energy in the context of a circular economy and energy recycling in Europe. *Energy*, *141*, 2013–2044.
8. Blair, J., & Mataraarachchi, S. (2021). A review of landfills, waste and the nearly forgotten nexus with climate change. *Environments*, *8*(8), 73.
9. Nandy, B., Sharma, G., Garg, S., Kumari, S., George, T., Sunanda, Y., & Sinha, B. (2015). Recovery of consumer waste in India—A mass flow analysis for paper, plastic and glass and the contribution of households and the informal sector. *Resources, Conservation and Recycling*, *101*, 167–181.
10. Clarke, C. J., Tu, W. C., Levers, O., Brohl, A., & Hallett, J. P. (2018). Green and sustainable solvents in chemical processes. *Chemical Reviews*, *118*(2), 747–800.
11. Ibrahim, M. I. M. (2016). Estimating the sustainability returns of recycling construction waste from building projects. *Sustainable Cities and Society*, *23*, 78–93.
12. Tsydenova, O., & Bengtsson, M. (2011). Chemical hazards associated with treatment of waste electrical and electronic equipment. *Waste Management*, *31*(1), 45–58.
13. Ahluwalia, I. J. (2019). Urban governance in India. *Journal of Urban Affairs*, *41*(1), 83–102.
14. Babazadeh, T., Nadrian, H., Mosaferi, M., & Allahverdipour, H. (2018). Identifying challenges and barriers to participating in the source separation of waste program in Tabriz, Northwest of Iran: A qualitative study from the citizens' perspective. *Resources*, *7*(3), 53.
15. Raghu, S. J., & Rodrigues, L. L. (2022). Solid waste management behavior among the student community: Integrating environmental knowledge and situational factors into the theories of planned behavior and value belief norm. *Journal of Environmental Planning and Management*, *65*(10), 1842–1874.

16. Otchere, F. (2018). Three Essays on Poverty Reduction and Socio-economic Development in Malawi. https://doi.org/10.17615/3mk0-9985
17. Kumar, R., Verma, A., Shome, A., Sinha, R., Sinha, S., Jha, P. K., & Vara Prasad, P. V. (2021). Impacts of plastic pollution on ecosystem services, sustainable development goals and need to focus on circular economy and policy interventions. *Sustainability*, *13*(17), 9963.
18. Wan, C., Shen, G. Q., & Choi, S. (2019). Waste management strategies for sustainable development. In *Encyclopedia of Sustainability in Higher Education*. Springer International Publishing: Cham, pp. 2020–2028.
19. Matheson, T. (2022). Disposal is not free: Fiscal instruments to internalize the environmental costs of solid waste. *International Tax and Public Finance*, *29*(4), 1047–1073.
20. Hao, H., Qiao, Q., Liu, Z., & Zhao, F. (2017). Impact of recycling on energy consumption and greenhouse gas emissions from electric vehicle production: The China 2025 case. *Resources, Conservation and Recycling*, *122*, 114–125.
21. Rawat, S., Verma, L., & Singh, J. (2020). Environmental hazards and management of E-waste. In *Environmental Concerns and Sustainable Development: Volume 2: Biodiversity, Soil and Waste Management*, pp. 381–398.
22. Chia, W. Y., Chew, K. W., Le, C. F., Lam, S. S., Chee, C. S. C., Ooi, M. S. L., & Show, P. L. (2020). Sustainable utilization of biowaste compost for renewable energy and soil amendments. *Environmental Pollution*, *267*, 115662.
23. Lohri, C. R., Camenzind, E. J., & Zurbrügg, C. (2014). Financial sustainability in municipal solid waste management—Costs and revenues in Bahir Dar, Ethiopia. *Waste Management*, *34*(2), 542–552.
24. Kang, H. Y., & Schoenung, J. M. (2006). Economic analysis of electronic waste recycling: Modeling the cost and revenue of a materials recovery facility in California. *Environmental Science & Technology*, *40*(5), 1672–1680.
25. Ezeah, C., Fazakerley, J. A., & Roberts, C. L. (2013). Emerging trends in informal sector recycling in developing and transition countries. *Waste Management*, *33*(11), 2509–2519.
26. Kala, K., & Bolia, N. B. (2020). Waste management communication policy for effective citizen awareness. *Journal of Policy Modeling*, *42*(3), 661–678.
27. Crociata, A., Agovino, M., & Sacco, P. L. (2015). Recycling waste: Does culture matter?. *Journal of Behavioral and Experimental Economics*, *55*, 40–47.
28. Gutberlet, J., & Uddin, S. M. N. (2017). Household waste and health risks affecting waste pickers and the environment in low-and middle-income countries. *International Journal of Occupational and Environmental Health*, *23*(4), 299–310.
29. De, S., & Debnath, B. (2016). Prevalence of health hazards associated with solid waste disposal-A case study of kolkata, India. *Procedia Environmental Sciences*, *35*, 201–208.
30. Knickmeyer, D. (2020). Social factors influencing household waste separation: A literature review on good practices to improve the recycling performance of urban areas. *Journal of Cleaner Production*, *245*, 118605.
31. Paul, J. G., Arce-Jaque, J., Ravena, N., & Villamor, S. P. (2012). Integration of the informal sector into municipal solid waste management in the Philippines—What does it need. *Waste Management*, *32*(11), 2018–2028.

32. Camacho, A. E., & Glicksman, R. L. (2021). Designing regulation across organizations: Assessing the functions and dimensions of governance. *Regulation & Governance*, *15*, S102–S122.
33. Thacker, S., Adshead, D., Fay, M., Hallegatte, S., Harvey, M., Meller, H., O'Regan, N., Rozenberg, J., Watkins, G., & Hall, J. W. (2019). Infrastructure for sustainable development. *Nature Sustainability*, *2*(4), 324–331.
34. Slekiene, J., Swan, N., & Kalina, M. (2024). Absorbent hygiene products disposal behaviour in informal settlements: Identifying determinants and underlying mechanisms in Durban, South Africa. *BMC Public Health*, *24*(1), 912.
35. Rahaman, M. A., Kalam, A., & Al-Mamun, M. (2023). Unplanned urbanization and health risks of Dhaka City in Bangladesh: Uncovering the associations between urban environment and public health. *Frontiers in Public Health*, *11*, 1269362.
36. Weißenbach, T., Pomberger, R., & Sarc, R. (2019, June). Composition of mixed commercial waste with focus on recyclable fractions. In *7th International Conference on Sustainable Solid Waste Management*, pp. 1–8.
37. Serranti, S., & Bonifazi, G. (2019). Techniques for separation of plastic wastes. In *Use of Recycled Plastics in Eco-efficient Concrete*. Woodhead Publishing, pp. 9–37.
38. Lin, L., Xu, F., Ge, X., & Li, Y. (2019). Biological treatment of organic materials for energy and nutrients production—Anaerobic digestion and composting. In *Advances in Bioenergy*. Elsevier, Vol. 4, pp. 121–181.
39. Paul, S., Choudhury, M., Deb, U., Pegu, R., Das, S., & Bhattacharya, S. S. (2019). Assessing the ecological impacts of ageing on hazard potential of solid waste landfills: A green approach through vermitechnology. *Journal of Cleaner Production*, *236*, 117643. https://doi.org/10.1016/j.jclepro.2019.117643
40. Choudhury, M., Jyethi, D. S., Dutta, J., Purkayastha, S. P., Deb, D., Das, R., Roy, G., Sen, T., & Bhattacharyya, K. G. (2022). Investigation of groundwater and soil quality near to a municipal waste disposal site in Silchar, Assam, India. *International Journal of Energy and Water Resources*, *6*(1), 37–47. https://doi.org/10.1007/s42108-021-00117-5
41. Mondal, T., Choudhury, M., Kundu, D., Dutta, D., & Samanta, P. (2023). Landfill: An eclectic review on structure, reactions and remediation approach. *Waste Management*, *164*, 127–142. https://doi.org/10.1016/j.wasman.2023.03.034
42. Choudhury, M., & Dutta, J. (2017). A comparative study of municipal solid waste management status for three major towns of Upper Assam—India. *International Journal of Waste Resources*, *7*(291), 2. https://doi.org/10.4172/2252-5211.1000291

PART III

Responsible Waste Management: Disposal, Treatment, and Recycling Strategies

CHAPTER 9

Managing Landfills: Design, Operation, and Environmental Consideration: A Case Study of Developed Countries

MURAD MUHAMMAD[1,2,3*], SANA ULLAH[4], and MUQADAS BATOOL[5]

[1]*State Key Laboratory of Desert and Oasis Ecology, Key Laboratory of Ecological Safety and Sustainable Development in Arid Lands, Xinjiang Institute of Ecology and Geography, Chinese Academy of Sciences, Urumqi, PR, China*

[2]*Xinjiang Key Laboratory of Biodiversity Conservation and Application in Arid Lands, Xinjiang Institute of Ecology and Geography, Chinese Academy of Sciences, Urumqi, China*

[3]*University of Chinese Academy of Sciences, Beijing, PR China*

[4]*Department of Agricultural Leadership, Education & Communication College of Agricultural and Environmental Sciences University of Georgia, Athens, GA, USA*

[5]*Department of Botany, Postgraduate College Layyah, Bahauddin Zakariya University, Multan, Pakistan*

*Corresponding author

ABSTRACT

This chapter delves into the multifaceted consequences of environmental considerations of landfills in developed countries. Sanitary landfilling is still the method of choice for final waste disposal; insufficient scientific evidence exists to reach definitive judgments regarding how trash impacts

Data Analytics and Smart Technologies in Solid Waste Management: Advancing Sustainability. Moharana Choudhury, Sushobhan Majumdar, and Avnish Chauhan (Eds.)

DOI: 10.1201/9781779640833-12

the environment and human health. Among the technological options available for managing and disposing of municipal solid waste, sanitary landfilling and open dumping are the preferred solutions in most countries because they are both affordable and require minimal technical expertise. Comparative evaluations of various waste management systems indicate this. The landfill sites are poorly designed and operated in most countries, including developed countries, which creates environmental and health concerns for the people living around these sites.

9.1 INTRODUCTION

Solid waste continues to accumulate at an alarming rate, making waste management a pressing concern on a worldwide scale. The per capita garbage production in affluent nations reached 800 kg in the 1990s, while in underdeveloped countries, it reached a maximum of 200 kg [1]. Two billion tons of solid trash were predicted to have been produced worldwide in 2010. For 50% of this total, we may thank the industrialized nations of North America, Oceania, Oceania, and Western Europe [2]. Nevertheless, it is said that as emerging nations' urban populations increase, this number will be revised. Additional estimates indicate that wealthy nations will generate less solid waste by 2035 than Asian and African nations, creating double as much garbage. Yet, efficient landfill use differs worldwide and is necessary for present and future waste outputs [3].

One uses a landfill, an artificial hole in the earth, to dispose of solid trash. Most of the garbage in landfills is either municipal solid waste (MSW), or commonplace single-use materials, including plastic, paper, metal, grass clippings, clothes, furniture, bottles, food scraps, newspapers, paint, and batteries. Different countries have different approaches to managing their municipal solid garbage. For example, although almost all MSW in Bulgaria ends up in landfills, none does in Sweden. A total of 41% of Europe's collected MSW was in landfills in 2014 [4]. Over half of the US garbage in 2015 was dumped in landfills, a practice that is both uncontrolled and unclean, in contrast to many Asian nations that dispose of their trash by open dumping or burning. Landfills are among the most frequent ways of trash disposal. However, there are other options as well. Most of the data pertain to garbage creation in the USA; hence, this evaluation will center on American landfill management.

According to the US Environmental Protection Agency, the USA produces 254 million tons of garbage annually. This amounts to around

2 kg of waste produced daily per individual. Composting accounts for 10.1% of US MSW treatment, incineration, or combustion, with energy recovery for 12.7%, recycling for 25%, and landfills for 52.1% [5]. The relative frequencies of different therapy modalities from 1960 to 2017. Landfills, which accounted for more than 52% of US trash disposal in 2017, are a more convenient option, according to the research [6]. The ever-increasing growth of our population has rendered us more reliant on landfills. The US population grew from 308.7 million in 2010 to 323.1 million in 2016, an increase of 281.4 million people [7]. Industrial landfills handle nonhazardous waste from businesses, while landfills dealing with toxic or hazardous waste are the most supervised and organized. However, MSW management facilities mainly collect home solid waste and other types of general solid waste [8]. Each of these landfills has developed its waste management system, which includes liners, gas collection, drainage systems, runoff control, leachate collection, and lots more. The following sections elaborate on the many forms of landfills. Landfills that process MSW receive the vast bulk of the garbage that our residences, institutions of higher learning, and businesses generate. Landfills for MSW can collect various materials, including hazardous waste, nonhazardous sludge, ISD, and construction debris, all of which pose risks to our air, groundwater, and wildlife. Most of these trash items are not biodegradable; therefore, they would not break down in landfills long [26].

Landfills have addressed the short-term requirement for trash disposal, but they are not the best option for managing trash over the long run, and they have a lot of dire consequences for people's health and the environment. However, not all solid waste landfills are effectively administered and constructed to protect the environment from harmful contaminants in the waste stream. Integrated waste management solutions are crucial in this context. Currently, most landfills continue to be a significant cause of groundwater contamination due to leachate, as well as emissions of greenhouse gases due to landfill gas [9]. Contaminated landfills may also make the land and soil around them useless. More people living in the country means more trash; therefore, Americans must figure out how to reduce and manage their waste properly so the environment does not worsen.

Municipal and industrial waste generation has increased quickly in the past few decades due to the ever-increasing industrial output and commerce in many nations across the globe. Developed nations' yearly per capita trash output was between 300 and 800 kg in the second part of the 1990s, whereas other countries' per capita waste production was below 200 kg [10]. Worldwide, practically every city is having trouble meeting its trash reduction goals, and by 2025, that number will have almost doubled to 2.2 billion metric tons [11]. Most countries' MSW is in landfills or open dumps worldwide.

Due to its economic benefits, landfilling is still a commonly recognized and employed option for ultimate MSW disposal. Research comparing several waste management options, including composting, incineration, and landfilling, has revealed that open dumping, also known as sanitary landfilling, is a popular and economical choice for treating and disposing of MSW in numerous nations. This method's low cost and ease of implementation make it a top choice [12–24].

The European Union directive on waste dumping reduces garbage disposal and sets strict landfilling and landfill site criteria. Even with this regulation, Europe is not uniform. Germany, Belgium, Switzerland, the Netherlands, Sweden, Austria, and Denmark landfill less than 5% of rubbish [13]. New member nations, candidate countries, and islands still employ landfilling for waste management. This figure is far higher compared to 37% in Poland, 38% in Slovakia, 64% in Bulgaria, and 50% in the Czech Republic (Yoder, 2003). Cholera and typhoid, watery diseases, have been related to pollution and health issues. Human actions, including ecosystem-altering technologies, can pollute the environment [14]. Energy sources, including light, heat, sound, and various chemical substances, can pollute. Gworek says pollution can be natural or caused by alien energy and substances. Urbanization and industry have released several soil organic carbons into the sky, each contributing to a different type of air pollution. All life on earth depends on the atmosphere [15]. The objectives of a chapter on landfill environmental impacts in industrialized nations are to cover several crucial points. First, the chapter describes landfill consumption and waste management patterns in industrialized countries to contextualize their environmental issues. Second, it examines landfill ecological impacts such as greenhouse gas emissions, leachate contamination, and habitat modification, focusing on ecosystems and human health. The chapter also evaluates industrialized countries' rules, policies, and technologies to reduce environmental damage and promote sustainable waste management. It also seeks landfill management trends, innovations, and best practices that promote environmental conservation and resource recovery [25, 26].

9.2 DESIGN AND REGULATIONS OF LANDFILLS

To safeguard human health and the environment, all landfills in the USA are required to fulfill minimum national requirements under the "Resource Conservation and Recovery Act (RCRA)" [16]. This includes both industrial and hazardous landfills, as well as MSW landfills. Under RCRA section D wastes, a landfill may be used to dispose of commercial solid waste and nonhazardous industrial trash. For licensed facilities, these regulations are found in part 264, and for interim status facilities, in part 265 [17]. Landfill owners and operators in the USA are generally required to adhere to these requirements. Each of the several landfill designs has its own unique set of procedures and features. There is a wide range of sustainability in landfill designs. This category includes open dumps and regulated, engineered, and sustainable landfills.

9.2.1 OPEN GARBAGE PITS

These open dumpsites have one thing in common: air pollution from unregulated burning of waste materials, public open defecation, water pollution from flooding and leaching, no containment of waste bodies, and poor or nonexistent regulation of waste kinds entering the site. Due to insufficient engineering, these exposed waste sites are unprotected from groundwater and have uncontrolled drainage [18]. Assessing the environmental dangers associated with these sites is necessary to decide whether to shut down the exposed waste sites or convert them to a controlled landfill. The location of the site (floodplains or groundwater), the depth and degree of compaction of the current open dumpsite, the variety of wastes within the site, and the possibility of mining decomposed organic elements should all be included in environmental impact assessments.

9.2.2 MANAGED DEPOSAL FACILITIES

Controlled landfills are the next level up from open dumps; they use standard waste management practices to consolidate and manage the overall body of waste, and they also use standard control mechanisms such as an on-site supervisor and restrictions on vehicle movement and access. Simple drainage management measures have been implemented to ensure that waste is not burned uncontrolled and that scavenging and foraging animals are kept to a minimum in these areas. Despite the increased regulation compared with open dump dumps, controlled landfills still fail to meet the basic requirements for garbage compaction and covering, rendering them unviable [9]. Restricting the working face area, implementing a litter barrier, and ensuring daily cover are all established operational protocols. The management of drainage systems, water quality, and waste volume is effectively regulated.

9.3 RECYCLED WASTE SITES AND SUSTAINABLE LANDFILLS

Engineered landfills are designed to regulate trash and prevent surface water from escaping the site through well-designed surface drainage. They are often referred to as sanitary landfills due to their high level of waste disposal. These landfills have a covered base for pile and compression, facilitating decomposition and preventing leachate seepage. Planning is crucial from the initial concept to the landfill's eventual decommissioning. Designed landfills reduce leachate seepage by preventing garbage from becoming saturated, resulting in slow waste degradation [19]. However, they can function as bioreactors, regulating fluid flow through garbage and speeding up the degrading process. This strategy is more advantageous for the long-term sustainability of airspace, processes, control, and product use due to its reduced negative impact on human and environmental health. Sustainable landfills consider factors that regulate chemical and biological processes, including water content, temperature, microflora, and compaction rates. Anaerobic bioreactors and aerobic cells are designed to enhance waste stabilization and optimize waste degradation rates [20]. Anaerobic bioreactors utilize an integrated system for collecting and recirculating leachate, including a geomembrane liner, a gas collection system, and a final cover. Conversely, aerobic biocell systems maximize the waste breakdown rate and the desirable gas carbon dioxide production through air circulation.

9.4 GROUND-BREAKING APPROACHES USED IN GARBAGE DISPOSAL

9.4.1 MINIMIZING GARBAGE DUMPING AND LIMITING THE RELEASE OF GASES

The problem of plastic waste is substantial, with a total of 6300 million metric tons of rubbish produced in 2015. Out of this, only 9% was recycled, 12% was burnt, and a staggering 79% was either dumped in landfills or released into the environment [21]. By 2050, an estimated 12,000 million metric tons of plastic trash will be deposited in landfills or released into the environment [22]. However, research suggests natural ways to dispose of trash, such as wax moth caterpillars or *Galleria mellonella*, which can decompose polyethylene into chemical bonds. This method could help reduce landfill trash and promote more comprehensive strategies for waste reduction. Another alternative is incorporating anaerobic bioreactors or aerobic biocells into landfills, which can reduce methane gas emissions, a greenhouse gas 25 times more harmful than carbon dioxide. A pilot trial in Denmark showed a significant decrease in methane output from 10 to 1 kg/h using a novel cover technology, which disperses gas through windows, converting methane into carbon dioxide [23]. This technology could be used globally to significantly reduce landfill methane emissions.

9.5 CONCLUSION

To reduce our environmental impact, we should honor the earth, reconsider our consumption habits, minimize trash and recycling, and reuse and recycle items whenever possible. The Annenberg Foundation reports that over 70% of trashes in landfills can be reused or recycled, alleviating economic, public health, and environmental problems. To reduce the amount and toxicity of trash sent to landfills, waste management firms should implement stringent landfill laws, and individuals should cut down on trash production. This research aimed to examine the health risks and environmental impacts of landfilling. Landfills are essential for waste management but also contribute to pollution, impacting aquatic life, biodiversity, flora, and wildlife. The presence of rodents and biocontaminant release near landfills negatively impacts air quality and groundwater quality. Landfills also pose health risks, including gas emissions into the atmosphere, lung cancer, respiratory health issues, and carcinogenic hazards. Children living near landfills face increased dangers due to environmental contamination. Some components in polluted groundwater have teratogenic effects. While landfills provide a convenient and effective solution to the large amount of trash generated, the negative impacts outweigh the positive ones. Greater technological use in garbage segregation, appropriate waste management, stricter enforcement of landfill rules, and establishing a larger concrete minimum distance for settlements are critical measures that must be evaluated and executed promptly.

KEYWORDS

- **landfills**
- **environment**
- **solid waste**
- **developed countries**

REFERENCES

1. Akter, N. (2000). Medical waste management: A review. Research Reports (1999): Health Studies, vol. XXVIII, pp. 370–394
2. Ayeleru, O. O., Dlova, S., Akinribide, O. J., Ntuli, F., Kupolati, W. K., Marina, P. F., Blencowe, A., & Olubambi, P. A. (2020). Challenges of plastic waste generation and management in sub-Saharan Africa: A review. *Waste Management*, *110*, 24–42.
3. Blair, J., & Mataraarachchi, S. (2021). A review of landfills, waste and the nearly forgotten nexus with climate change. *Environments*, *8*(8), 73.
4. Ghosh, S. K. (2016). Sustainable SWM in developing countries focusing on faster growing economies, India and China. *Procedia Environmental Sciences*, *35*, 176–184.
5. Grasso, D. (2017). *Hazardous Waste Site Remediation*. Routledge: New York.
6. Hashisho, J., & El-Fadel, M. (2014). Determinants of optimal aerobic bioreactor landfilling for the treatment of the organic fraction of municipal waste. *Critical Reviews in Environmental Science and Technology*, *44*(16), 1865–1891.
7. Hassan, S. H., Halim, A. A., Yusoff, M. S., Wang, L. K., & Wang, M.-H. S. (2021). Legislation for solid waste management. *Solid Waste Engineering and Management: Volume 1*, 85–141.
8. Kaplan, G. G., & Ng, S. C. (2017). Understanding and preventing the global increase of inflammatory bowel disease. *Gastroenterology*, *152*(2), 313–321. e312.
9. Khan, S., Anjum, R., Raza, S. T., Bazai, N. A., & Ihtisham, M. (2022). Technologies for municipal solid waste management: Current status, challenges, and future perspectives. *Chemosphere*, *288*, 132403.
10. Kumar, A., & Agrawal, A. (2020). Recent trends in solid waste management status, challenges, and potential for the future Indian cities—A review. *Current Research in Environmental Sustainability*, *2*, 100011.
11. Kunce, M. Perception and Reality: United States mortality.
12. Lawrence, B. D. (2013). *Investigation of Synergistic NOx reduction from Cofiring and Air Staged Combustion of Coal and Low Ash Dairy Biomass in a 30 KiloWatt low NOx furnace*.
13. Malinauskaite, J., Jouhara, H., Czajczyńska, D., Stanchev, P., Katsou, E., Rostkowski, P., Thorne, R. J., Colon, J., Ponsá, S., & Al-Mansour, F. (2017). Municipal solid waste

management and waste-to-energy in the context of a circular economy and energy recycling in Europe. *Energy*, *141*, 2013–2044.
14. Meegoda, J. N., Hettiarachchi, H., & Hettiaratchi, P. (2016). Landfill design and operation. In *Sustainable Solid Waste Management*, pp. 577–604.
15. Nanda, S., & Berruti, F. (2021). Municipal solid waste management and landfilling technologies: A review. *Environmental Chemistry Letters*, *19*(2), 1433–1456.
16. Owen, N. E. (2009). *Effects of Increased Kerbside Provision and Materials Recovery Facility Development on Recycling Rates in a Rural Community*. Cardiff University: United Kingdom.
17. Ozbay, G., Jones, M., Gadde, M., Isah, S., & Attarwala, T. (2021). Design and operation of effective landfills with minimal effects on the environment and human health. *Journal of Environmental and Public Health, 2021*.
18. Papakonstantinou, E., Mitsis, T., Dragoumani, K., Efthimiadou, A., Bacopoulou, F., Chrousos, G. P., Eliopoulos, E., & Vlachakis, D. (2021). Materials of biological origin and biofuels: Small environmental footprint and epigenetic impact. *International Journal of Epigenetics*, *1*(3), 1–8.
19. Schaub, E. O. (2023). *Year of No Garbage: Recycling Lies, Plastic Problems, and One Woman's Trashy Journey to Zero Waste*. Simon and Schuster.
20. Singh, S., & Kumar, R. (2022). Air pollution and its associated impacts on atmosphere and biota health. In *Extremes in Atmospheric Processes and Phenomenon: Assessment, Impacts and Mitigation*. Springer, pp. 29–58.
21. Soni, A., Das, P. K., & Kumar, P. (2023). A review on the municipal solid waste management status, challenges and potential for the future Indian cities. *Environment, Development and Sustainability*, *25*(12), 13755–13803.
22. Tojo, N., Alexander, N., & Ingo, B. (2008). Waste management policies and policy instruments in Europe: An overview.
23. Wojnowska-Baryła, I., Bernat, K., & Zaborowska, M. (2022). Plastic waste degradation in landfill conditions: The problem with microplastics, and their direct and indirect environmental effects. *International Journal of Environmental Research and Public Health*, *19*(20), 13223.
24. Mondal, T., Choudhury, M., Kundu, D., Dutta, D., & Samanta, P. (2023). Landfill: An eclectic review on structure, reactions and remediation approach. *Waste Management*, *164*, 127–142. https://doi.org/10.1016/j.wasman.2023.03.034
25. Paul, S., Choudhury, M., Deb, U., Pegu, R., Das, S., & Bhattacharya, S. S. (2019). Assessing the ecological impacts of ageing on hazard potential of solid waste landfills: A green approach through vermitechnology. *Journal of Cleaner Production*, *236*, 117643. https://doi.org/10.1016/j.jclepro.2019.117643
26. Choudhury, M., & Dutta, J. (2017). A comparative study of municipal solid waste management status for three major towns of Upper Assam—India. *International Journal of Waste Resources*, *7*(291), 2. https://doi.org/10.4172/2252-5211.1000291

CHAPTER 10

Anaerobic Digestion and Composting: A Sustainable Solution for Organic Waste Management

RATUL PEGU,[1*] SARMISTHA PAUL,[2] BISHAL PAUL,[1] AKASH CHETRY,[1] and ELKANA HEZRON[3]

[1]*Department of Environmental Sciences, Tezpur University, Tezpur, Assam, India*

[2]*Department of Botany, Visva Bharati University, Santiniketan, Bolpur, Birbhum, West Bengal, India*

[3]*Department of Sustainable Agriculture, Biodiversity and Ecosystem Management, School of Life Sciences and Bioengineering, The Nelson Mandela African Institution of Science and Technology, Arusha, Tanzania*

*Corresponding author

ABSTRACT

Solid waste production is growing quickly on a global scale every day, which has an influence on the environment in many ways. Organic and inorganic wastes make up solid waste. Due to its putrefying nature, organic waste creates a nuisance in the environment. Nowadays, one of the biggest issues faced in all developing nations is the handling of organic waste. Organic waste originates from various sources such as schools, hospitals, markets, workplaces, restaurants, households, and wastewater treatment plants. The economy, ecology, and public health are all seriously threatened by the rapidly rising volume of organic waste. Organic wastes

Data Analytics and Smart Technologies in Solid Waste Management: Advancing Sustainability. Moharana Choudhury, Sushobhan Majumdar, and Avnish Chauhan (Eds.)

DOI: 10.1201/9781779640833-13

are biodegradable and can be converted to value-added products using suitable techniques. To manage organic waste sustainably, the cause and characteristics of its creation must be identified, followed by appropriate management measures. This chapter discusses the role and effectiveness of anaerobic digestion and composting as long-term sustainable organic waste management solutions. Understanding how organic waste impacts the ecosystem's health, how anaerobic digestion and composting can manage organic waste, and the benefits of these techniques will be highlighted in this chapter.

10.1 INTRODUCTION

The population of the world is predicted to increase to 8.6 billion in 2030 and 9.8 billion in 2050 [1]. One of the inevitable effects of human existence is waste production, which rises in tandem with urbanization, industrialization, and population expansion. Appropriate waste management practices and societal economic standing influence the amount of waste produced. The uncontrolled decomposition of waste in open dumping can worsen epidemic diseases, multiply foul odors, and contribute to climate change. The unplanned and excessive disposal of solid waste (SW) in developing countries' large cities has been connected to several health issues and negative environmental effects [2, 3].

It is estimated that global waste production will rise from 2 billion tons in 2016 to 3.4 billion tons in 2050, with most of this output coming from countries in Asia and Africa [1, 4]. Of the total SW generated worldwide, organic waste (OW) accounts for 46% [5]. It includes animal manure, food processing waste, crop residues, municipal biosolids, kitchen waste, and vegetable waste [6]. OWs are typically byproducts of municipal, industrial, and farming operations; they are referred to as "wastes" because they are undesirable or useless goods [7]. OW production is rising and is predicted to do so for some time to come, which has become a global problem. Food, energy, and water supplies may be impacted if the massive amount of OW is improperly managed and the quality of the water, air, and land is compromised [8].

To address this waste generation tendency, the current waste management system must incorporate new concepts, procedures, and technology, particularly for solid organic waste (SOW). Now, most SOW are disposed using traditional methods including burning and landfilling. However, landfills and incineration are not sustainable ways of SW disposal, due to the limited space and negative impact on the environment [1, 2]. Landfill releases gases such as CH_4, CO_2, and N_2O play a major role in global warming. Leaching and runoff of heavy metals and other contaminants from *landfill* areas also pollute groundwater [1, 9, 43]. Furthermore, the burning of SW releases harmful chemicals such as dibenzofurans and chlorinated dibenzo-*p*-dioxins as well as greenhouse gases [1]. Therefore, we need some sustainable ways of disposing of SOW which valorize our generated SOW.

Composting (CP) and anaerobic digestion (AD) are the most used biological valorization processes. Both the processes degraded OW through biological processes, but take place in anaerobic and aerobic environments, respectively. Through this chapter, we discussed the relationship between population growth and OW generation, global and Indian scenarios of OW management, CP and AD process, and their role in sustainable OW management strategies.

10.2 POPULATION AND WASTE GENERATION

The world's population currently stands at 8 billion, and estimates indicate that by 2080, it will likely reach 10.4 billion [10]. This is the result of advances in living conditions and health rates, especially in children. The alteration of lifestyle and household consumption habits brought about by population growth and rural and urban development leads to a rise in waste generation and other issues in contemporary society. Waste characteristics, waste content, and waste volume have all altered as a result of changes in household consumption habits [11–13]. In addition to raising living standards in developing nations, economic growth and urbanization have increased the amount and complexity of municipal solid waste (MSW).

Global waste production is a result of growing population density, rural-to-urban migration, and industrialization. These factors provide significant economic, social, and sociodemographic concerns [14, 15]. The increasing amount of OW produced is a consequence of population growth, economic development, and human activity [16]. Global MSW generation is now expected to reach 2.01 billion tons per year. It is expected that by 2050, the figure will climb to 3.40 billion tons. This is more than double the pace of population growth at the time [15]. In addition, from that total MSW generation, at least 33% is not managed in an environment-friendly manner.

Everyone on the planet creates between 0.11 and 4.54 kg of waste every day, with an average of 0.74 kg [17]. The amount of waste generated and income level have a strong positive correlation [18]. As a result, high-income nations produce more waste than low-income nations. In high-income countries, daily per capita waste generation is anticipated to increase by 19% by 2050, whereas it is expected to increase by 40% or more in low and middle-income countries [17]. Table 10.1 shows the 10 top waste producers in the world with their waste generation, recycling, incineration, landfill, open dumping, and uncounted waste amount.

TABLE 10.1 Top 10 waste producers of the world

Rank	Country	Estimated Annual MSW Generation (Million Tonnes)	Per Capita Waste Generation (kg/day)
1	United States	292	2.2–2.5
2	China	235	1.0–1.2
3	India	170	0.4–0.5
4	Brazil	80	1.1
5	Indonesia	65	0.7
6	Russia	60	1.1
7	Mexico	54	1
8	Japan	52	1.0–1.1
9	Germany	50	1.4–1.6
10	Nigeria	42	0.5

Source: Global Waste Index 2025) [19].

10.3 EFFECT OF ORGANIC WASTE ON THE ENVIRONMENT

A large volume of OW generation is not only a social or human issue; it is also a big issue for the environment. Air pollution and climate change are intimately linked to inadequate or nonscientific waste management practices, which have an impact on numerous ecosystems and species. OWs contaminated with various toxic pollutants such as pesticides [20], pharmaceuticals [20], and heavy metals [21] pose serious risks to surface and groundwater, soil, and food chains. Landfills are used to dispose of approximately 37% of the waste produced worldwide and others are disposed of by recycling, open dumping, incineration, and CP. Additionally, there are many proposals about the environmental effects of each disposal frame.

OWs are collected from different areas and finally disposed of in landfill sites. There is a significant environmental emission of particulate matter and carbon dioxide (CO_2) during the collection, transportation, and processing of OW. Landfill dumping contaminates soil, surface, and subsurface water, causes land area loss, and emits methane into the atmosphere. One other significant area of worry is the potential climatic impact resulting from fugitive methane emissions from the natural gas supply chain. These releases are mostly responsible for short-term global warming since methane (CH_4) is 26 times more hazardous than CO_2. Fugitive CH_4 emissions also trigger landfill fires. This is a result of waste burning spontaneously when CH_4 and oxygen (O_2) come into contact.

Leachate is the source of contamination in groundwater and surface waters. Leachate is a foul-smelling liquid that leaks from landfills as OW degrades anaerobically. Leachate comprises a variety of harmful organic contaminants [20], heavy metals [21], ammonia nitrogen compounds, and other more complicated components than residential sewage. Since groundwater is used as a source of drinking water, leachate contaminated with groundwater and other natural sources causes risk to the population.

Open dumping results in vegetation loss, slope instability, and unintentional fire hazards. Open burning and poorly managed incineration send harmful chemicals and pollutants into the air, raising the risk of birth abnormalities, cancer, and other health problems. Concerns about human health are also noted at every stage of the handling, treating, and disposing of garbage. Health issues are associated with both direct and indirect exposure to waste. Indirect exposure happens when polluted water, soil, or food is consumed. Direct exposure happens when one is exposed to dangerous materials, parasite worms or insects, and waste odors. The principal routes of exposure are ingestion of tainted landfill leachate water, inhalation of gas emissions from incinerators and landfills, and ingestion of food tainted with bacteria, viruses, and persistent organic pollutants from waste [22].

10.4 GLOBAL AND INDIAN CONTEXTS FOR ORGANIC WASTE MANAGEMENT

Waste management is the process of handling, observing, and controlling waste materials as well as their transportation, collection, processing, recycling, treatment, and disposal. The term describes the technique for dealing with waste materials left over by human activity and decreasing waste's impacts on the environment and public health. Waste management is an essential service for protecting the environment and saving costs or reducing expenses for disposal. The first step of waste management is to provide reliable collection services to all citizens and eliminate uncontrolled dumping and open burning. Currently, the world is progressing toward this target.

Managing the production of SW is a challenging task that faces challenges globally due to a lack of social consciousness, accountability, and innovative solutions [23]. Human activity generates waste when energy and resources are used inefficiently [24, 25]. To control or manage immense volumes of waste, proper waste treatment and disposal practices are needed. Improving waste management practices helps in the reduction of raw material generation and by recycling we can reduce the production of new resources. Without a proper waste management system, this growth in waste generation leads to pollution of the rivers, lakes, and the ocean ecosystem.

The global production of municipal solid garbage is predicted to increase from 2.3 billion tons in 2023 to 3.8 billion tons by 2050 [26]. As we mentioned above, 45% of total generated SW comprises OW. The generation of OW and its management is a widely discussed topic in global energy and resource recovery because it is necessary to recover energy and materials from OW to execute the circular economy. To manage or handle this amount of waste needs modern techniques and instruments. Also, for handling or managing organic parts, CP and conversion are the most popular and secure methods. Due to the low economy, lack of social consciousness, accountability, and innovative solutions, developing countries cannot handle them through effective ways. Therefore, the effects of unsustainable waste management are more pronounced in impoverished metropolitan areas as compared to industrialized and developing countries. Around 90% of waste in low-income countries is burned in the open or dumped in uncontrolled sites, which poses major risks to public health, safety, and the environment [18]. Poor waste management practices provide a breeding ground for disease-carrying insects, create urban cruelty, and produce methane, which accelerates climate change. However, the collection, treatment, and disposal of waste in developed nations are well organized and waste is appropriately collected by a curb system and separated for recycling. In Europe, less than 40 million tons of OW, or biowaste, are collected separately and processed into high-quality compost and digestate. This is only 17% of the total generated MSW. Every year, between 118 and 138 million tons of biowaste are produced across the European Union; of this, only around 40% are recycled through high-quality compost and digestate [27]. This is because developed countries prefer open dumping and disposal since it is less expensive and requires less maintenance [28].

Every year, India generates 62 million tons of waste. From that, 70% of waste is collected, of which 12 million are processed and 31 million are disposed of in landfills [29]. MSW generation is expected to exceed 165 million tons by 2030, due to changing consumption patterns and strong economic expansion. As a result, the majority of landfills in India have surpassed both their 20-m height restriction and capacity [29]. It is estimated that these locations encompass over 10,000 ha of metropolitan land.

In India, collected, transported, disposed of, and segregated SW are under the jurisdiction of the urban local bodies. As per reports, Indian MSW constitutes 51% of OW, followed by 32% of inert and non-OW. Glass, paper, and plastics make up about 17% of waste that is categorized as recyclable [30]. From the total amount of generated waste, 81% of the waste is disposed of in landfills without treatment putting human health at risk and creating major environmental issues [30]. India can recycle this waste and generate 5.4 MT of city compost per year. There are 700 CP facilities in operation in India, with an annual capacity of 18.9 million MT [31]. However, only 0.2 million MT are generated from the city compost because most of the plants are underutilized.

To handle sewage and MSW management, the Ministry of Housing and Urban Affairs launched "Swachh Bharat Mission Urban" in 2014. To improve the international waste management hierarchy and the integrated SW management system, the Ministry of Environment Forests and Climate Change published new Solid Waste Management Rules in 2016. The OW management system still confronts certain difficulties despite the government's numerous endeavors, including improper source segregation, contamination with unrelated materials, odor problems, unsafe use, and is expensive in comparison to other products. Furthermore, neither the locals nor the government is very knowledgeable about the effective compost policy tools.

10.5 SUSTAINABLE WASTE MANAGEMENT PRACTICES AND THEIR IMPORTANCE AND BENEFITS

The World Commission on Environment and Development coined the term "sustainable" in its Brundtland Report. According to them, sustainable development is "development that satisfies current needs without jeopardizing future generations' ability to satiate their own needs." Considering this idea, current activities should not endanger a society's culture or level of living.

Countries vary in terms of their size, wealth, living standards, cultures, political and administrative structures, and ability to achieve sustainable development. While wealthy and technologically advanced nations may find it easier to pursue sustainable development, this is not always the case.

Moving away from the traditional make-use-dispose production model toward a more circular economy is the main component of sustainable waste management (SWM) practices. SWM aims to reduce the quantity of SW burned or disposed of in landfills while increasing the useful life of resources [44]. Figure 10.1 depicts the SWM hierarchy and practically all waste management practices fall under this hierarchy. This hierarchy provides a framework where waste management options are ranked to help decision-makers decide "how to deal with waste." The option that is most beneficial to the environment is waste avoidance, even if disposal is the least preferred option in the hierarchy.

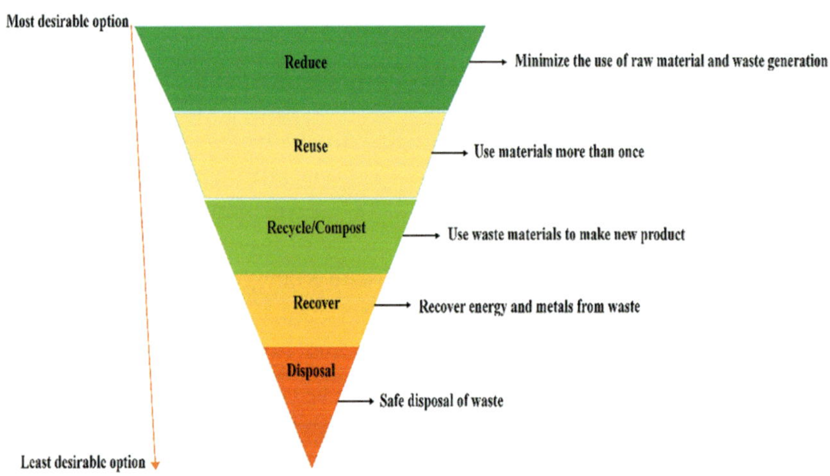

FIGURE 10.1 Waste management hierarchy.

One sentence is insufficient to fully express the importance or benefits of SWM practices. SWM practices keep the environment fresh and neat; save the earth and conserve energy; reduce waste disposal costs; cut back on greenhouse gas emissions; encourage energy conservation; reduce the production of SW annually; increase economic efficiency in the production and consumption of goods that can be achieved by developing markets for recycled materials, treating, and disposing of resources properly, and using them to improve economic efficiency; decrease in the need for more landfill space; reduces the odor and visual intrusion of landfills; improved corporate reporting and green badges for businesses; enables more effective and efficient waste management decisions to be made during the event; supports collaboration between stakeholders (such as residents, private facility owners, state, local, tribal, and territorial governments, and owners of storage, treatment, and disposal facilities) in order to prevent accidents; and protects natural resources by enabling the company to substitute waste products in their manufacturing process.

10.6 COMPOSTING AND ANAEROBIC DIGESTION OF ORGANIC WASTE

The OW management system has traditionally been based on CP and AD, and sometimes a combination of both [32, 33]. CP is a controlled, aerobic technique used to treat SOW with little cost, where SOW is converted into a nutrient-rich and biologically stable soil amendment. Three steps comprise the self-heating process of CP: mesophilic, thermophilic, and maturity [33]. Its primary objective is to manage OW [34]. CP takes advantage of the aerobic metabolism (less than 6% v/v of oxygen) of various bacteria that can use OW as a source of carbon and energy. They mostly turn OW into CO_2, water, compost, and heat due to their metabolisms [35]. During the active phase, a thermophilic environment (above 55°C) is required to inactivate pathogens and weed seeds [34]. CP is a desirable soil supplement because it is an organic-rich material mostly made up of stable and resistant organic components [36]. CP has been used widely for many years to manage OW because of several advantages: (1) agriculture's ability to recover plant nutrients and stable organic matter; (2) maintain product cleanliness; (3) reduce volume owing to considerable moisture loss during the active phase; (4) demand minimal initial outlay of funds; and (5) perform simple tasks without the need for professional labor.

Microorganisms use a set of biological processes called AD to break down biodegradable materials in the absence of oxygen [37]. AD anaerobically breaks down OW to generate digestate and biogas in four sequential steps: hydrolysis, acidogenesis, acetogenesis, and methanogenesis [38]. AD can function in three different temperature regimes: mesophilic (15–20°C, 35–40°C, and 55–60°C, respectively); the last two are better for breaking down organic matter and producing biogas. In addition, AD types can be categorized as semidry, dry, or wet. In both mono- and multistage AD, whether the procedure stages are completed in a single digester or isolated in two or more reactors, the process is based on adding up to solid substances within the reactor.

AD is used to break down a variety of waste materials such as sewage sludge, animal dung, food waste, and lignocellulosic biomass [39, 40]. AD has been utilized increasingly often in the last 20 years to handle OW from municipalities and industry, both liquid and solid. Government laws that offer financial incentives play a significant part in driving the process.

10.7 ROLE OF ANAEROBIC DIGESTION AND COMPOSTING IN SUSTAINABLE ORGANIC WASTE MANAGEMENT STRATEGIES

When resources are scarce, waste or residues especially OW should be seen as sources of secondary raw materials for civilization. SWM techniques provide alternate items for use in addition to reducing the amount of waste and landfill space. As of right now, the liquid phase of the digestate's biogas, compost, and nutrients only partially recovers the "value" of OW. The creation of renewable energy is often used to subsidize them due to their low economic value [33, 41].

As noted in the previous section, OW accounts for 45% of the MSW produced globally. A proper waste strategy is necessary to sustainably manage such a huge amount of OW. The most practical methods for managing OW sustainably are AD and CP. AD and CP are two efficient waste management techniques that can deal with this problem by recycling OW and producing useful goods.

Waste minimization or prevention, material reuse, recycling, CP, energy recovery, and landfill disposal are all included in the hierarchy of SWM. AD and CP systems meet the requirements of all three strategies by reducing waste output through recycling and producing valuable goods. Landfilling lessens the quantity of waste that is composted.

Waste is aerobically broken down, releasing CO_2 and heat as it forms compost [42]. Compost is a nutrient-rich product used as a soil amendment for food production in agriculture. Additionally, AD produces digestate and biogas from organic matter. The main ingredient in biogas is methane, which may be utilized to generate electricity sustainably and with less water usage than fossil fuels [43]. Because of its high nutrient content, digestate may be added to the soil [44]. Thus, using AD can improve the security of food, energy, and water while simultaneously reducing the environmental dangers related to OW.

10.8 CONCLUSION AND FUTURE RECOMMENDATION

The production of solid organic garbage is rising daily due to population expansion worldwide. The overproduction of SOW must be controlled with SWM techniques. The two most popular management techniques for lowering SOW are AD and CP systems. They each have unique processes of degradation and unique benefits. Compared to AD, CP has higher rates of microbial development and diversification, which can result in a higher organic breakdown efficiency. However, the largest annoyance of composing is the smell generated, which is mostly caused by NH_3, H_2S, and VOC emissions. In contrast, AD generates biogas as a sustainable energy source, which benefits the environment by emitting fewer greenhouse gases and odors. Additionally, depending on plant size and final product value, AD may be more profitable than CP, but CP is more profitable at smaller scales. While a considerable number of farmers are thought to be able to use animal manure on their farms by implementing on-farm AD, small-scale enterprises cannot benefit from AD due to its high cost. As a result, AD may be advantageous for centralized treatment of sewage sludge or food waste. Decentralized treatment, like on-farm animal maturity, may benefit from CP. However, because animal dung contains a lot of ammonia, emissions should be closely regulated. For this reason, current research on AD has prioritized process optimization and product value, while research on CP has concentrated on odor and greenhouse gas emissions reduction as well as CP time reduction. In AD, inoculation is essential for process improvement, which can be accomplished by adding certain hydrolytic microorganisms or recirculating digestate. In AD, micro-aeration can also be utilized to boost hydrolysis and encourage facultative hydrolytic bacteria.

In earlier research, the pretreatment of lignocellulosic biomass for AD has been thoroughly examined and debated. However, because of their high cost or slow reaction time, they are not currently practical to apply in a full-scale operation. AD's profitability may increase with the upgrading of biogas for use as a fuel for transportation and the alternative use of digestate. Aeration is essential to CP and needs to be maximized. Nonetheless, no single aeration plan can effectively deliver oxygen and reduce emissions at the same time. CP could be improved by multistage inoculation; however, this would complicate the process. Although pretreatment of substrates shows promise in reducing the duration of CP, it is not feasible for use. In CP, the addition of microbes to lower emissions has been tried, with varying degrees of success. Furthermore, it has not been discovered that a single addition works well to concurrently lower GHG emissions and odor. Additionally, there were significant differences in the reported greenhouse gas emissions from AD and CP between studies, which were mostly caused by differences in the study's borders and underlying assumptions. Future studies in the areas of economic and environmental analysis, process integration and optimization, and online monitoring and control should be carried out. Furthermore, more research is needed to comprehend the mechanisms of improvement and the interactions between the provided microorganisms and the existing bacteria to decrease emissions during CP or increase performance in AD.

KEYWORDS

- **organic waste**
- **anaerobic digestion**
- **composting**

REFERENCES

1. Chen, T., Zhang, S., & Yuan, Z. (2020). Adoption of solid organic waste composting products: A critical review. *Journal of Cleaner Production*, *272*, 122712. https://doi.org/10.1016/j.jclepro.2020.122712.
2. Lou, X. F., & Nair, J. (2009). The impact of landfilling and composting on greenhouse gas emissions—A review. *Bioresource Technology*, *100*, 3792–3798. https://doi.org/10.1016/j.biortech.2008.12.006.

3. Viéitez, E. R., & Ghosh, S. (1999). Biogasification of solid wastes by two-phase anaerobic fermentation. *Biomass and Bioenergy*, *16*, 299–309. https://doi.org/10.1016/S0961-9534(99)00002-1.
4. Karim, A. A., Kumar, M., Mohapatra, S., & Singh, S. K. (2019). Nutrient rich biomass and effluent sludge wastes co-utilization for production of biochar fertilizer through different thermal treatments. *Journal of Cleaner Production*, *228*, 570–579. https://doi.org/10.1016/j.jclepro.2019.04.330.
5. Pace, S. A., Yazdani, R., Kendall, A., Simmons, C. W., & VanderGheynst, J. S. (2018). Impact of organic waste composition on life cycle energy production, global warming and Water use for treatment by anaerobic digestion followed by composting. *Resources, Conservation and Recycling*, *137*, 126–135. https://doi.org/10.1016/j.resconrec.2018.05.030.
6. Mata-Alvarez, J., Macé, S., & Llabrés, P. (2000). Anaerobic digestion of organic solid wastes. An overview of research achievements and perspectives. *Bioresource Technology*, *74*, 3–16. https://doi.org/10.1016/S0960-8524(00)00023-7.
7. Westerman, P. W., & Bicudo, J. R. (2005). Management considerations for organic waste use in agriculture. *Bioresource Technology*, *96*, 215–221. https://doi.org/10.1016/j.biortech.2004.05.011.
8. Lin, L., Xu, F., Ge, X., & Li, Y. (2018). Improving the sustainability of organic waste management practices in the food-energy-water nexus: A comparative review of anaerobic digestion and composting. *Renewable and Sustainable Energy Reviews*, *89*, 151–167. https://doi.org/10.1016/j.rser.2018.03.025.
9. Mor, S., Ravindra, K., Dahiya, R. P., & Chandra, A. (2006). Leachate characterization and assessment of groundwater pollution near municipal solid waste landfill site. *Environmental Monitoring and Assessment*, *118* 435–456. https://doi.org/10.1007/s10661-006-1505-7.
10. Foflonker, F. (2024). Population growth. Retrieved from: https://www.britannica.com/science/population-growth.
11. Aziz, S. Q., Aziz, H. A., Bashir, M. J. K., & Yusoff, M. S. (2011). Appraisal of domestic solid waste generation, components, and the feasibility of recycling in Erbil, Iraq. *Waste Management & Research: The Journal for a Sustainable Circular Economy*, *29*, 880–887.
12. Widyaningsih, N., Tjiptoherijanto, P., Widanarko, S., & Seda, F. S. S. F. (2015). Linkage model between sustainable consumption and household waste management. *Procedia Environmental Sciences*, *28*, 195–203.
13. Darban Astane, A. R., & Hajilo, M. (2017). Factors affecting the rural domestic waste generation. *Global Journal of Environmental Science and Management*, *3*, 417–426.
14. Moya, D., Aldás, C., López, G., & Kaparaju, P. (2017). Municipal solid waste as a valuable renewable energy resource: A worldwide opportunity of energy recovery by using Waste-To-Energy Technologies. *Energy Procedia*, *134*, 286–295.
15. Prajapati, P., Varjani, S., Singhania, R. R., Patel, A. K., Awasthi, M. K., Sindhu, R., Zhang, Z., Binod, P., Awasthi, S. K., & Chaturvedi, P. (2021). Critical review on technological advancements for effective waste management of municipal solid waste—Updates and way forward. *Environmental Technology & Innovation*, *23*, 101749.

16. Twagirayezu, G., Huang, K., & Xia, H. (2023). Effects of bio-contaminants in organic waste products on the soil environment, In: *Fate of Biological Contaminants During Recycling of Organic Wastes*. Elsevier, pp. 187–212.
17. World Bank Group. (2022). Solid waste management. Retrieved from: https://www.worldbank.org/en/topic/urbandevelopment/brief/solid-waste-management.
18. Kumari, T., & Raghubanshi, A. S. (2023). Waste management practices in the developing nations: Challenges and opportunities. *Waste Management and Resource Recycling in the Developing World*, 773–797. https://doi.org/10.1016/B978-0-323-90463-6.00017-8.
19. G.W. Index, These are the biggest waste polluters worldwide, (2025). 382 https://sensoneo.com/global-waste-index/.
20. Pegu, R., Paul, S., Bhattacharyya, P., Prakash, A., & Bhattacharya, S. S. (2023). Exorbitant signatures of pesticides and pharmaceuticals in municipal solid wastes (MSWs): Novel insights through risk analysis, dissolution dynamics, and model-based source identification. *Science of The Total Environment*, *900*, 165855. https://doi.org/10.1016/j.scitotenv.2023.165855.
21. Paul, S., Choudhury, M., Deb, U., Pegu, R., Das, S., & Bhattacharya, S. S. (2019). Assessing the ecological impacts of ageing on hazard potential of solid waste landfills: A green approach through vermitechnology. *Journal of Cleaner Production*, *236*, 117643. https://doi.org/10.1016/j.jclepro.2019.117643.
22. Kiyasudeen, K., Ibrahim, M. H., Quaik, S., & Ismail, S. A. (2016). Organic waste management practices and their impact on human health. In: *Prospects of Organic Waste Management and the Significance of Earthworms*. pp. 245–252. https://doi.org/10.1007/978-3-319-24708-3_11.
23. Nimita Jebaranjitham, J., Selvan Christyraj, J. D., Prasannan, A., Rajagopalan, K., Chelladurai, K. S., & Gnanaraja, J. K. J. S. (2022). Current scenario of solid waste management techniques and challenges in Covid-19—A review. Heliyon. *8*, e09855. https://doi.org/10.1016/j.heliyon.2022.e09855.
24. Sharma, K. D., Jain, S. (2020). Municipal solid waste generation, composition, and management: the global scenario. *Social Responsibility Journal*, *16*, 917–948.
25. Laohalidanond, K., Chaiyawong, P., Kerdsuwan, S. (2015). Municipal solid waste characteristics and green and clean energy recovery in Asian megacities. *Energy Procedia*, *79*, 391–396.
26. UNEP. (2024). Global-waste-management-outlook-2024. Retrieved from: https://www.unep.org/resources/global-waste-management-outlook-2024.
27. Gottschall, R., Thelen-Jüngling, M., Kranert, M., & Kehres, B. (2023). Suitability of biowaste and green waste composts for organic farming in Germany and the resulting utilization potentials. *Agriculture*, *13*, 740. https://doi.org/10.3390/agriculture13030740.
28. Renou, S., Givaudan, J. G., Poulain, S., Dirassouyan, F., & Moulin, P. (2008). Landfill leachate treatment: Review and opportunity. *Journal of Hazardous Materials*, *150*, 468–493. https://doi.org/10.1016/j.jhazmat.2007.09.077.
29. Thamarai, M., & Naresh, V. S. (2023). Smart self-power generating garbage management system using deep learning for smart cities. *Microprocessors and Microsystems*, *98*, 104816. https://doi.org/10.1016/j.micpro.2023.104816.

30. Mani, S., & Singh, S. (2016). Sustainable municipal solid waste management in India: A policy agenda. *Procedia Environmental Sciences*, *35*, 150–157. https://doi.org/10.1016/j.proenv.2016.07.064.
31. Management of Organic Waste in India. (2021). Retrieved from: https://tuewas-asia.org/wp-content/uploads/2021/10/2021_FS_ExI_8_India-Organic-Waste_EN-1.pdf.
32. Ma, Y., & Liu, Y. (2019). Turning food waste to energy and resources towards a great environmental and economic sustainability: An innovative integrated biological approach. *Biotechnology Advances*, *37*, 107414. https://doi.org/10.1016/j.biotechadv.2019.06.013.
33. Alibardi, L., Astrup, T. F., Asunis, F., Clarke, W. P., De Gioannis, G., Dessì, P., Lens, P. N. L., Lavagnolo, M. C., Lombardi, L., Muntoni, A., Pivato, A., Polettini, A., Pomi, R., Rossi, A., Spagni, A., & Spiga, D. (2020). Organic waste biorefineries: Looking towards implementation. *Waste Management*, *114*, 274–286. https://doi.org/10.1016/j.wasman.2020.07.010.
34. Cerda, A., Artola, A., Font, X., Barrena, R., Gea, T., & Sánchez, A. (2018). Composting of food wastes: Status and challenges. *Bioresource Technology*, *248*, 57–67. https://doi.org/10.1016/j.biortech.2017.06.133.
35. Wang, S., & Zeng, Y. (2018). Ammonia emission mitigation in food waste composting: A review. *Bioresource Technology*, *248*, 13–19. https://doi.org/10.1016/j.biortech.2017.07.050.
36. Cucina, M., Tacconi, C., Sordi, S., Pezzolla, D., Gigliotti, G., & Zadra, C. (2018). Valorization of a pharmaceutical organic sludge through different composting treatments. *Waste Management*, *74*, 203–212. https://doi.org/10.1016/j.wasman.2017.12.017.
37. Zhou, S. X., Dong, Y. P., & Zhang, Y. L. (2012). Solid-state anaerobic digestion for methane production from corn stalks with stack-pretreated. *Materials Science Forum*, 697–698, 326–330. https://doi.org/10.4028/www.scientific.net/MSF.697-698.326.
38. Wang, P., Wang, H., Qiu, Y., Ren, L., & Jiang, B. (2018). Microbial characteristics in anaerobic digestion process of food waste for methane production—A review. *Bioresource Technology*, *248*, 29–36. https://doi.org/10.1016/j.biortech.2017.06.152.
39. Mata-Alvarez, J., Dosta, J., Romero-Güiza, M. S., Fonoll, X., Peces, M., & Astals, S. (2014). A critical review on anaerobic co-digestion achievements between 2010 and 2013. *Renewable and Sustainable Energy Reviews*, *36*, 412–427.
40. Lin, L., Xu, F., Ge, X., & Li, Y. (2019). Biological treatment of organic materials for energy and nutrients production—Anaerobic digestion and composting. *Advances in Bioenergy*, *4*, 121–181.
41. Clarke, W. P. (2018). The uptake of anaerobic digestion for the organic fraction of municipal solid waste—Push versus pull factors. *Bioresource Technology*, *249*, 1040–1043. https://doi.org/10.1016/j.biortech.2017.10.086.
42. Tuomela, M., Vikman, M., Hatakka, A., & Itävaara, M. (2000). Biodegradation of lignin in a compost environment: A review. *Bioresource Technology*, *72*, 169–183. https://doi.org/10.1016/S0960-8524(99)00104-2.
43. Yang, L., Ge, X., Wan, C., Yu, F., & Li, Y. (2014). Progress and perspectives in converting biogas to transportation fuels. *Renewable and Sustainable Energy Reviews*, *40*, 1133–1152. https://doi.org/10.1016/j.rser.2014.08.008.

44. Sheets, J. P., Yang, L., Ge, X., Wang, Z., & Li, Y. (2015). Beyond land application: Emerging technologies for the treatment and reuse of anaerobically digested agricultural and food waste. *Waste Management*, *44,* 94–115. https://doi.org/10.1016/j.wasman.2015.07.037.

CHAPTER 11

Insights into Hazardous and Special Wastes: A Case Study of Latin America

PRANGYA RATH[1], FARHANA RAHMAN[2], HARSHITA JAIN[1], LAXMI KANT BHARDWAJ[3*], PANKAJ KANTI JODDER[4], BHAWANA SHARMA[5], MOHARANA CHOUDHURY[6], CÍNTIA SOARES[7], and NATAN PADOIN[8]

[1]*Amity Institute of Environmental Sciences (AIES), Amity University, Noida, Uttar Pradesh, India*

[2]*University of Technology Sydney (UTS), Ultimo, Sydney, NSW, Australia*

[3]*Amity Institute of Environmental Toxicology, Safety & Management (AIETSM), Amity University, Noida, Uttar Pradesh, India*

[4]*Department of Geography and Earth Science, University of North Carolina at Charlotte, Charlotte, NC, USA*

[5]*Department of Bioscience and Biotechnology, Banasthali Vidyapith, Vanasthali, Rajasthan, India*

[6]*Environmental Research and Management Division, Voice of Environment (VoE), Guwahati, Assam, India*

[7]*Laboratory of Materials and Scientific Computing (LabMAC), Chemical and Food Engineering Department, Federal University of Santa Catarina, Florianópolis, Santa Catarina, Brazil*

[8]*Department of Chemical and Food Engineering, Federal University of Santa Catarina, Florianópolis, Santa Catarina, Brazil*

*Corresponding author

Data Analytics and Smart Technologies in Solid Waste Management: Advancing Sustainability. Moharana Choudhury, Sushobhan Majumdar, and Avnish Chauhan (Eds.)

DOI: 10.1201/9781779640833-14

ABSTRACT

In recent decades, waste has emerged as a pressing global environmental concern, spurred by urbanization and population growth. In developing nations like Latin America, waste poses significant human health risks. Waste composition has evolved notably since the 20th-century technological revolution, including pollutants such as pesticides, pharmaceuticals, plastics, and nonmaterials, complicating waste management efforts worldwide. This chapter deals with the escalating health and environmental challenges stemming from toxic and special waste in Latin America. It examines the rapid emergence of electronic waste (e-waste), pesticides, medical, mining, and industrial waste due to technological advancements and population growth. Particular attention is paid to the sometimes-disregarded predicament of garbage pickers, constantly exposed to dangerous products, highlighting the necessity of protective regulations. The research illustrates the potential and difficulties in trash management by examining the complex link between garbage creation and technological advancement. To guarantee a sustainable future for waste management in the area, emphasis is focused on the significance of international collaboration and the creation of creative policies. This chapter aims to offer a thorough and innovative viewpoint on handling hazardous and unusual trash in Latin America.

11.1 INTRODUCTION

South America, Central America, Mexico, and the Caribbean islands are all included in the region known as Latin America [1]. For many years, the area has struggled with the management of hazardous and special waste (H&SW) [2]. Environmental conditions are becoming an increasingly pressing issue in many contexts, from thriving urban centers to isolated wilderness locations. H&SW includes a broad spectrum of materials that, because of their intrinsic qualities, might be hazardous to the environment and human health. Varshney et al. [3] stated that H&SWs exhibit toxicity, flammability, corrosivity, or reactivity or contain substances listed as hazardous by regulatory authorities. Various nations may enforce distinct regulations and classifications for perilous and exceptional wastes, although some broad classifications include the following:

- Industrial wastes: They are generated from manufacturing processes [4], for example, textiles, plastics, microplastics, chemicals, and so on.
- Electronic wastes (e-wastes): These wastes are discarded electronic devices that may contain hazardous heavy metals like lead (Pb), mercury (Hg), and cadmium (Cd) [5, 6], for example, computers, ovens, mobile phones, television, and so on.
- Radioactive wastes: These wastes include radioactive substances that are generated from nuclear power plants, and research laboratories [7], for example, nuclear fuel, contaminated tools, and so on.
- Medical or infectious wastes: These wastes are generated from hospitals and medical laboratories that contain hazardous materials [8, 9], for example, medical caps and shoes, used needles, contaminated gloves, and so on.
- Mining wastes: These wastes are generated from mines and can contaminate the soil [10, 11], for example, waste rock, mine water, and so on.
- Construction and demolition wastes: These wastes are generated during the construction or demolition of buildings [12], for example, concrete, wood, plastics, metal, and so on.
- Household hazardous wastes: These wastes are generated from household activities [13], for example, plastic waste, municipal solid waste, electronics, tires, and so on.
- Oil and petroleum-related wastes: These wastes are generated from oil industries, oil refining industries, and petrol pumps [14], for example, hazardous chemicals, spent oil, oil-contaminated soils, some persistent organic pollutants, and so on.
- Special wastes: These wastes include those materials that do not fit into standard waste categories due to their unique characteristics [15], for example, inks, liquid from industries, cooling fluids, and so on.
- Fluorescent bulbs and lamps: They contain Hg particles and therefore these wastes require special handling and disposal methods [16], for example, bulbs, compact fluorescent bulbs (CFLs), and so on.
- Pesticides and agrochemical wastes: They are generated from agricultural activities and can contaminate soil. Owing to its volatile nature, most of the chemicals are transported worldwide through the long-range atmospheric transport mechanism (LRATM) [17–23], for example, obsolete pesticides and fertilizers, and so on.

- Batteries: They are generated from industrial and household activities [24], for example, plastics, heavy metals, and so on.

Galvão et al. [25] have emphasized the intricate link between environmental health and hazardous waste. The need for an integrated approach is straightforward: one that combines innovative waste management technologies, public health initiatives, and so on [26]. Hoornweg and Bhada-Tata [27] studied the H&SWs and reported different types of wastes such as organic waste (54%), cardboard and paper waste (16%), plastic waste (12%), glass (4%), metals (2%), and other wastes (12%) (Figure 11.1). Images of different types of waste generated in Latin America are shown in Figure 11.2.

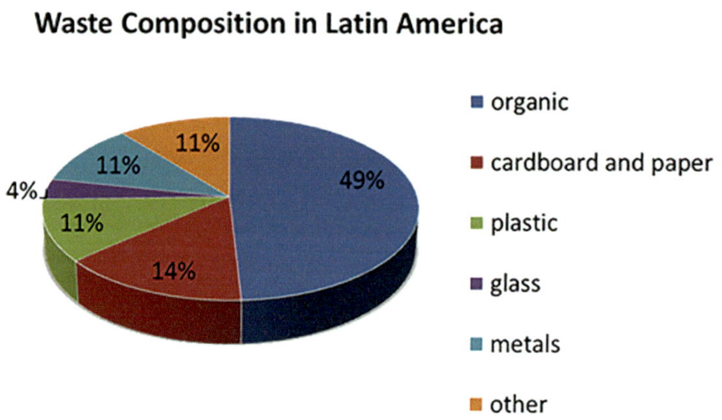

FIGURE 11.1 Different types of waste composition in Latin America.

Source: Adapted from Ref. [27]

The primary aim of this chapter is to offer an overview of the health issues associated with H&SWs in Latin America. It will explore the origins of this crisis, its effects on both individuals and the environment, and the suggested approaches to tackle it.

11.2 A TOXIC LEGACY: HOW THE WASTE CRISIS IN LATIN AMERICA DEVELOPS INEQUALITY

Hazardous waste is a widespread problem in Latin America, although its impact is not equally felt. Some areas and localities bear a disproportionate

FIGURE 11.2 Photographs depict a brief overview of different types of waste generated in Latin America. (a) The wastes dumped at the dumping site; (b) dumping sites for different types of wastes; (c, d) the height of the accumulated garbage.

Source: Cruvinel et al. [8].

amount of the hazards and fallout. Think of the following categories as being especially at risk.

The urban poor: around overflowing dumps and contaminated rivers in expanding megacities, informal colonies frequently spring up. The issue is exacerbated by inadequate garbage collection and sanitation, which provides a haven for the spread of illness and environmental dangers [8].

- Indigenous villages—deep within the Amazon: indigenous groups have historically coexisted peacefully with their surroundings in the Andean highlands and rainforest. However, hazardous waste from mining, logging, and agriculture is frequently dumped onto fragile ecosystems. Higher incidence of cancer problems, birth defects, and respiratory ailments are among the effects [28, 29].
- Rural laborers: the widespread usage of pesticides persists across extensive agricultural regions. The majority of those affected by health difficulties, including as neurological illnesses and persistent respiratory problems, are farmworkers and their families [30, 31].
- Women and children: hazardous waste has a gendered effect that disproportionately affects women and children. Children are especially susceptible to toxins because of their developing bodies and immune systems [32].

11.3 FACTORS AFFECT THE PRODUCTION OF HAZARDOUS AND SPECIAL WASTES IN LATIN AMERICA

Waste management is becoming a more significant issue for the developing metropolitan centers of Latin America. Owing to the region's expanding population, increasing consumption, and quickening technological developments, waste production is outpacing current infrastructure and endangering the ecosystem. Growing urbanization and new technological advancements fuel increased consumerism, which increases packaging waste [33]. This covers not just commonplace goods but also the expanding issue of electronic garbage, or e-waste, as technology ages [10]. In addition, the process of urbanization itself encourages a shift in lifestyle toward handy single-use products that add significantly to waste streams [34]. Projects involving construction and demolition also greatly contribute to the total trash picture [35].

These concerns are made worse by the improper handling of plastic garbage. Single-use plastics in particular should be disposed of properly to avoid polluting landfills and harming wildlife and coastal areas [36, 37–54]. Greater waste generation is produced by larger populations, and if waste management systems are not developed enough to keep up with this expansion, damage may result in threats to public health and the environment [38, 55].

The solution to this issue must take a diverse approach. The infrastructure for waste management, which includes facilities for disposal, recycling, and transportation, has to be strengthened [39]. Public education on proper recycling and trash disposal methods is also essential [40]. Strict laws and policies that promote recycling and decrease the negative environmental effects of trash disposal are ultimately necessary to effectively manage the growing waste stream in Latin America [41].

11.4 EFFECTS OF HAZARDOUS AND SPECIAL WASTES ON THE ENVIRONMENT AND HUMAN HEALTH

Environmental risks are enormous in Latin America due to a variety of H&SWs that come from industrial, electronic, and medicinal sources [26]. Ecological imbalances are exacerbated by air, water, and soil pollution brought on by inappropriate waste disposal. Hazardous pollutants contaminate ecosystems and threaten biodiversity [26]. There is an urgent need to address the causes of public health concerns. A study conducted by Moraes et al. [41] focused on the hazards inevitably encountered by garbage pickers and highlighted the critical need for improved safety protocols and legislative changes. Waste pickers confront socioeconomic difficulties such as unstable working conditions, and physical risks [42]. Physical risks are common in waste pickers throughout their work such as sharp items, and dangerous substances can harm them [13].

The agricultural sector in Latin America confronts severe dangers of prolonged pesticide use, leading to a spectrum of health issues [43]. A study by Olivero-Verbel et al. [44] focused on the exposure of Pb and Hg due to H&SW in the Colombian and Brazilian Amazon. Inadequate garbage collection and the predicament of indigenous people were brought to light by this circumstance, underscoring the pressing need for efficient waste management and public health initiatives to protect vulnerable groups. Children in these areas are more sensitive and exposed to toxins in the air they breathe, as highlighted by Laborde et al. [45]. Only 55% of urban solid waste is disposed of in permitted landfills, according to Lopez-Yamunaqué and Iannacone [46], highlighting the inadequate reach of final disposal services.

11.5 CURRENT STRATEGIES AND POLICIES FOR THE MANAGEMENT OF HAZARDOUS AND SPECIAL WASTE

The countries of the Latin American region have several challenges in hazardous waste management. Sustainable waste management practices and circular economy principles are being increasingly emphasized to mitigate the environmental impact of growing urban areas in the region [35]. Recent studies have shown how important it is for Latin America to have effective garbage management systems to protect the environment and the well-being of individuals. To save the environment, Margallo et al. [26] emphasized the need to employ sophisticated waste management technologies. Hettiarachchi et al. [47] mentioned how many individuals in the region employ unlawful means to reduce debris and urge for waste recovery solutions. According to Lethbridge [48], municipal solid waste (MSW) management services are crucial for safeguarding public health and the environment. Terleeva [43] studied MSW and stated that several challenges in the management of MSW affect the transition to a circular economy.

Hoornweg and Giannelli [49] looked into the role that the private industry plays in collecting garbage. In their investigation of the challenges of managing urban solid waste, Lopez-Yamunaqué and Iannacone [46] focused on the insufficient supply of disposal services. Qian et al. [50] addressed how important is to create risk management plans for individuals and governments. Margallo et al. [26] emphasized the requirement for advanced waste management technologies to mitigate the environmental impacts of H&SWs [51].

11.6 CHALLENGES IN THE MANAGEMENT OF HAZARDOUS AND SPECIAL WASTE IN LATIN AMERICA

Recent studies emphasized the emerging strategy of collaboration between academia and industry in Latin America for hazardous waste management [52].

- waste management procedures are greatly impacted by businesses that produce hazardous waste, including manufacturing, mining, and chemical industries.
- challenges in hazardous waste management in Latin America include inadequate funding, lack of technical expertise, insufficient monitoring and enforcement mechanisms, and issues related to illegal dumping.

The waste management situation in Latin America and the Caribbean (LA&C) and its current difficulties and potential have been examined. Moreover, considering the imperative to harmonize these policies with the concept of a circular economy and the anticipated greenhouse gas (GHG) emissions reduction strategies outlined in the Paris Agreement, it is plausible to speculate that the LA&C industry could experience substantial changes from 2020 to 2030 [15].

The management of MSW is of utmost importance due to its significant impact on GHG emissions. Costa Rica has made remarkable advancements in the management of hazardous waste [53]. The collective responsibility (CR) concept, involving the separate collection of biowaste and recyclables, contrasts with business as usual (BAU) and market individualism (MI) scenarios, which highlight the necessity of sustainable waste management strategies, supported by frameworks like the Basel Convention and Rio Summit [46].

11.7 SUSTAINABLE AND LONG-TERM SOLUTIONS FOR HAZARDOUS AND SPECIAL WASTES

Managing H&SWs sustainably requires a comprehensive approach that addresses their generation, handling, treatment, and disposal. Here are some sustainable and long-term solutions:

- the most efficient approach is to implement and adopt cleaner production methods.
- promote the recycling and reuse of hazardous materials whenever possible and decrease the negative effects on the environment.
- utilize advanced treatment technologies to neutralize, detoxify, or render hazardous wastes less harmful, which may include biological, chemical, thermal, and physical processes.
- strengthen and enforce regulations governing the management of H&SWs. Strict compliance ensures businesses adhere to responsible waste management practices.
- increase public awareness about the hazards and the importance of proper disposal. Education can promote responsible consumer behavior, encouraging safe disposal practices.
- collaborate with international organizations to develop and implement standardized guidelines managing H&SWs.
- invest in research and development to discover new technologies for managing hazardous waste.

11.8 RECOMMENDATIONS

The author suggests several recommendations for the minimization of the production of H&SW:

To establish equitable and sustainable waste management techniques with the cooperation between the public, corporate sectors, and nongovernmental groups. Local governments and educational institutions should work together to enforce efficient waste management regulations. There is a need for a comprehensive method that includes cutting-edge technology and improved governance for solving the problem of H&SW. To adopt advanced and sustainable technologies like waste-to-energy and environmentally sound landfill systems for mitigating environmental impacts. To ensure adequate resources and robust enforcement mechanisms for deterring illegal dumping, promoting compliance, and upholding the integrity of waste management systems.

11.9 CONCLUSION

Many communities in Latin America are dealing with the consequences of H&SWs. Ignoring these consequences will only continue the cycle of injustice and environmental degradation. By focusing on these voices and recognizing the vulnerabilities of regions and communities, we can develop a more nuanced understanding of the waste crisis. This understanding can help us advocate for solutions prioritizing justice and ensuring fairness in distributing responsibility and resources. Then, it may be possible to truly heal the wounds caused by this legacy on planet Earth, securing a healthier future for all habitats in Latin America. These hazardous wastes need specialized handling, treatment, and disposal techniques. Existing policies and regulations often need to address the complexities of dangerous waste management effectively. Latin American countries can improve hazardous waste management through regulatory updates, capacity building, investment in infrastructure, and fostering collaboration between the government, industries, and communities. This approach should prioritize the utilization of cutting-edge technologies, the enhancement of operational proficiency, and an increase in the involvement of the business sector.

ACKNOWLEDGMENTS

The authors are thankful to their respective organizations for their support and cooperation in doing this study.

KEYWORDS

- **environmental contaminants**
- **health hazards**
- **resource management**
- **sustainable waste management**

REFERENCES

1. Adams, J., Pendlebury, D., Potter, R., & Szomszor, M. (2021). Latin America: South and Central America, Mexico and the Caribbean.
2. Cruvinel, V. R. N., Marques, C. P., Cardoso, V., Novaes, M. R. C. G., Araújo, W. N., Angulo-Tuesta, A., & Da Silva, E. N. (2019). Health conditions and occupational risks in a novel group: waste pickers in the largest open garbage dump in Latin America. *BMC Public Health*, *19*, 1–15.
3. Varshney, R., Singh, P., & Yadav, D. (2022). Hazardous wastes treatment, storage, and disposal facilities. In: *Hazardous Waste Management*. Elsevier, pp. 33–64.
4. Gupta, P., Sharma, A., & Bhardwaj, L. K. (2023). Solid waste management (SWM) and its effect on environment & human health. 2023. https://doi.org/10.20944/preprints202309.0384.v1
5. Bhardwaj, L. K., Rath, P., & Choudhury, M. (2023). Assessment of E-waste pollution level in soil and water and its impact on human health: A review. Preprints, 111714. https://doi.org/10.20944/preprints202311.1714.v1.
6. Bhardwaj, L. K., Rath, P., & Tokas, R. (2024). E-waste management in developing countries current practices challenges disposal and impact on human health and environment. In: *Sustainable Solutions for E-Waste and Development*. IGI Global Scientific Publishing, pp. 90–102. . https://doi.org/10.4018/979-8-3693-1018-2.ch007
7. Deng, D., Zhang, L., Dong, M., Samuel, R. E., Ofori-Boadu, A., & Lamssali, M. (2020). Radioactive waste: A review. *Water Environment Research*, *92*(10), 1818–1825.
8. Cruvinel, V. R. N., Marques, C. P., Cardoso, V., Novaes, M. R. C. G., Araújo, W. N., Angulo-Tuesta, A., Escalda, P. M. F., Galato, D., Brito, P., & Da Silva, E. N. (2019). Health conditions and occupational risks in a novel group: Waste pickers in the largest open garbage dump in Latin America. *BMC Public Health*, *19*(1), 581.
9. Bhardwaj, L. K., Rath, P., Bajpai, S., Upadhyay, D., Jain, H., Kumar, N., & Sinha, S. (2023). COVID-19 and the interplay with antibacterial drug resistance. . https://doi.org/10.20944/preprints202312.0842.v1
10. Andeobu, L., Wibowo, S., & Grandhi, S. (2023). Environmental and health consequences of E-waste dumping and recycling carried out by selected countries in Asia and Latin America. *Sustainability*, *15*(13), 10405.
11. Choudhury, M., Jyethi, D. S., Dutta, J., Purkayastha, S. P., Deb, D., Das, R., Roy, G., Sen, T., & Bhattacharyya, K. G. (2022). Investigation of groundwater and soil quality near to a municipal waste disposal site in Silchar, Assam, India. *International Journal of Energy and Water Resources*, *6*(1): 37–47. https://doi.org/10.1007/s42108-021-00117-5
12. Berry, T. A., Belluso, E., Vigliaturo, R., Gieré, R., Emmett, E. A., Testa, J. R., Steinhorn, G., & Wallis, S. L. (2022). Asbestos and other hazardous fibrous minerals: Potential exposure pathways and associated health risks. *International Journal of Environmental Research and Public Health*, *19*(7), 4031.
13. El-Khateeb, M. (2022). Household hazardous waste: Handling, precaution and hazard reduction. *Egyptian Journal of Chemistry*, *65*(8), 625–642.
14. Kutralam-Muniasamy, G., Pérez-Guevara, F., Elizalde-Martínez, I., & Shruti, V. C. (2020). Review of current trends, advances and analytical challenges for microplastics contamination in Latin America. *Environmental Pollution*, *267*, 115463.

15. Orona-Návar, C., García-Morales, R., Loge, F. J., Mahlknecht, J., Aguilar-Hernández, I., & Ornelas-Soto, N. (2022). Microplastics in Latin America and the Caribbean: A review on current status and perspectives. *Journal of Environmental Management*, *309*, 114698.
16. Cayumil, R., Khanna, R., Konyukhov, Y., Burmistrov, I., Kargin, J. B., & Mukherjee, P. S. (2021). An overview on solid waste generation and management: Current status in Chile. *Sustainability*, *13*(21), 11644.
17. Bhardwaj, L. K., Sharma, S., Ranjan, A., & Jindal, T. (2019). Persistent organic pollutants in lakes of Broknes peninsula at Larsemann Hills area, East Antarctica. *Ecotoxicology*, *28*, 589–596.
18. Bhardwaj, L. K., & Jindal, T. (2020). Persistent organic pollutants in lakes of Grovnes Peninsula at Larsemann Hill area, East Antarctica. *Earth Systems and Environment*, *4*, 349–358.
19. Bhardwaj, L. K., Sharma, S., & Jindal, T. (2021). Occurrence of polycyclic aromatic hydrocarbons (PAHs) in the Lake water at Grovnes Peninsula Over East Antarctica. *Chemistry Africa, 4*, 965–980.
20. Bhardwaj, L. K., & Jindal, T. (2022). Polar ecotoxicology: Sources and toxic effects of pollutants. In; *New frontiers in environmental Toxicology* . pp. 9–14.
21. Mairaj, S., Nagar, R. D., Bhardwaj, L. K., Rehman, F., Punnakal, A., Chauhan, A., & Jindal, T. (2022). Antioxidant responses of arsenite-induced oxidative stress in rice (Oryza sativa L.) and its modulation by eugenol (extracted from *Ocimum sanctum*). *Indian Journal of Agricultural Research*, *56*(4), 396–400.
22. Alam, S., Bhardwaj, L. K., Mallick, R., & Rai, S. (2023). Estimation of heavy metals and fluoride ion in vegetables grown nearby the stretch of River Yamuna, Delhi (NCR), India. *Indian Journal of Environmental Protection, 43*(1), 64–73.
23. Bhardwaj, L. K., Sharma, S., & Jindal, T. (2023). Estimation of physico-chemical and heavy metals in the lakes of Grovnes & Broknes peninsula, Larsemann Hill, East Antarctica. *Chemistry Africa*, *6*, 2677–2694.
24. Schismenos, S., Chalaris, M., & Stevens, G. (2021). Battery hazards and safety: A scoping review for lead acid and silver-zinc batteries. *Safety Science*, *140*, 105290.
25. Galvao, L. A. C., Câmara, V., & Buss, D. (2020). Environmental health in Latin American Countries. In: *Oxford Research Encyclopedia of Global Public Health*..
26. Margallo, M., Ziegler-Rodriguez, K., Vázquez-Rowe, I., Aldaco, R., Irabien, Á., & Kahhat, R. (2019). Enhancing waste management strategies in Latin America under a holistic environmental assessment perspective: A review for policy support. *Science of the Total Environment*, *689*, 1255–1275.
27. Hoornweg, D., & Bhada-Tata, P. (2012). What a waste: A global review of solid waste management. Retrieved from: www.wds.worldbank.org/external/default/WDSContentServer/WDSP/IB/2012/07/25/000333037_20120725004131/Rendered/PDF/681350WP0REVIS0at0a0Waste20120Final.pdf.
28. Quispe-Jofré, A., Philimon, P. P., & Alfaro-Lira, S. (2021). Socio-environmental conflict over abandoned mining waste in Copaquilla, Chile. *Environmental Science and Pollution Research*, *28*, 68773–68791.
29. Mestanza-Ramón, C., Ordoñez-Alcivar, R., Arguello-Guadalupe, C., Carrera-Silva, K., D'Orio, G., & Straface, S. (2022). History, socioeconomic problems and environmental impacts of gold mining in the Andean Region of Ecuador. *International Journal of Environmental Research and Public Health*, *19*(3), 1190.

30. Bundschuh, J., Schneider, J., Alam, M. A., Niazi, N. K., Herath, I., Parvez, F., Tomaszewska, B., Guilherme, L. R. G., Maity, J. P., López, D. L., Cirelli, A. F., Pérez-Carrera, A., Morales-Simfors, N., Alarcón-Herrera, M. T., Baisch, P., Mohan, D., & Mukherjee, A. (2021). Seven potential sources of arsenic pollution in Latin America and their environmental and health impacts. *Science of the Total Environment*, *780*, 146274.
31. de Morias Lima, P., de Morais, M. F., Constantino, M. A., Paulo, P. L., & Magralhães Filho, F. J. C. (2021). Environmental assessment of waste handling in rural Brazil: Improvements towards circular economy. *Cleaner Environmental Systems*, *2*, 100013.
32. Marques, C. P., Zolnikov, T. R., Noronha, J. M. D., Angulo-Tuesta, A., Bashashi, M., & Cruvinel, V. R. N. (2021). Social vulnerabilities of female waste pickers in Brasília, Brazil. *Archives of Environmental & Occupational Health*, *76*(3), 173–180.
33. Ncube, L. K., Ude, A. U., Ogunmuyiwa, E. N., Zulkifli, R., & Beas, I. N. (2021). An overview of plastic waste generation and management in food packaging industries. *Recycling*, *6*(1), 12.
34. Hardoy, J. E., Mitlin, D., & Satterthwaite, D. (2021). The rural, regional and global impacts of cities in Africa, Asia and Latin America. In: *The Earthscan Reader in Sustainable Cities*. Routledge, pp. 426–462.
35. Ferronato, N., Moresco, L., Guisbert Lizarazu, G. E., Gorritty Portillo, M. A., Conti, F., & Torretta, V. (2023). Comparison of environmental impacts related to municipal solid waste and construction and demolition waste management and recycling in a Latin American developing city. *Environmental Science and Pollution Research*, *30*(4), 8548–8562.
36. Mihai, F. C., Gündoğdu, S., Markley, L. A., Olivelli, A., Khan, F. R., Gwinnett, C., Gutberlet, J., Reyna-Bensusan, N., Llanquileo-Melgarejo, P., & Meidiana, C. (2021). Plastic pollution, waste management issues, and circular economy opportunities in rural communities. *Sustainability*, *14*(1), 20.
37. Paul, S., Choudhury, M., Deb, U., Pegu, R., Das, S., & Bhattacharya, S. S. (2019). Assessing the ecological impacts of ageing on hazard potential of solid waste landfills: A green approach through vermitechnology. *Journal of Cleaner Production*, *236*, 117643. https://doi.org/10.1016/j.jclepro.2019.117643.
38. Colorado, H. A., & Echeverri-Lopera, G. I. (2020). The solid waste in Colombia analyzed via gross domestic product: towards a sustainable economy. *Revista Facultad de Ingenieria Universidad de Antioquia*, *96*, 51–63.
39. Ferronato, N., Portillo, M. A. G., Lizarazu, G. E. G., & Torretta, V. (2021). Formal and informal waste selective collection in developing megacities: Analysis of residents' involvement in Bolivia. *Waste Management & Research*, *39*(1), 108–121.
40. Castro, F. D., Júnior, A. B. B., Bassin, J. P., Tenório, J., Cutaia, L., Vaccari, M., & Espinosa, D. (2023). E-waste policies and implementation: A global perspective. In: *Waste Management and Resource Recycling in the Developing World*. Elsevier, pp. 271–307.
41. Moraes, D. M., Silva, P. H., Rampasso, I. S., Anholon, R., Quelhas, O. L., Farias Filho, J. R., & Toledo, R. F. (2021). Environmentally responsible corporate: Actions analysis of Latin American pulp and paper industry. *Environmental Challenges*, *4*, 100153.

42. Shittu, O. S., Williams, I. D., & Shaw, P. J. (2021). Global E-waste management: Can WEEE make a difference? A review of e-waste trends, legislation, contemporary issues and future challenges. *Waste Management*, *120*, 549–563.
43. Terleeva, A. V. (2021). Circulation of organic municipal solid waste in Latin American Countries: The current state and problems of industry development. *Regionalistica*, *8*(6), 116–131.
44. Olivero-Verbel, J., Alvarez-Ortega, N., Alcala-Orozco, M., & Caballero-Gallardo, K. (2021). Population exposure to lead and mercury in Latin America. *Current Opinion in Toxicology*, *27*, 27–37.
45. Laborde, A., Tomasina, F., Bianchi, F., Bruné, M. N., Buka, I., Comba, P., Corra, L., Cori, L., Duffert, C. M., Harari, R., Iavarone, I., McDiarmid, M. A., Gray, K. A., Sly, P. D., Soares, A., Suk, W. A., & Landrigan, P. J. (2015). Children's health in Latin America: The influence of environmental exposures. *Environmental Health Perspectives*, *123*(3), 201–209.
46. Lopez-Yamunaqué, A., & Iannacone, J. A. (2021). La gestión integral de residuossólidosurbanosen América Latina. *Paideia XXI*, *11*(2), 453–474.
47. Hettiarachchi, H., Ryu, S., Caucci, S., & Silva, R. (2018). Municipal solid waste management in Latin America and the Caribbean: Issues and potential solutions from the governance perspective. *Recycling*, *3*(2), 19.
48. Lethbridge, J. (2017). Municipal solid waste management services in Latin America.
49. Hoornweg, D., & Giannelli, N.(2007). Managing municipal solid waste in Latin America and the Caribbean: Integrating the private sector, harnessing incentives.
50. Qian, G., Duanmu, C., Ali, N., Khan, A., Malik, S., Yang, Y., & Bilal, M. (2021). Hazardous wastes, adverse impacts, and management strategies: A way forward to environmental sustainability. *Environment, Development and Sustainability*, *24*, 9731–9756.
51. Segovia-Hernández, J. G., Hernández, S., Cossío-Vargas, E., & Sánchez-Ramírez, E. Tackling sustainability challenges in Latin America and Caribbean from the chemical engineering perspective: A literature review in the last 25 years. *Chemical Engineering Research and Design*, 2022, *188*, 483–527.
52. Romero-Esquivel, L. G., & Acuña-Piedra, A. N. D. R. E. A. (2020). Hazardous waste management in Costa Rica: An academic–small company collaboration. *WIT Transactions on Ecology and the Environment*, *247*, 161–170.
53. Lara-Topete, G. O., Yebra-Montes, C., Orozco-Nunnelly, D. A., Robles-Rodríguez, C. E., & Gradilla-Hernández, M. S. (2022). An integrated environmental assessment of MSW management in a large city of a developing country: Taking the first steps towards a circular economy model. *Frontiers in Environmental Science*, *10*, 838542.
54. Mondal, T., Choudhury, M., Kundu, D., Dutta, D., & Samanta, P. (2023). Landfill: An eclectic review on structure, reactions and remediation approach. *Waste Management*, *164*, 127–142. https://doi.org/10.1016/j.wasman.2023.03.034
55. Choudhary, R., Kumar, A., Narayan, S., Mishchenko, I., Choudhury, M., Bhardwaj, L. K., ... & Boiko, O. Integrating fuzzy logic into life cycle assessment for sustainable municipal solid waste management in Metro Cities. Ecological Engineering & Environmental Technology, 26(7).

CHAPTER 12

Burning Bright: Incineration Processes, Energy Recovery, and Emission Control

KIRAN BALA[1*], ANSHUL[2], SAKSHI VERMA[3], ARUSHI DOGRA[4], PANKAJ KANTI JODDER[5], and ABHISHEK NALLURI[6]

[1]*Department of Zoology, Deshbandhu College, University of Delhi, New Delhi, Delhi, India*

[2]*Institute of Environmental Studies, Kurukshetra University, Kurukshetra, Haryana, India*

[3]*Department of Nutrition Biology, Central University of Haryana, Mahendergarh, Haryana, India*

[4]*Faculty of Life Sciences and Biotechnology, South Asian University, New Delhi, Delhi, India*

[5]*Department of Geography and Earth Science, University of North Carolina at Charlotte, Charlotte, NC, USA*

[6]*Huazhong University of Science and Technology, Wuhan, China*

*Corresponding author

ABSTRACT

Solid waste management remains a significant global challenge, necessitating efficient treatment and disposal methods and comprehensive garbage identification, reduction, recycling, storage, collection, transfer, and transportation systems. Energy from waste (EfW) emerges as a pivotal solution, harnessing waste-derived energy to address sustainability concerns. Modern approaches prioritize sustainable green development, diminishing

Data Analytics and Smart Technologies in Solid Waste Management: Advancing Sustainability. Moharana Choudhury, Sushobhan Majumdar, and Avnish Chauhan (Eds.)

DOI: 10.1201/9781779640833-15

the reliance on landfills and instead focusing on EfW conversion. This strategy manages solid waste and mitigates the impacts of global warming by utilizing waste as an energy source for power generation. Moreover, solid waste recycling offers diverse opportunities, including creating construction materials, biofuels, and biosorbents for various applications, contributing to resource conservation. EfW systems play a vital role in transforming waste into valuable energy forms like biohydrogen, biogas, and bioalcohol, promoting sustainability in energy production. However, environmental considerations are paramount, with the need to address and mitigate the impacts of thermal treatment techniques, such as incineration, on air, water, and solid residues. This chapter provides an overview of EfW, emphasizing its technological diversity, environmental implications, economic aspects, and prospects within the renewable energy landscape. Strategies for emission control, including catalytic converters and scrubbers, are highlighted to enhance the social acceptability of incineration.

12.1 INTRODUCTION

The escalation of waste production worldwide is a cause for concern. Since 2016, urban areas have generated over 2 billion tons of solid waste each year. Rapid population growth and urbanization are expected to cause the total to increase by 70%, potentially reaching a total of 3.40 billion tons by the year 2050 [1]. Potentially valuable resources like components of rare earth and noble metals, waste often contain hazardous substances like heavy metals. Because of the absence of natural resources, the excessive production and improper handling of waste might endanger the environment and human health and destabilize this sector's technical industry [2]. An emerging focus in India pertains to the process of obtaining energy from municipal solid waste (MSW), owing to its potential to furnish valuable energy resources, diminish waste volume, and mitigate certain health and safety risks inherent in prevailing waste handling methodologies. The Asia–Pacific area is witnessing unparalleled growth in the waste-to-energy (WtE) sector, largely attributed to developments in India and China [2]. The growth in these countries that has recently occurred is propelled by heightened cognizance of the dangers and ecological repercussions linked with MSW, alongside escalating energy demands and land utilization prerequisites [3].

12.2 CLASSIFICATION OF SOLID WASTE

12.2.1 MUNICIPAL WASTE

The MSW is a term that covers the variety of waste products generated by various activities such as street cleaning, construction, and demolition [4]. There is no segregation practice for various types of waste, including e-waste, commercial, industrial, and household plastic, in India [5]. Landfilling untreated MSW causes chemical and biological degradation of soil and water resources, as well as air pollution from gas and particulate emissions [6, 45]. The Central Pollution Control Board (CPCB) states that Indian cities generated more than 62 million tons (Mt) of MSW during the year 2015, translating into an average per capita generation rate of approximately 450 g/day [7]. Moreover, only about a quarter, that is, 28% or 14 Mt, underwent proper handling whereas the remaining three-fourths, that is, 72% or 36 Mt, were disposed of openly without any treatment [7, 8]. In 2001, the disposal of waste necessitated 240 km^2, comprising 50% of Mumbai's total area, for unsanitary disposal. By the conclusion of 2011, this area had expanded to 380 km^2, encompassing 90% of Chennai's area. Projections suggest that by 2021, waste production will demand 590 km^2 of land, a size comparable to that of Hyderabad, India's largest city [8].

12.2.2 INDUSTRIAL WASTE

It is challenging to quantify waste precisely, but 43,936 industrial units in India generate an estimated 51.1 million metric tons (MMT) of rubbish annually, of which 7.46 MMT are classified as toxic waste [9]. Of this hazardous waste, 46% (3.41 MMT) is disposed of in landfills, 9% (0.69 MMT) is subjected to incineration, and 45% (3.35 million metric tons) is recycled. In terms of producing hazardous trash, Gujarat leads the country, contributing 28.76% of the total. It is noteworthy that hazardous trash makes up around 10–15% of industrial waste, and the amount of solid hazardous waste is witnessing an annual growth rate of 2–5%. Notably, more than 70% of all industrial toxic waste is made up of coal ash from thermal power plants [9].

12.2.3 AGRICULTURAL WASTE

Agriculture, being fundamental to sustaining life on Earth, assumes a pivotal role in providing sustenance for both human and animal populations while significantly contributing to economic prosperity. However, this advancement is accompanied by a substantial environmental dilemma, with an annual production of more than 1.3 billion tons of agricultural waste worldwide, representing approximately a third of the entire agricultural output. Fruits, vegetables, roots, and tubers account for 40–50% of the worldwide quantitative loss, amounting to 520–650 million tons annually [10]. This agricultural waste encompasses residues from the processing of raw agricultural products, dust, leftovers from crops, and damaged or decayed crops [11]. Various types of agro-industrial waste, characterized by differing chemical compositions, are generated by sectors such as dairy, poultry, meat, seafood, grains, and legumes processing. In addition, toxic and hazardous chemical waste is produced when pesticides, herbicides, and insecticides are used in farming activities [9] (Figure 12.1).

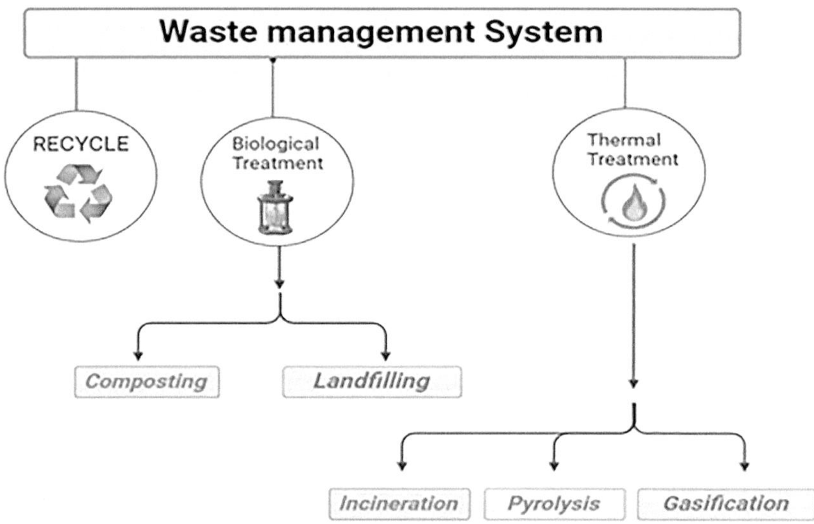

FIGURE 12.1 Waste management system.

12.3 THERMAL PROCESSES IN WASTE MANAGEMENT

In India, improper waste disposal challenges can be addressed through effective material recovery and source separation, concentrating on thermal recovery. Thermal recovery, which employs recovered waste as fuel, is a crucial determinant. When it comes to thermal processes, combustion-based WtE plants dominate and provide combined heat and power. Combining maximum recycling with thermal methods can significantly cut down landfilling [44]. Furthermore, there is increased interest in thermal processes for unsegregated, low-calorific value waste, and further alternatives are available, such as gasification and pyrolysis [12].

12.3.1 INCINERATION

Incineration is a heat treatment method where waste is burned at high temperatures with excess air. This process oxidizes and breaks down toxic substances in the waste, making it safer. It is an established technology for treating hazardous waste, developed over more than 20 years, with various types of incinerators like rotary kilns, liquid jets, pyrolysis, fluidized beds, and multilayer incinerators available for industrial use. This approach maximizes the harmlessness, reduction, and potential recycling of hazardous waste [13] (Table 12.1).

TABLE 12.1 Types of Incinerators Used in Waste Management

Technology	Description	Advantages	References
Rotary Kiln incinerators	It has two chambers: a recombustion chamber for full burning with secondary air and an angled primary chamber for thermal disintegration	No presorting required High thermal efficiency (up to 85%) Effective destruction of hazardous chemicals	Rani et al. [46]
Fluidized bed incinerators	Sludge, which has a high moisture content but a low calorific value, may be burned efficiently in this flexible incinerator together with other different kinds of trash. Because carbonates and char layers are removed during turbulent mixing in the bed material, combustion efficiency is increased	When compared with alternative techniques, combustion at lower temperatures with great efficiency lowers NO_x and CO emissions, making it more ecologically friendly	Van Caneghem et al. [47]
Moving grate incinerators	The waste is subsequently burned and dumped into an ash pit for disposal. The grate—which is usually composed of metal—allows the primary and secondary air to mix thoroughly, providing an abundance of oxygen for full combustion	Flexibility is provided by the many grate designs, which include forward, backward, double, rocking, and rolling movements Achieves thermal efficiency of up to 85%	Rani et al. [46]

TABLE 12.1 *(Continued)*

Technology	Description	Advantages	References
Multiple hearth	A vertical steel shell with many firebrick hearths and a refractory	MHFs ensure high solids and gas retention	Winchell et al. [48]
Incinerators	Liner is called a multiple hearth furnace (MHF). The process of evaporation happens first, then the start of combustion	Due to mechanical mixing and gas flow	
Catalytic combustion	A catalyst is used by the waste treatment system to increase the efficiency of incineration and lower emissions of nitrogen oxides. It works well for managing waste materials that are complicated and need to be destroyed by heat	Limits formation of undesirable by-products	Abah et al. [49]
Waste-gas flare	VOCs, methane, hydrogen, and carbon monoxide are among the waste gases that can be burned using waste-gas flare (a high-temperature oxidation method)	The US EPA (2015) estimates combustion efficiency at approximately 96.5% and destruction efficiency at around 98% for waste gas flares	US EPA [21]
Fixed grate/ Direct flame	This type of incinerator introduces garbage through a throat and directs it down a grate into an ash pit	Organic toxin breakdown, cost- effective solution for waste treatment	Speight [50]
Liquid injection incinerators	Waste is injected, vaporized, and burned in high-temperature combustion chambers to produce innocuous gases and ash	Specialized facilities used for burning liquid wastes like hazardous chemicals or medical waste	Zhang [54]

12.3.2 PYROLYSIS

The process of pyrolysis is divided into three principal stages: the first is feeding and dosing the source material, the second is converting the mass, and the third is collecting and separating the end products bio-oil, gases, and coke. Pressure, heating rate, raw materials composition, and final temperature determine how the products are distributed [14]. Pyrolysis technology converts wastes and gives an insight into how different factors may impact the final product's composition. Pyrolysis may be characterized by its benefits, incredible efficiency, with up to 80% energy recovery, MSW is converted into bio-oil, biochar, and pyrolysis gas with minimal implementation requirements, reduced emissions of NO_x and SO_2 the opportunity to process diverse waste types, including plastics, and processing of all waste plastic sorts without the necessity of preliminary cutting effort [15].

12.3.3 GASIFICATION

Gasification waste residue is produced by pyrolysis and gasification, which is a complex thermochemical process, carried out at high temperatures [16]. The properties of the waste, the kind of gasifier that is used, and the oxidant that is selected all affect how effective gasification is. Common oxidants include pure oxygen, gas steam, and oxygen-enriched air. The choice of oxidant has a significant impact on the calorific properties of the resulting syngas [17]. Commercial gasifiers may be generically categorized into three types: fluidized bed frameworks, downdraft type, and updraft type. Pressure and temperature serve as pivotal parameters, with pressurized gasifiers demonstrating superior productivity and augmented generation of hydrogen [18]. Fixed-bed gasifiers with updraft present drawbacks owing to issues related to tar concentration, and the incorporation of filters proves to be both costly and intricate. In these gasifiers, the material is injected from the top while the oxidant is supplied from the bottom, syngas is collected from the upper section, and particles are kept above the bed until discharge [19].

12.4 ENERGY FROM WASTE MECHANISMS

Given the increasing volume of waste, municipal solid waste incineration (MSWI) is essential for its treatment and the supply of the generated energy as an alternative source. MSWI plays multiple roles in the field of waste treatment and energy recovery that can complement or even substitute its counterparts [20, 21]. More than 80% of the global energy is based on fossil sources [22, 23]. In this context, MSWI is central to fossil energy reduction, increase in renewable share, and compatible and efficient waste volume treatment [24]. WtE is a multimillion dollar industry with an annual revenue of more than $20 billion [25]. To facilitate the procedure, the energy-from-waste EfS incineration plant uses a boiler system to recover the thermal energy emerging during cooling [26]. Thermal energy is recuperated as steam in the boiler, with the flue gas undergoing cooling through indirect water heating in water pipes. A significant portion of the steam is directed to a steam turbine, generating electricity. Alternatively, steam and heated water discharged through the turbine present additional options as heat sources. Notably, as steam temperature and pressure increase, the efficiency of energy recovery improves (Figure 12.2).

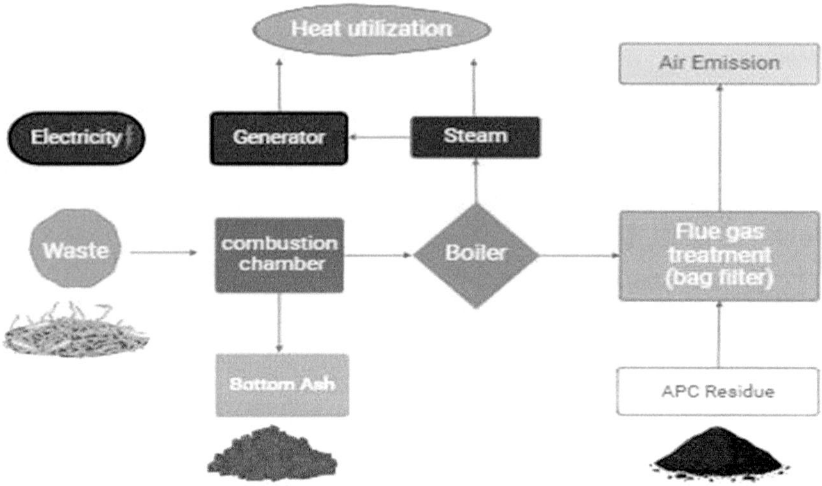

FIGURE 12.2 Energy from waste mechanism.

12.5 ENVIRONMENTAL IMPLICATIONS

The incineration process converts waste into different alternatives, and toxic ash, as well as pollutants, scatter into the air and water, are among the most severe environmental implications [27]. Incineration does not destroy some toxic elements such as per- and poly-fluoroalkyl substances, resulting in their discharge into the air and incorporation into ash, from which contamination of land and aqua results. Water consequences include the garbage disposal of lakes and the uncontrolled dumping of industrial waste by the lakes [27, 28]. Urban and industrial activities introduce fertilizers, drugs, pesticides, and high-risk pollutants accompanying the earth with their effluent of wastewater to the groundwater. Agricultural and nonagricultural pollutants in soil and water sources threatened with contamination [29, 30]. Industrial emissions, which also cause acid rain, are also responsible for water self-pollution and harm to the environment and health of both humans and nature [30]. Incineration is also a cause of soil pollution with heavy metals, which results in pollution and therefore contamination of groundwater, transfer into water bodies, dust inhalation, and bioaccumulation in the neighboring ecosystems. A brief description of some pollutants emitted from the incinerator stack is given in Table 12.2.

TABLE 12.2 Air Pollutants Caused Due to Incineration Process

Air Pollutants	Description	References
Acid chlorides and fluorides	• Plastic debris, such as PVC pipes, and salty food are the main sources of chlorine in MSW • Fluorinated fabrics and polymers like polytetrafluoroethylene (PTFE) are sources of fluorine	Karmakar et al. [51]
Volatile organic compounds (VOCs)	• Consists of carbon chains or rings and is mainly formed due to incomplete combustion	Hu et al. [52]
Mercury (Hg)	• Made from common home items such as button cells, thermometers, alkaline batteries, and fluorescent lights	Hu et al. [52]
Dioxins and furans	• Formed as by-products in trace amounts when H_2, O_2, C, and Cl are exposed to heat	McKay [53]

12.6 RULES AND REGULATIONS REGARDING EMISSIONS

An improved urban environment and a decreased risk of disease transmission are the results of effectively handling MSW. The efficacy of the MSW management system serves as a gauge for effective governance, thereby bolstering the credibility of local authorities and increasing the likelihood of attracting external investments [31]. However, in India, regulations governing emissions from incinerators, particularly MSWI plants, encounter challenges and discrepancies. Currently, the Environmental Protection Act's national ambient air-quality standards set emission limitations for stack emissions in traditional incinerators; however, MSW incinerators are not subject to any particular national emission regulations. Instead, WtE plants adhere to operational certificate agreements issued by individual state governments, with emission standards in several states notably relaxed. India's WtE plants monitor only a limited range of air emission indicators, including carbon dioxide (CO_2), oxygen (O_2), sulfur oxides (SO_x), nitrogen oxides (NO_x), and particulate matter (PM). Prominently, inadequate monitoring exists for important pollutants such as mercury, heavy metals, volatile organic compounds (VOCs), and persistent organic pollutants (POPs). There is a facility located in New Delhi called Ecopolis that maintains PM emission limits 10 times greater than those seen in the United Kingdom. The emission data that have been provided exceed the emissions of plants situated in the United Kingdom by a factor of a hundred. Residents residing near WtE facilities have reported instances of emission limit violations and data manipulation [32].

12.7 HOW TO CONTROL EMISSIONS FROM INCINERATORS

Particulate matter is removed from the air using a variety of techniques. Physical mechanisms are used by mechanical collectors, such as cyclone separators. By using an electric charge to attract and remove particulate matter, electrostatic precipitators (ESPs) are an efficient way to remove dust and PM. Wet scrubbers also remove impurities from the gas stream by capturing and using liquids, such as water. For the removal of acid gases, a variety of wet scrubbers are used, such as spray towers, packed bed scrubbers, venture scrubbers, and dry scrubbers [33].

12.7.1 CYCLONES

A cyclone works by generating a vortex inside a chamber. Through centrifugal force, particulates are thrown toward the chamber walls where their momentum surpasses the drag force. The cyclone is fundamentally a rectangular and conical section design. Cyclones are fabricated from high-temperature materials and corrosive substances. Interior surfaces are kept smooth to enhance particulate migration toward the dust hopper [34]. Dry cyclones are reasonably capable of separating sand particles, materials like starch, fly ash, and powders that can form finer particulates may require supplementary pollution control equipment [35].

12.7.2 BAG FILTERS

Bag filters, commonly denominated as "bag houses," are utilized as cylindrical bags of varied fabrics across sectors to extract particles from gas streams. Fabrics are divided into rows, splitting the gas flow, and allowing particles to accumulate on the fabric surface. These particles are then periodically dislodged by filter shaking. Sodium bicarbonate precoat is a technique used to eliminate gaseous sulfur dioxide from the exhaust emissions of power boilers. Precoat compounding with activated carbon must be performed in cases where hazardous dioxins are present [34].

12.7.3 ELECTROSTATIC PRECIPITATOR

Electrostatic precipitators generate an electric field, then execute the corona discharge and gas stream ionization, followed by the charging of particles and the enrolment of charged particles into the special collection plates. In the case of ESP, the work of the discharge electrodes leads to the creation of corona discharge caused by a high-voltage application, which allows the gas molecules that were neutral earlier to become positively and negatively charged and hit the suspended particles [34]. Eliminating dispersed particulate matter from the collecting electrode is the main goal of ESPs [36] (Figure 12.3).

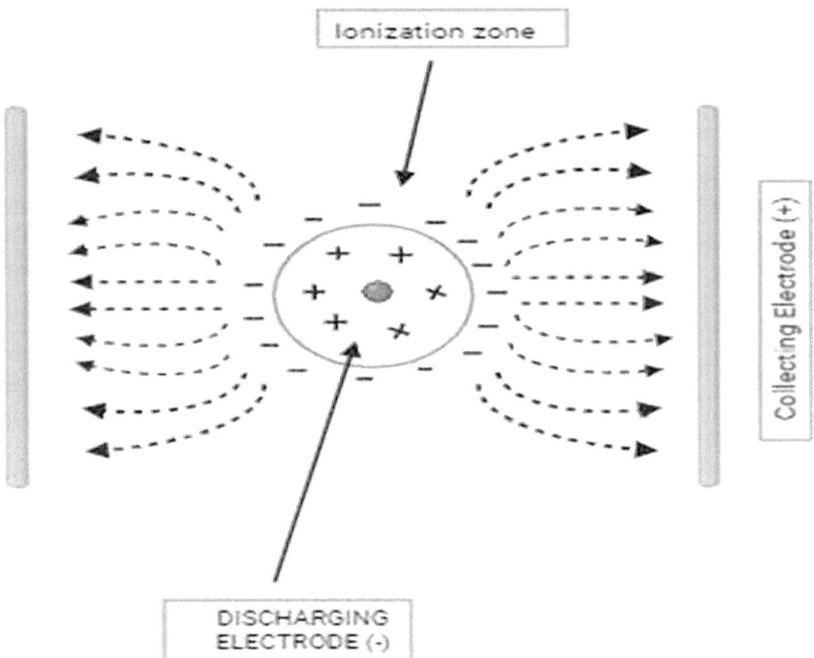

FIGURE 12.3 Working of electrostatic precipitator.

12.7.4 SCRUBBER

Scrubbers work based on inertial impaction and Brownian motion concepts. They are devices that gather particles by spraying them with a liquid, usually water [34]. Their mechanism involves sprinkling liquid-laden gas streams with particles, allowing the particles to become wet, and then extracting them using a collecting surface. Wet scrubbers come in a variety of forms; the most effective are venturi scrubbers, which may achieve up to 98% effectiveness for particles as small as 0.5 μm.

12.7.5 DRY SORBENT INJECTION

A finely powdered substance, such as sodium bicarbonate ($NaHCO_3$) or trona ($Na_2CO_3 \cdot NaHCO_3 \cdot 2H_2O$), is injected into the flue gas duct within the framework of dry sorbent injection (DSI) systems. The sodium sorbent then turns into porous sodium carbonate ($NaCO_3$) following calcination. This porous substance interacts with HF, HCl, SO_2, and SO_3 acid gasses. Several factors affect the effectiveness of a DSI system: the temperature at the injection site, the particulate control device used (ESP or bag house), the sorbent particle size, the sorbent's penetration and mixing with the flue gas, and the sorbent's residence time in the flue gas stream [37].

12.7.6 CATALYTIC REDUCTION

The catalytic converter is made up of a metal shell that encloses a ceramic honeycomb structure with thin-walled passages coated with aluminum oxide; it allows precious metal reactions involving palladium, rhodium, and platinum to convert pollutants through oxidation and reduction processes [38]. Additionally, thermal oxidizers burn volatile organic compounds (VOCs) at high temperatures; adsorption systems use activated carbon or other absorbents to remove VOCs, thereby reducing emissions by an order of magnitude and improving the air quality substantially [33].

12.8 SCOPE OF ENERGY FROM WASTE

A fascinating option for trash management and sustainable development is to use waste to generate energy [39]. This may involve generating ecologically benign bioenergy in various forms of hydrogen energy, biogas, and bioethanol, which can stimulate economic growth by generating employment possibilities [40]. Generating combined heat and power (CHP) is another use for bioenergy [41]. By tackling environmental degradation, energy generation, and waste management simultaneously, the WtE nexus is essential to achieving a circular economy [13]. While developed countries have made strides in utilizing waste for energy production, developing nations encounter challenges in waste collection and disposal [42]. Recent reports highlight emerging trends in WtE initiatives, concentrating on factors that are technological, environmental, economic, and social [43]. An analytical assessment of these reports can offer insights into future developments and recommendations for tackling issues in this quickly expanding industry, with a focus on biochemical routes for the conversion of waste to energy [13].

12.9 CONCLUSION

The waste management issue is critical in today's era of development, as mismanagement poses risks to both humans and the environment. EfW offers a sustainable solution by converting waste materials into valuable energy resources, addressing waste volume challenges, and promoting environmentally sustainable energy production. Thermal treatment methods like incineration and pyrolysis are commonly used for EfW, with incineration favored for its high thermal efficiency and energy recovery potential. However, incineration poses environmental implications, necessitating strict regulations and emission control techniques like catalytic converters and scrubbers. EfW offers affordable waste management and generates income through by-products like electricity or heat. Overall, EfW presents a comprehensive and sustainable solution to the waste problem, requiring waste segregation, recycling, efficient WtE conversion technologies, public awareness campaigns, and regulatory policies for effective implementation and further research to ensure long-term sustainability.

KEYWORDS

- **thermal treatment techniques**
- **refused derived fuel**
- **green development**
- **sustainability concerns**

REFERENCES

1. Voukkali, I., Papamichael, I., Loizia, P., & Zorpas, A. A. (2024). Urbanization and solid waste production: Prospects and challenges. *Environmental Science and Pollution Research*, *31*(12), 17678–17689.
2. World Energy Council. (2011). World energy resources: Waste to energy. Retrieved from http://www.worldenergy.org/wpcontent/uploads/2013/10/WER_2013_7b_Waste_to_Energy.pdf
3. Yang, N., Zhang, H., Shao, L., Lü, F., & He, P. (2013). Greenhouse gas emissions during MSW landfilling in China: Influence of waste characteristics and LFG treatment measures. *Journal of Environmental Management*, *129*, 510–521.
4. World Bank. (2019). Solid waste management. Retrieved from https://www.worldbank.org/en/topic/urbandevelopment/brief/solid-waste-management
5. Leray, L., Sahakian, M., & Erkman, S. (2016). Understanding household food metabolism: Relating micro-level material flow analysis to consumption practices. *Journal of Cleaner Production*, *125*, 44–55.
6. Cremiato, R., Mastellone, M. L., Tagliaferri, C., Zaccariello, L., & Lettieri, P. (2018). Environmental impact of municipal solid waste management using life cycle assessment: The effect of anaerobic digestion, materials recovery and secondary fuels production. *Renewable Energy*, *124*, 180–188.
7. CPCB (Central Pollution Control Board). (2016). Central pollution control board (CPCB) bulletin, Government of India. Retrieved from http://cpcb.nic.in/openpdffile.php?id=TGF0ZXN0RmlsZS9MYXRlc3RfMTIzX1NVTU1BUllfQk9PS19GUy5wZGY
8. MNRE (Ministry of New and Renewable Energy). (2016). Power generation from municipal solid waste. Retrieved from http://164.100.47.193/lsscommittee/Energy/16_Energy_20.pdf
9. The Associated Chambers of Commerce of India (ASSOCHAM) & PricewaterhouseCoopers (PwC). (2017). India would need to bring 88 km2 land under waste disposal by 2050: Study. Retrieved from http://www.assocham.org/newsdetail.php?id=6350
10. Ravindran, R., Hassan, S. S., Williams, G. A., & Jaiswal, A. K. (2018). A review on bioconversion of agro-industrial wastes to industrially important enzymes. *Bioengineering*, *5*(4), 93.
11. Obi, F. O., Ugwuishiwu, B. O., & Nwakaire, J. N. (2016). Agricultural waste concept, generation, utilization and management. *Nigerian Journal of Technology*, *35*(4), 957–964.
12. Kumar, S., Smith, S. R., Fowler, G., Velis, C., Kumar, S. J., Arya, S. R., Kumar, R., & Cheeseman, C. (2017). Challenges and opportunities associated with waste management in India. *Royal Society Open Science*, *4*(2), 160764. https://doi.org/10.1098/rsos.160764
13. Qiu, B., Xu, L., Chen, J., & Qin, J. (2016). A brief discussion on the incineration technology of hazardous waste rotary kiln [Abstract]. *Jiangxi Chemical Industry*, 51–53.
14. Miandad, R., Barakat, M., Aburiazaiza, A. S., Rehan, M., Ismail, I., & Nizami, A. (2017). Effect of plastic waste types on pyrolysis liquid oil. *International Biodeterioration & Biodegradation*, *119*, 239–252.

15. Ławińska, O., Korombel, A., & Zajemska, M. (2022). Pyrolysis-based municipal solid waste management in Poland—SWOT analysis. *Energies*, *15*(2), 510. https://doi.org/10.3390/en15020510
16. Kantarelis, E., & Zabaniotou, A. (2009). Valorization of cotton stalks by fast pyrolysis and fixed bed air gasification for syngas production as precursor of second generation biofuels and sustainable agriculture. *Bioresource Technology*, *100*(2), 942–947. https://doi.org/10.1016/j.biortech.2008.07.
17. Zhang, R., Yin, R., & Luo, Y. (2018). Inhibition of CCl formation during the combustion of MSW gasification syngas: An experimental study on the synergism and competition between oxidation and chlorination. *Waste Management*, *76*, 472–482. https://doi.org/10.1016/J.WASMAN.2018.03.027
18. Watson, J., Zhang, Y., Si, B., Chen, W.-T., & de Souza, R. (2018). Gasification of biowaste: A critical review and outlooks. *Renewable and Sustainable Energy Reviews*, *83*, 1–17. https://doi.org/10.1016/J.RSER.2017.10.003
19. Ramos, A., Monteiro, E., Silva, V., & Rouboa, A. (2018). Co-gasification and recent developments on waste-to-energy conversion: A review. *Renewable and Sustainable Energy Reviews*, *81*, 380–398. https://doi.org/10.1016/J.RSER.2017.07.025
20. Cucchiella, F., et al. (2017). Sustainable waste management: Waste to energy plant as an alternative to landfill. *Energy Conversion and Management*, *131*, 18–31.
21. EPA. (2016). Energy recovery from the combustion of municipal solid waste (MSW). Retrieved from https://www.epa.gov/smm/energy-recovery-combustion-municipal-solidwaste-msw
22. BP. (2016). *BP Statistical Review of World Energy June 2016*, 65th edn, BP. https://www.bp.com/content/dam/bp/pdf/energy-economics/statistical-review-2016/bp-statistical-review- of-world-energy-2016-full-report.pdf
23. IEA. (2016). *World Energy Outlook Report 2016*. International Energy Agency. Retrieved from https://www.iea.org/media/publications/weo/WEO2016Chapter1.pdf
24. Tester, J. W., et al. (2012). *Sustainable Energy: Choosing Among Options*. The MIT Press.
25. World Energy Council. (2013). Waste-to-energy. Retrieved from http://www.worldenergy.org/wp-content/uploads/2013/10/WER_2013_7b_Waste_to_Energy.pdf
26. Schifftner, K. C. (2021). *Air Pollution Control Equipment Selection Guide*. CRC Press.
27. Akhtar, N., Syakir Ishak, M. I., Bhawani, S. A., & Umar, K. (2021). Various natural and anthropogenic factors responsible for water quality degradation: A review. *Water*, *13*, 2660. https://doi.org/10.3390/w13192660
28. Government Newsletter. (1992). Ground water issue. U.S. environmental protection agency, technology innovation office, office of solid waste and emergency response, Washington, DC 20460. EPA/540/S-92/001. *Remediation Journal*, 3, 136–137.
29. Parris, K. (2011). Impact of agriculture on water pollution in OECD countries: Recent trends and future prospects. *International Journal of Water Resources Development*, *27*, 33–52. https://doi.org/10.1080/07900627.2010.541937
30. Masi, F., Rizzo, A., & Regelsberger, M. (2017). The role of constructed wetlands in a new circular economy, resource oriented, and ecosystem services paradigm. *Journal of Environmental Management*, *216*, 275–284. https://doi.org/10.1016/j.jenvman.2017.01.045

31. Wilson, D. C., Rodic, L., Modak, P., Soos, R., Carpintero Rogero, A., Velis, C., Iyer, M., & Simonett, O. (2015). *Global Waste Management Outlook Report*. UNEP: Nairobi, Kenya.
32. Nixon, J. D., Dey, P. K., & Ghosh, S. K. (2017). Energy recovery from waste in India: An evidence- based analysis. *Sustainable Energy Technologies and Assessments*, *21*, 23–32. http://dx.doi.org/10.1016/j.seta.2017.04.003
33. Schiavon, M., Ravina, M., Zanetti, M., & Panepinto, D. (2024). State-of-the-art and recent advances in the abatement of gaseous pollutants from waste-to-energy. *Energies*, *17*, 552. https://doi.org/10.3390/en17030552
34. Sahu, M., Malyan, V., & Mayya, Y. S. (2021). Technologies for controlling particulate matter emissions from industries. In *Pollution Control Technologies: Current Status and Future Prospects*, pp. 253–290.
35. Okedere, O. B., Sonibare, J. A., Fakinle, B. S., & Jimoda, L. A. (2013). Usefulness of particulate cyclone in air pollution control. *Management of Environmental Quality: An International Journal*, *24*(6), 771–781.
36. Parker, K. R. (1997). *Applied Electrostatic Precipitation*. Blackie Academic Professional.
37. Kong, Y., & Davidson, H. (2010, January). Dry sorbent injection of sodium sorbents for SO2, HCl, and mercury mitigation. In *North American Waste-to-Energy Conference*, Vol. 43932, pp. 317–320.
38. Dey, S., & Mehta, N. S. (2020). Automobile pollution control using catalysis. *Resources, Environment and Sustainability*, *2*, 100006.
39. Srivastava, R. K., Shetti, N. P., Reddy, K. R., & Aminabhavi, T. M. (2020). Sustainable energy from waste organic matters via efficient microbial processes. *Science of the Total Environment*, *722*, 137927. https://doi.org/10.1016/j.scitotenv.2020.137927
40. Banerjee, S., Kaushik, S., & Tomar, R. S. (2019). *Global Scenario of Biofuel Production: Past, Present and Future*. Springer International Publishing. https://doi.org/10.1007/978-3-030-14463-0_18
41. Ren21. (2019). Renewables 2019: Global status report. REN21, the Renewable Energy Policy Network for the 21st Century. http://www.ren21.net/wp- content/uploads/2019/05/gsr_2019_full_report_en.pdf
42. Moya, D., Aldas, C., López, G., & Kaparaju, P. (2017). Municipal solid waste as a valuable renewable energy resource: A worldwide opportunity of energy recovery by using waste-to-energy technologies. *Energy Procedia*, *134*, 286–295. https://doi.org/10.1016/j.egypro.2017.09.618
43. Kamali, M. (2020). An opinion on multi-criteria decision-making analysis for sustainability- based spatial planning practices. Time to improve. *Journal of Settlements and Spatial Planning*, 1–3.
44. Mondal, T., Choudhury, M., Kundu, D., Dutta, D., & Samanta, P. (2023). Landfill: An eclectic review on structure, reactions and remediation approach. *Waste Management*, *164*, 127–142. https://doi.org/10.1016/j.wasman.2023.03.034
45. Paul, S., Choudhury, M., Deb, U., Pegu, R., Das, S., & Bhattacharya, S. S. (2019). Assessing the ecological impacts of ageing on hazard potential of solid waste landfills: A green approach through vermitechnology. *Journal of Cleaner Production*, *236*, 117643. https://doi.org/10.1016/j.jclepro.2019.117643

46. Rani, R. U., Banu, J. R., Tsang, D. C., & Lay, C. H. (2020). Thermochemical conversion of food waste for bioenergy generation. In M. Aslam, & R. Kumar (Eds.), *Food Waste to Valuable Resources* (pp. 97–118). Academic Press.
47. Van Caneghem, J., Brems, A., Lievens, P., Block, C., Billen, P., Vermeulen, I., Dewil, R., Baeyens, J., ...& Vandecasteele, C. (2012). Fluidized bed waste incinerators: Design, operational and environmental issues. *Progress in Energy and Combustion Science*, *38*(4), 551–582. https://doi.org/10.1016/J.PECS.2012.02.002
48. Winchell, L. J., Ross, J. J., Wells, M. J., Fonoll, X., Norton Jr, J. W., & Bell, K. Y. (2021). Per- and polyfluoroalkyl substances thermal destruction at water resource recovery facilities: A state of the science review. *Water Environment Research*, *93*(6), 826–843.
49. Abah, E. O., Ahamed, T., & Noguchi, R. (2021). Catalytic temperature effects on conversion efficiency of PM2.5 and gaseous emissions from rice husk combustion. *Energies*, *14*(19), 6131. https://doi.org/10.3390/en14196131
50. Speight, J. G. (2020). Production of fuels from nonfossil fuel feedstocks. In J. G. Speight (Ed.), *The Refinery of the Future* (pp. 391—-426).
51. Karmakar, A., Daftari, T., Sivagami, K., Chandan, M. R., Shaik, A. H., Kiran, B., & Chakraborty, S. (2023). A comprehensive insight into wWaste to eEnergy conversion strategies in India and its associated air pollution hazard. *Environmental Technology & Innovation*, *29*, 103017.
52. Hu, Y., Cheng, H., & Tao, S. (2018). The growing importance of waste-to-energy (WTE) incineration in China's anthropogenic mercury emissions: Emission inventories and reduction strategies. *Renewable and Sustainable Energy Reviews*, 97, 119–137. http://dx.doi.org/10.1016/j.rser.2018.08.026
53 McKay, G., 2002. Dioxin characterisation, formation and minimisation during municipal solid waste (MSW) incineration: review. *Chemical. Engineering. Journal.* 86, 343–368. http://dx.doi.org/10.1016/S1385-8947(01)00228-5.
54. Zhang, E. (2023). Research on urban solid waste treatment methods. In *E3S Web of Conferences* (Vol. 424, p. 03010). EDP Sciences.

PART IV

Shaping the Future: Policies, Regulations, and Outlook in Environmental Management

CHAPTER 13

Revolutionizing Waste Management: Cutting-Edge Technologies Leading the Charge Toward Sustainable Futures

RATAN CHOWDHURY[1*], GISANDU K. MALUNGUJA[2], and AMBARISH BHUYAN[3]

[1]*Department of Botany, Rangapara College, Rangapara, Assam, India*

[2]*Department of Natural Sciences, Mbeya University of Science and Technology, Mbeya, Tanzania*

[3]*Department of Botany, D.H.S. Kanoi College, Dibrugarh, Assam, India*

*Corresponding author

ABSTRACT

As global concerns over environmental sustainability intensify, the demand for innovative waste management solutions has never been more pressing. This chapter explores the transformative role of cutting-edge technologies in reshaping traditional waste management practices, paving the way for a more sustainable future. It emphasizes the integration of Internet of Things, sensors, data analytics, and machine learning in real-time monitoring. Smart bins and collection systems, equipped with radio-frequency identification technology and GPS-enabled routing, are examined for their efficiency in waste tracking and route optimization. Waste-to-energy innovations are also discussed, including biomass and bioenergy technologies, and the extraction of power from landfill gas and organic waste. Through the lens of case studies, the chapter highlights successful implementations of these technologies worldwide, emphasizing their positive impact

Data Analytics and Smart Technologies in Solid Waste Management: Advancing Sustainability.
Moharana Choudhury, Sushobhan Majumdar, and Avnish Chauhan (Eds.)

DOI: 10.1201/9781779640833-17

and key lessons learned. Future trends and emerging technologies offer a glimpse into potential game-changers on the horizon, along with the continued evolution toward a circular economy. Challenges and considerations are addressed by examining the ethical and environmental implications of technology adoption, stressing the need to balance innovation with affordability and accessibility. A collaborative effort among stakeholders and decision-makers is essential to drive the transition toward sustainable waste management practices.

13.1 INTRODUCTION

In today's world, waste management stands at a critical juncture. With increasing urbanization and consumption patterns, the volume of waste generated globally has surged, placing immense pressure on our environment and natural resources [1]. The current state of recycling programs is characterized via both challenges and opportunities. Despite advancements in recycling and waste reduction endeavors, a considerable quantity of waste still finds its way into dumping grounds or combustion facilities, adding to ecosystem deterioration and the release of greenhouse gases [2]. In many regions, inadequate infrastructure and lack of awareness further exacerbate the waste management crisis, leading to pollution of land, water, and air [3, 4].

There is an unprecedented imperative to discover sustainable solutions for waste. As the global population continues to grow, projections indicate that waste generation will increase exponentially in the coming decades [5]. Due to which it not just poses ecosystem concerns, but it also threatens public health and safety. Sustainable waste management practices are essential to mitigate these risks, preserve natural resources, and create a healthier and more resilient future for all [6].

Technology has emerged as a game-changer in waste management, driving innovations that promise to revolutionize the industry. High-tech solutions, such as Internet of Things (IoT), robotics, and artificial intelligence, are being deployed to optimize waste collection, sorting, and recycling processes [7]. Additionally, innovative trash-to-energy systems are transforming trash into valuable resources, reducing landfill dependence and contributing to renewable energy production [8, 9]. The integration of technology not only enhances efficiency but also opens up new avenues for sustainable waste management practices, marking a paradigm shift in how we perceive and manage waste.

13.2 TRADITIONAL CHALLENGES IN WASTE MANAGEMENT

Waste management has long been grappling with a range of challenges that have persisted over the years. These traditional hurdles have underscored the need for innovative and sustainable solutions to address the growing waste crisis.

13.2.1 LANDFILL OVERLOAD AND ENVIRONMENTAL IMPACTS

Landfill overload remains a significant challenge in waste management. As waste generation continues to rise, landfills are reaching their capacity, posing environmental risks and health hazards [10]. Improperly managed landfills can contaminate soil, groundwater, and air with toxic substances, leading to long-term environmental degradation and public health issues [11–50]. Additionally, the methane gas emitted from decomposing organic waste in landfills contributes to climate change, highlighting the urgent need for alternative waste disposal methods [12–49].

13.2.2 LIMITED EFFECTIVENESS OF CONVENTIONAL RECYCLING METHODS

Conventional recycling methods, although beneficial, have shown limited effectiveness in addressing the mounting waste problem. While recycling aims to divert waste from landfills and conserve resources, it often faces challenges such as contamination, inefficient sorting, and lack of infrastructure [13]. Contaminated recyclables can compromise the quality of recycled products, making them less marketable and reducing their economic viability [14]. Moreover, the high costs associated with collection, processing, and transportation further impede the scalability and efficiency of traditional recycling methods [15].

13.2.3 GROWING GLOBAL CONCERNS AND CALLS FOR CHANGE

The growing global concerns over waste management have led to widespread calls for change. Public awareness about the environmental and social impacts of unsustainable waste practices has increased, prompting demands for more responsible and innovative waste management solutions [16]. Communities, organizations, and governments around the world are advocating for policies and initiatives that prioritize waste reduction, recycling, and sustainable disposal methods [17]. The rise of environmental movements and grassroots campaigns reflects a global shift toward greater environmental consciousness and a collective commitment to addressing the waste crisis [18].

Traditional waste management practices are fraught with challenges that necessitate a paradigm shift toward more sustainable and effective solutions. As we navigate through these challenges, it becomes increasingly clear that innovation and technology will play a pivotal role in revolutionizing waste management and paving the way for a cleaner, healthier, and more sustainable future.

13.3 THE RISE OF TECHNOLOGICAL SOLUTIONS

In response to the traditional challenges faced by waste management, technological solutions have emerged as transformative tools that promise to revolutionize the industry. These technological advancements are reshaping how we manage, monitor, and process waste, paving the way for more sustainable and efficient practices. Technological advancements in trash management are ushering in a new era of innovation and efficiency. From smart bins to advanced sorting systems, technology is being leveraged to enhance processing efficiency, waste collection, and recycling processes [19]. These advancements not only improve operational efficiency but also enhance environmental sustainability by reducing landfill dependence and resource consumption [20]. As we delve deeper into the realm of technological solutions, it becomes evident that these innovations are key drivers in transforming waste management practices for the better (Figure 13.1).

13.3.1 INCORPORATING IOT AND SENSOR TECHNOLOGIES FOR INSTANTANEOUS MONITORING

The integration of IoT and sensors has revolutionized real-time monitoring in waste management. Sensor-enabled waste receptacles outfitted are capable of discerning their fill levels, optimize collection routes, and alert

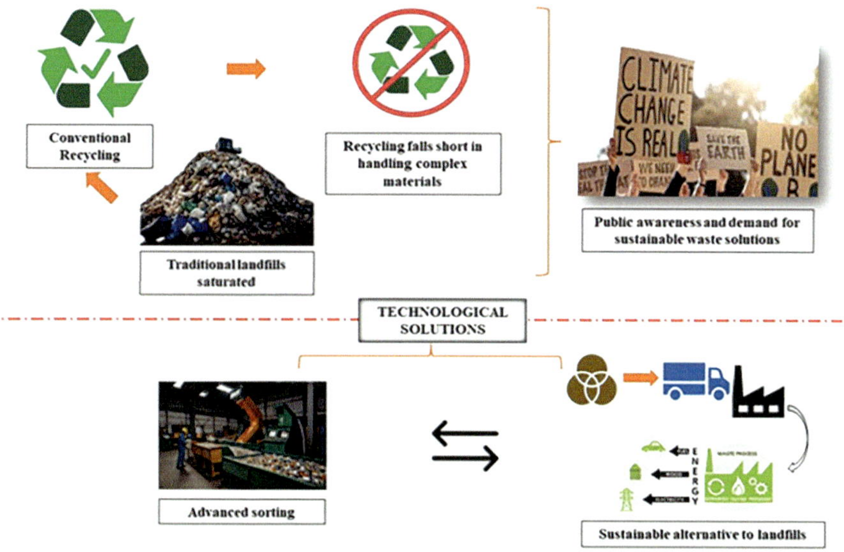

FIGURE 13.1 Challenges in waste management and technological solutions.

authorities when maintenance is required [21]. This real-time monitoring capability enables more efficient waste collection, reduces operational costs, and minimizes environmental impact by preventing overflowing bins and reducing unnecessary trips [22]. Moreover, IoT-enabled sensors provide valuable data insights that can inform decision-making and help municipalities and organizations implement data-driven waste management strategies [23].

13.3.2 DATA ANALYTICS AND MACHINE LEARNING IN WASTE SORTING AND RECYCLING

Data analytics and machine learning are playing an increasingly important role in waste sorting and recycling processes. Advanced algorithms can analyze vast amounts of data to optimize waste sorting, improve recycling rates, and reduce contamination [24]. Machine learning models can learn from past sorting mistakes, adapt to new waste streams, and enhance sorting accuracy over time [25]. Furthermore, predictive analytics can forecast waste generation patterns, enabling better resource allocation and planning for future waste management needs [26].

13.4 SMART BINS AND COLLECTION SYSTEMS

Smart bins and collection systems represent a significant advancement in waste management technology. Smart waste bins are equipped with sensors and IoT technology to provide real-time monitoring and management of waste collection. These bins can detect fill levels, compaction rates, and even identify the type of waste deposited [27]. By leveraging this data, smart bins can optimize waste collection schedules, reduce collection costs, and minimize environmental impact by preventing overflowing bins and reducing unnecessary trips [22].

13.4.1 RADIO-FREQUENCY IDENTIFICATION TECHNOLOGY FOR EFFICIENT WASTE TRACKING

Radio-frequency identification (RFID) technology has been increasingly adopted in waste management for efficient waste tracking and identification. RFID tags can be attached to waste bins or individual waste items to track their movement, monitor sorting processes, and manage inventory in recycling facilities [28]. This technology enables better traceability of waste streams, reduces contamination, and improves recycling rates by ensuring that recyclables are properly sorted and processed [29]. Moreover, RFID technology facilitates automated billing and invoicing, streamlining the waste management process, and improving operational efficiency [30].

13.4.2 GPS-ENABLED ROUTING FOR OPTIMIZED COLLECTION ROUTES

GPS-enabled routing systems are employed to optimize waste collection routes, reducing fuel consumption, and minimizing collection times. These systems utilize GPS technology to track collection vehicles in real-time, allowing for dynamic route adjustments based on traffic conditions, bin fill levels, and collection priorities [31]. By optimizing collection routes, GPS-enabled systems can significantly reduce operational costs, improve service efficiency, and lower carbon emissions associated with waste collection [32]. Furthermore, these systems provide valuable data insights that can inform future route planning and optimization strategies, contributing to continuous improvement in waste management practices [33].

13.5 WASTE-TO-ENERGY INNOVATIONS

Waste-to-energy (WtE) innovations represent a promising approach to managing waste while generating clean and renewable energy. These technologies convert various forms of waste into energy, reducing the volume of waste sent to landfills and contributing to a more sustainable energy mix. As WtE technologies continue to evolve and become more integrated into waste management and energy systems, they hold the potential to play a significant role in transitioning toward a circular economy and a cleaner, greener future [34].

13.5.1 EMERGING TECHNOLOGIES IN BIOMASS AND BIOENERGY

Advanced conversion technologies such as anaerobic digestion and gasification can convert organic waste, agricultural residues, and energy crops into biogas, biofuels, and heat [35]. These technologies offer a sustainable alternative to fossil fuels, reducing carbon emissions and mitigating climate change [36]. Moreover, bioenergy production from biomass can provide decentralized energy solutions, supporting rural development and enhancing energy access in underserved areas [37].

13.5.2 HARNESSING POWER FROM LANDFILL GAS AND ORGANIC WASTE

Landfill gas, primarily composed of methane and carbon dioxide, can be captured and utilized as a renewable energy source for electricity generation or heat production [38]. Advanced landfill gas recovery systems can optimize gas collection efficiency, reduce methane emissions, and generate revenue through the sale of electricity or renewable energy credits [39]. Additionally, organic waste materials such as food waste and agricultural residues can be processed through anaerobic digestion or composting to produce biogas or compost, respectively [40].

13.6 CASE STUDIES: SUCCESSFUL IMPLEMENTATIONS

Examining real-world applications of innovative waste management technologies and strategies provides valuable insights into their effectiveness, challenges, and potential for widespread adoption. Here, we present case studies of successful implementations from around the globe, highlighting the positive impact of technology adoption and drawing lessons from leading innovators in sustainable waste management. Recycle Track System (RTS) is a technology-driven waste management company in the United States that leverages IoT, data analytics, and mobile technology to optimize waste collection and recycling services [19]. RTS's innovative approach has enabled it to increase recycling rates, reduce contamination, and improve operational efficiency. By providing real-time tracking, transparent reporting, and customized waste solutions, RTS demonstrates the positive impact of technology adoption in waste management [41]. The case studies of successful implementations in sustainable waste management demonstrate the transformative power of innovation, technology adoption, and collaborative efforts. By learning from these leading innovators and adopting best practices, we can accelerate progress toward achieving a more sustainable, efficient, and environmentally responsible waste management ecosystem globally.

13.7 THE CONTINUED EVOLUTION TOWARD A CIRCULAR ECONOMY

The transition toward a circular economy, where resources are reused, recycled, and regenerated, will continue to gain momentum, supported by technological advancements, policy initiatives, and public awareness. Governments and regulators will increasingly implement extended producer responsibility (EPR) schemes, requiring producers to take responsibility for the entire lifecycle of their products, from design to disposal, promoting product stewardship and waste reduction [42]. The rise of collaborative and sharing economy models, where resources, products, and services are shared, rented, or leased, will contribute to reducing consumption, waste generation, and environmental impact, fostering a more sustainable and resource-efficient economy [43]. Businesses will continue to innovate and adopt circular business models, such as product-as-a-service, sharing platforms, and closed-loop systems, which prioritize resource efficiency, waste reduction, and sustainability, while creating value and driving economic growth [44].

13.8 CHALLENGES AND CONSIDERATIONS

As we navigate the transformative journey toward revolutionizing waste management through cutting-edge technologies, it is essential to recognize and address the challenges and considerations that arise. While technology holds immense potential to drive innovation and sustainability in waste management, its successful implementation requires careful planning, strategic considerations, and stakeholder engagement. By addressing technological, ethical, environmental, and affordability considerations, stakeholders can foster a more inclusive, equitable, and sustainable approach to waste management, harnessing the full potential of technological innovations to create a cleaner, greener, and more resilient future for all.

One of the significant challenges in implementing new technologies in waste management is ensuring compatibility and seamless integration with existing systems and infrastructure. As waste management systems vary across regions and jurisdictions, achieving interoperability and integration can be complex and require customized solutions [45]. With the proliferation of sensors, IoT devices, and data analytics platforms in waste management, managing and safeguarding data privacy and security becomes paramount. Establishing robust data management practices, encryption, and compliance with data protection regulations are crucial to protect sensitive information and build trust among stakeholders [46].

13.9 ETHICAL AND ENVIRONMENTAL CONSIDERATIONS IN TECH-DRIVEN WASTE SOLUTIONS

While technology offers promising solutions to improve waste management efficiency, it is essential to conduct comprehensive sustainability and lifecycle assessments to evaluate the environmental footprint, resource use, and long-term impacts of tech-driven waste solutions. Adopting a holistic approach that considers environmental, social, and economic factors is crucial to ensure that technology adoption aligns with sustainability goals and principles [47]. Addressing equity and social inclusion concerns is vital to ensure that tech-driven waste solutions benefit all members of society, including marginalized and underserved communities. Fostering inclusive decision-making processes, promoting access to technology, and addressing digital divides can contribute to equitable and fair waste management practices [48].

13.10 CONCLUSION

As we conclude this exploration into the transformative potential of cutting-edge technologies in revolutionizing waste management, it is evident that we stand at a pivotal juncture in shaping the future of waste management practices and sustainability. The convergence of innovation, technology, and sustainability offers unprecedented opportunities to address the challenges of waste generation, environmental degradation, and resource depletion. The journey toward revolutionizing waste management through technological innovations is both exciting and challenging. By embracing innovation, fostering collaboration, prioritizing sustainability, and empowering stakeholders, we can pave the way for a cleaner, greener, and more resilient future. Let us seize this opportunity, take collective action, and work together toward creating a sustainable and prosperous future for all.

KEYWORDS

- **environmental sustainability**
- **waste management solutions**
- **innovative technologies**
- **recycling technologies**

REFERENCES

1. Oberle, B., Bringezu, S., Hatfield-Dodds, S., Hellweg, S., Schandl, H., & Clement, J. (2019). *Global Resources Outlook: 2019*. International Resource Panel, United Nations Environment Programme: Paris, France.
2. Yang, H., Ma, M., Thompson, J. R., & Flower, R. J. (2018). Waste management, informal recycling, environmental pollution and public health. *Journal of Epidemiology and Community Health*, *72*(3), 237–243.

3. Godfrey, L., Ahmed, M. T., Gebremedhin, K. G., Katima, J. H., Oelofse, S., Osibanjo, O., Richter, U. H., & Yonli, A. H. (2019). Solid waste management in Africa: Governance failure or development opportunity. *Regional development in Africa, 235.* doi:10.5772/intechopen.86974 .
4. Hardoy, J. E., & Satterthwaite, D. (1991). Environmental problems of Third World cities: A global issue ignored. *Public Administration and Development, 11*(4), 341–361.
5. World Bank. (2018). *WHAT A WASTE 2.0: A Global Snapshot of Solid Waste Management to 2050.* World Bank Group: Washington, DC.
6. Koval, V., Mikhno, I., Udovychenko, I., Gordiichuk, Y., & Kalina, I. (2021). Sustainable natural resource management to ensure strategic environmental development. *TEM Journal, 10*(3), 1022–1030.
7. Sinthiya, N. J., Chowdhury, T. A., & Haque, A. B. (2022). Artificial intelligence based Smart Waste Management—a systematic review. In *Computational Intelligence Techniques for Green Smart Cities*, pp. 67–92.
8. Bosmans, A., Vanderreydt, I., Geysen, D., & Helsen, L. (2013). The crucial role of Waste-to-Energy technologies in enhanced landfill mining: A technology review. *Journal of Cleaner Production, 55*, 10–23.
9. Bishoge, O. K., Huang, X., Zhang, L., Ma, H., & Danyo, C. (2019). The adaptation of waste-to-energy technologies: Towards the conversion of municipal solid waste into a renewable energy resource. *Environmental Reviews, 27*(4), 435–446.
10. Chen, D. M. C., Bodirsky, B. L., Krueger, T., Mishra, A., & Popp, A. (2020). The world's growing municipal solid waste: Trends and impacts. *Environmental Research Letters, 15*(7), 074021.
11. Siddiqua, A., Hahladakis, J. N., & Al-Attiya, W. A. K. (2022). An overview of the environmental pollution and health effects associated with waste landfilling and open dumping. *Environmental Science and Pollution Research, 29*(39), 58514–58536.
12. Gautam, M., & Agrawal, M. (2021). Greenhouse gas emissions from municipal solid waste management: A review of global scenario. In *Carbon Footprint Case Studies: Municipal Solid Waste Management, Sustainable Road Transport and Carbon Sequestration*, pp. 123–160.
13. Lead, C., Adedipe, N. O., Sridhar, M. K. C., & Verma, M. (2005). Waste management, processing, and detoxification. In *Ecosystems and Human Well being: Policy Responses*, pp. 313–334.
14. Ferreira, B., Monedero, J., Martí, J. L., Aliaga, C., Hortal, M., & López, A. D. (2008). The economic aspects of recycling. In *Post-Consumer Waste Recycling and Optimal Production*, p. 99.
15. Munir, M. T., Li, B., & Naqvi, M. (2023). Revolutionizing municipal solid waste management (MSWM) with machine learning as a clean resource: Opportunities, challenges and solutions. *Fuel, 348*, 128548.
16. Wan, C., Shen, G. Q., & Choi, S. (2019). Waste management strategies for sustainable development. In *Encyclopedia of Sustainability in Higher Education.* Springer International Publishing: Cham, pp. 2020–2028.
17. Song, Q., Li, J., & Zeng, X. (2015). Minimizing the increasing solid waste through zero waste strategy. *Journal of Cleaner Production, 104*, 199–210.

18. Rootes, C. (2013). From local conflict to national issue: When and how environmental campaigns succeed in transcending the local. *Environmental Politics*, *22*(1), 95–114.
19. Salem, K. S., Clayson, K., Salas, M., Haque, N., Rao, R., Agate, S., Singh, A., Levis, J. W., Mittal, A., Yarbrough, J. M., Venditti, R., Jameel, H., Lucia, L., & Pal, L. (2023). A critical review of existing and emerging technologies and systems to optimize solid waste management for feedstocks and energy conversion. *Matter*, *6*(10), 3348–3377.
20. Nanda, S., & Berruti, F. (2021). Municipal solid waste management and landfilling technologies: A review. *Environmental Chemistry Letters*, *19*(2), 1433–1456.
21. Henaien, A., Elhadj, H. B., & Fourati, L. C. (2024). A sustainable smart IoT-based solid waste management system. *Future Generation Computer Systems*, *157*, 587–602.
22. de Morais, C. S., Ramos, T. R. P., Lopes, M., & Barbosa-Póvoa, A. P. (2024). A data-driven optimization approach to plan smart waste collection operations. *International Transactions in Operational Research*, *31*(4), 2178–2208.
23. Bibri, S. E. (2018). The IoT for smart sustainable cities of the future: An analytical framework for sensor-based big data applications for environmental sustainability. *Sustainable Cities and Society*, *38*, 230–253.
24. Xia, W., Jiang, Y., Chen, X., & Zhao, R. (2022). Application of machine learning algorithms in municipal solid waste management: A mini review. *Waste Management & Research*, *40*(6), 609–624.
25. Erkinay Ozdemir, M., Ali, Z., Subeshan, B., & Asmatulu, E. (2021). Applying machine learning approach in recycling. *Journal of Material Cycles and Waste Management*, *23*, 855–871.
26. Dyson, B., & Chang, N. B. (2005). Forecasting municipal solid waste generation in a fast-growing urban region with system dynamics modeling. *Waste Management*, *25*(7), 669–679.
27. Arthur, M. P., Shoba, S., & Pandey, A. (2024). A survey of smart dustbin systems using the IoT and deep learning. *Artificial Intelligence Review*, *57*(3), 56.
28. Hannan, M. A., Al Mamun, M. A., Hussain, A., Basri, H., & Begum, R. A. (2015). A review on technologies and their usage in solid waste monitoring and management systems: Issues and challenges. *Waste Management*, *43*, 509–523.
29. Denuwara, N., Maijala, J., & Hakovirta, M. (2019). Sustainability benefits of RFID technology in the apparel industry. *Sustainability*, *11*(22), 6477.
30. Condemi, A., Cucchiella, F., & Schettini, D. (2019). Circular economy and E-waste: An opportunity from RFID TAGs. *Applied Sciences*, *9*(16), 3422.
31. Alwabli, A. S. (2021). *Dynamic Route Optimization for Waste Collection in Smart City*, PhD dissertation..
32. Watling, S. D., & Connor, J. B. (2010). GPS/GIS technology to provide sustainable solutions. In *World Environmental and Water Resources Congress 2010: Challenges of Change*, pp. 208–217.
33. Anjum, M., Sarosh Umar, M., & Shahab, S. (2022, February). Analysis of IoT and communication technologies to develop waste management service framework for smart city. In *Proceedings of Second International Conference on Sustainable Expert Systems: ICSES 2021*. Springer Nature Singapore: Singapore, pp. 691–707.

34. Kumar, A., & Samadder, S. R. (2017). A review on technological options of waste to energy for effective management of municipal solid waste. *Waste Management, 69*, 407–422.
35. Lee, S. Y., Sankaran, R., Chew, K. W., Tan, C. H., Krishnamoorthy, R., Chu, D. T., & Show, P. L. (2019). Waste to bioenergy: A review on the recent conversion technologies. *BMC Energy*, *1*(1), 1–22.
36. Ali, J., Rasheed, T., Afreen, M., Anwar, M. T., Nawaz, Z., Anwar, H., & Rizwan, K. (2020). Modalities for conversion of waste to energy—Challenges and perspectives. *Science of the Total Environment*, *727*, 138610.
37. Mohammed, Y. S., Mokhtar, A. S., Bashir, N., & Saidur, R. (2013). An overview of agricultural biomass for decentralized rural energy in Ghana. *Renewable and Sustainable Energy Reviews*, *20*, 15–25.
38. Themelis, N. J., & Ulloa, P. A. (2007). Methane generation in landfills. *Renewable Energy*, *32*(7), 1243–1257.
39. Ahmed, S. I., Johari, A., Hashim, H., Mat, R., Lim, J. S., Ngadi, N., & Ali, A. (2015). Optimal landfill gas utilization for renewable energy production. *Environmental Progress & Sustainable Energy*, *34*(1), 289–296.
40. Zupančič, G. D., & Grilc, V. (2012). Anaerobic treatment and biogas production from organic waste. In *Management of Organic Waste*, p. 2.
41. Rao, A. S., Radanovic, M., Liu, Y., Hu, S., Fang, Y., Khoshelham, K., Palaniswami, M., & Ngo, T. (2022). Real-time monitoring of construction sites: Sensors, methods, and applications. *Automation in Construction*, *136*, 104099.
42. Shooshtarian, S., Maqsood, T., Wong, P. S., Khalfan, M., & Yang, R. J. (2021). Extended producer responsibility in the Australian construction industry. *Sustainability*, *13*(2), 620.
43. Fernandes, S., Lucas, J., Madeira, M. J., Cruchinho, A., & Honório, I. D. (2019). Circular and collaborative economies as a propulsion of environmental sustainability in the new fashion business models. In *Innovation, Engineering and Entrepreneurship*. Springer International Publishing, pp. 925–932.
44. Lacy, P., Long, J., Spindler, W., Lacy, P., Long, J., & Spindler, W. (2020). The circular business models. *The Circular Economy Handbook: Realizing the Circular Advantage*, 17–42.
45. McDougall, F. R., White, P. R., Franke, M., & Hindle, P. (2008). *Integrated Solid Waste Management: A Life Cycle Inventory*. John Wiley & Sons.
46. Adekugbe, A. P., & Ibeh, C. V. (2024). Navigating ethical challenges in data management for US program development: Best practices and recommendations. *International Journal of Management & Entrepreneurship Research*, *6*(4), 1023–1033.
47. Dharmayanti, N., Ismail, T., Hanifah, I. A., & Taqi, M. (2023). Exploring sustainability management control system and eco-innovation matter sustainable financial performance: The role of supply chain management and digital adaptability in Indonesian context. *Journal of Open Innovation: Technology, Market, and Complexity*, *9*(3), 100119.
48. Ellis, S. C. (2005). Meaningful consideration? A review of traditional knowledge in environmental decision making. *Arctic*, 58(1), 66–77.

49. Mondal, T., Choudhury, M., Kundu, D., Dutta, D., & Samanta, P. (2023). Landfill: An eclectic review on structure, reactions and remediation approach. *Waste Management*, *164*, 127–142. https://doi.org/10.1016/j.wasman.2023.03.034
50. Paul, S., Choudhury, M., Deb, U., Pegu, R., Das, S., & Bhattacharya, S. S. (2019). Assessing the ecological impacts of ageing on hazard potential of solid waste landfills: A green approach through vermitechnology. *Journal of Cleaner Production*, *236*, 117643. https://doi.org/10.1016/j.jclepro.2019.117643

CHAPTER 14

Innovative Approaches to Waste Management for Sustainable Practices Through Data-Driven Solutions

HEM CHANDRA JOSHI,[1] ABBAS SHIEH,[2] ANSHUL JAISWAL,[3] LALITHA SHANMUGASUNDARAM,[4] MARIANA FLORES GARCIA,[5] and SUSHOBHAN MAJUMDAR[6*]

[1]*Faculty of Natural Resource Management, Far Western University, Bhimdatta, Nepal*

[2]*Department of Planning, Policy, and Design, Azad University, Australia*

[3]*Department of Botany, Banaras Hindu University, Varanasi, Uttar Pradesh, India*

[4]*George Washington University, Washington, DC, USA*

[5]*Faculty of Architecture, Universidad La Salle Bajío, León de los Aldama, Mexico*

[6]*Department of Geography, Pandit Raghunath Murmu Smriti Mahavidyaluya, Bankura, West Bengal, India*

*Corresponding author

ABSTRACT

This chapter explores innovative approaches to waste management aimed at fostering sustainable practices through the integration of data-driven solutions. With the ever-increasing global waste production posing significant environmental challenges, there is a growing need for effective waste management strategies. Traditional methods often fall short in addressing

Data Analytics and Smart Technologies in Solid Waste Management: Advancing Sustainability. Moharana Choudhury, Sushobhan Majumdar, and Avnish Chauhan (Eds.)

DOI: 10.1201/9781779640833-18

the complexities of modern waste streams and fail to maximize resource recovery while minimizing environmental impact. In response, this chapter advocates for the adoption of data-driven solutions to enhance waste management practices. Leveraging advanced technologies such as Internet of Things sensors, machine learning algorithms, and big data analytics, waste management systems can be optimized for improved efficiency, waste diversion, and recycling rates. By harnessing real-time data insights, stakeholders can make informed decisions regarding waste collection, sorting, treatment, and disposal, leading to more sustainable outcomes. Furthermore, the integration of data-driven approaches enables the development of predictive models and optimization algorithms to anticipate future waste trends and optimize resource allocation. Through case studies and examples, this chapter demonstrates the potential of data-driven solutions to revolutionize waste management practices and contribute to a more sustainable and circular economy.

14.1 INTRODUCTION

Waste management has been a subject of great concern nowadays. Previously, the population was less with limited production of nondegradable solid and industrial wastes, but with the rapid urbanization and industrialization, the waste production has been major environmental issue. The waste management became a very organized, specialized, and complex activity due to risk associated and its impact on human health and environment [15–52]. Four primary methods of waste management are common in practice: landfill, incineration, recycling, and composting [33–50]. Various challenges are associated with waste management. Some of the major challenges are financial constraint, ineffective recycling, conflict with localities, rapid urbanization, and limited resources.

When addressing waste management, prioritizing financial analysis is essential. On one hand, effective waste management and recycling prove cost-effective in the long term, yet commands significant upfront investment. The expenses associated with waste management can vary based on factors such as disposal/recycling location, waste type, and other considerations [16]. Also, we need to focus seriously on reuse and recycle to the maximum extent. Ineffective recycling has led to new production, which needs large quantity of natural resources, human and financial capital as well as production of different pollutants in the environment.

High-quality materials, such as plastic, paper, and metal, are disposed into landfills, despite their potential for recovery and reuse. Retrieving such valuable materials not only diminishes landfill waste but also mitigates the detrimental effects of manufacturing new plastics from scratch [45–49]. Reuse and recycle are equivalent to production and this thought should be appreciated by the respective stakeholders for sustainable resource use [11]. Though landfill have been largely taken as the easy, efficient and the commonly used means of solid waste management across the globe, there is always conflict with the localities with regards to the environmental, health, and infrastructure issues [46]. Solid waste management presents considerable challenges, exacerbated by rapid urbanization and overpopulation in major cities. Densely populated areas naturally generate substantial waste volumes. However, the disposal of waste occurs in both sanctioned and unsanctioned sites, compounding the issue of waste management in urban regions [3].

Sustainable waste management involves activities and approaches of reduce, reuse and recycle to the maximum potential which in turns focuses on sustainable resource utilization [27]. Recycling and reuse initiatives not only creates employment opportunities by requiring additional manpower for collection, sorting, and processing of recyclable wastes, but they also bolster job creation through reuse centers [4]. Applying 3R principle significantly reduces air pollution and reduces climate emissions, such as CO_2 and greenhouse gas (GHG) emissions [20]. In addition, as more items are reduced or reused and recycled, the amount of waste that needs to go to the landfill or incinerator is also reduced, in this way we can save the costs of landfill and incineration as well as we can save landfill space waste management, overlooking citizen concerns, such as inadequate modeling of odor dispersion, during the planning and operation of waste facilities near residential zones, can lead to direct issues for residents [46]. Geographic Information System (GIS) and remote sensing tools are used to select the suitable location for landfill site considering the soil, geology, road network, residential areas, waterbodies, wetlands, streams, etc. [37].

14.2 TRADITIONAL WASTE MANAGEMENT PRACTICES AND LIMITATIONS

Landfilling is one of the major traditional waste management approaches that involves the deposition, compaction, and containment of waste at designated sites. Currently, it offers a convenient, adaptable, and cost-effective disposal option compared to other methods. It stands out as the sole comprehensive solution for all types of waste disposal. Despite a decline in solid waste landfilling, it is anticipated to persist as a vital component within integrated solid waste management systems worldwide [46]. Landfills can significantly degrade the quality of surface and groundwater, primarily through the release of landfill leachate [51]. The composition of leachate is influenced by various factors such as waste density, composition, climate conditions, and moisture content [9]. Additionally, the age of the landfill plays a crucial role in determining leachate characteristics. In its initial stages, a young landfill undergoes an acidogenic phase, characterized by markedly high levels of biochemical oxygen demand (BOD) and chemical oxygen demand (COD) content in the leachate. However, as the landfill matures, the concentrations of BOD and COD in the leachate diminish gradually. Similarly, landfill site deteriorates the soil quality by the contamination of heavy metals [40] and microorganisms such as Coliform, *Clostridia perfringens* causing eco-toxicity, which degrade the biological, chemical, and physical properties of the soils [32]. Similarly various harmful gases like methane, toluene, ammonia, xylene, carbon monoxide, hydrogen sulfide, nitrogen dioxide, formaldehyde, and ethylbenzene are released in the landfill causing serious harm to surrounding people health.

Incineration reduces waste volume and toxicity, but emissions pose health concerns. Finding suitable disposal sites for incinerator ash is challenging, surpassing plant placement. To gain public ownership and acceptance, strict government regulations governing ash disposal must be established [44]. Incineration produces particulate matters, acids and other gases, heavy metals, and carbon compounds in the environment [41]. Combustion and incineration systems are engineered to enhance the likelihood of complete reactions, yet they invariably fall short of achieving perfection. Instead, they may yield small quantities of various byproducts, influenced by the waste's chemical makeup and the encountered combustion conditions. These byproducts, alongside potentially unreacted waste components, constitute the emissions from the incinerator [38].

Reduce, reuse, and recycle are the components of sustainable waste management. Though recycling has multiple benefits, still there is very limited recycling activities done in the case of solid waste management. Some of the major challenges of recycling are awareness of the recycling, policy provision for promoting recycling, market demand, workers safety, and skilled manpower [39]. To boost recycling rates, it is imperative for local governments to stimulate demand for recycled materials within markets and bolster the workforce within recycling companies [35]. Further critical considerations involve providing financial support for various recycling projects to support the infrastructure of the recycling companies in the country is essential.

Innovative technologies for treating Municipal Solid Waste (MSW) include biogas recovery, composting, anaerobic digestion, and thermal processes like incineration, gasification, and pyrolysis. The utilization of these technologies is contingent upon the unique combination of local, regional, and national priorities driving both waste management strategies and GHG mitigation efforts [27]. Biomethanation or anaerobic digestion emerges as a promising alternative for MSW treatment, harnessing renewable energy and leveraging the recycling potential of biodegradable organic waste from various sources [7]. Furthermore, MSW pyrolysis and gasification technologies stand out as environmentally favorable methods for waste treatment, boasting lower pollution emissions compared to conventional approaches, and holding the promise of enhanced energy production efficiency [30]. In addition, advancements in artificial intelligence (AI) and technology offer promising avenues [1] for enhancing efficiency and safeguarding human health in solid waste management.

14.3 ROLE OF DATA IN WASTE MANAGEMENT

Data has a critical role in effective waste management. In waste management, it is essential to monitor data such as waste typologies (e.g., urban or domestic, industrial or hospital, etc.) and waste locations, landfill site selection, collection optimization, and recycling are necessary information for waste management. GIS and geospatial analysis techniques could effectively improve waste management [8].

New data-driven approaches based on data analytics and machine learning have emerged and applied in waste management in recent years. For example, a Finnish study has introduced a data-driven approach that applies a self-organizing map and k-means algorithm to create waste generation profiles using container-level waste weight data [36]. These types of methods emphasize container-level waste monitoring data collection, which can facilitate the application of advanced analytics in waste management, leading to understanding waste generation patterns on various scales for better decision-making in combination with public sector data sources.

Data-driven waste management enhances efficiency and accuracy through real-time data collection, identification methods, and communication techniques, enhancing waste tracking and monitoring. Combining AI models with other technologies, such as Internet of Things (IoT) and data science, can increase their applicability and improve the quality of models by providing high-quality training data and lead to better solutions for waste management challenges [14].

14.4 INNOVATIVE TECHNOLOGIES FOR WASTE MANAGEMENT

With the advancement of technology, a range of new intelligent methods can be applied to waste collection. Most waste management systems can employ sensors, cameras, and GIS systems to collect data on the location and level of trash cans and take action to collect them [24]. There are also suggestions for using smart trash cans, wireless sensors to check the level of trash in the trash can, creating a user program as a monitoring platform for waste management, and creating awareness about waste for people [24].

IoT can also be used to check the amount of waste produced and whether it can be recycled or not, check the level of wastebins and the performance of the municipality in waste collecting, and even help the forces according to the data collected by the relevant software. It can be applied to measure the bins' location, find the optimal collection path, and track the recycling steps [10, 43].

AI can increase waste management efficiency in different ways. Cities' waste management can be improved by integrating the IoT and data access networks, combinatorial optimization, and GIS [19]. AI can play a critical role in waste data collection and statistical analysis regarding the rate of trash cans being filled and finding the route to the places that are the priority for garbage collection. Automating the waste management process by machine learning is considered an excellent use of technology to prevent human interference and related vulnerability to human health. It can happen through familiarizing the machines with possible scenarios and taking action in waste management, separation, and recycling [19].

Blockchain technology enhances waste management by providing a secure, transparent, and transparent process for data storage and processing. It offers opportunities for efficient waste resource management, reliable channelization, document protection, efficient resource management, and transparency in waste collection [5].

14.5 DATA-DRIVEN STRATEGIES FOR WASTE REDUCTION AND RECYCLING

Data-driven solutions optimize resource allocation, aid decision-making, and manage waste efficiently. Key strategies include predictive modeling, waste collection routes, sorting technologies, and monitoring systems. Predictive modeling, particularly using artificial neural networks (ANNs), enhances efficiency and sustainability in waste management. ANNs, trained on historical data, can accurately predict waste generation, enabling better resource allocation and effective waste collection strategies. In Poland, ANN models forecasted a 2% increase in waste generation by 2024 based on socioeconomic data [25]. Coupling ANN with particle swarm optimization further improves predictive accuracy and cost efficiency, as demonstrated in a study that outperformed conventional ANN models [13]. Accurate predictive modeling is particularly important, as it allows for stakeholders to target and develop management strategies for specific areas. Data in essence provides an avenue for stakeholders to explore other strategies, including the optimization of waste collection routes.

Optimizing waste collection routes is crucial for efficiency and environmental sustainability. Technologies like GIS, Backtracking Search Algorithms (BSA), and Ant Colony Systems (ACS) are effective for this purpose. GIS-based smart collection systems have shown to reduce operation costs by 19%, carbon dioxide emissions by 5–22%, and collection trip times by 18–42% [2]. BSA integrated into a capacitated vehicle routing problem model reduced route distances by 36.80% and improved collection efficiency by 36.78%, alongside significant reductions in fuel consumption and CO_2 emissions [6]. ACS algorithms optimize MSW collection by considering various dynamic factors, significantly reducing tour lengths and operational costs [22].

Optimizing waste collection routes can reduce costs and environmental impact. However, efficient sorting technologies are crucial for proper disposal. Advanced sorting technologies, including direct methods using magnetic properties and indirect methods using sensors, are essential for efficient waste management [18]. These technologies enable precise separation of recyclables, improving the quality of recycled products and reducing contamination.

Predictive modeling, waste route optimization, and sorting technologies enhance waste management efficiency. Proper monitoring and feedback systems are crucial for effective management. Data-driven solutions like IoT sensors can enhance these systems. [12]. Common sensors like DHT for basic data logging and DS18B20 for precise temperature measurement are widely used. Expanding research to include parameters like gas composition and light intensity can further enhance monitoring practices [12]. In solid waste management, smart bin solutions with a three-layer IoT architecture include sensors to measure waste levels, Wi-Fi, and GSM technologies for data transmission, and cloud services for data storage. These systems notify cleaning services when bins reach a threshold and display bin locations on a map, reducing human effort, time, and costs, while promoting a greener environment [26].

These real-time monitoring systems illustrate the potential of IoT technologies to improve waste management by providing timely data and efficient feedback mechanisms, thus supporting sustainability and a circular economy. These examples are few compared to the numerous technologies and data-driven solutions out there; however, all these technologies demonstrate the innate power of data in resolving waste management issues. Each of these technologies were far more efficient than their traditional methods. Thus, it is imperative to foster and develop new technologies with a data focus in order to bring about more effective waste management strategies.

14.6 CHALLENGES AND BARRIERS TO ADOPTION

14.6.1 DATA PRIVACY AND SECURITY CONCERN

The energy system's vast datasets pose a threat to cyberattacks, necessitating privacy and security measures. Enhancements in IT infrastructure are needed for smart energy systems, ensuring customer data security. Industry self-management, technological advancements, and strengthened laws can enhance data-intensive smart systems' security and privacy.

14.6.2 INFRASTRUCTURE AND TECHNOLOGY REQUIREMENT

The rise of energy-related big data has challenged conventional infrastructure, necessitating infrastructure improvements in network transmission, storage, processing, transfer, visualization, and collaboration. Initiatives aim to build smart urban computing systems for sustainable urban planning and operational management. However, barriers include lack of infrastructure for storage, transportation, treatment, disposal technologies, inadequate technology for Restriction of Hazardous Substances policies, and poor waste forecasting and planning.

14.6.3 RESISTANCE TO CHANGE FROM TRADITIONAL WASTE MANAGEMENT PRACTICES

Traditional waste management practices face resistance due to perceived reliability, comfort, fear of unknowns, and significant infrastructure expenditures. This resistance hinders the adoption of innovative sustainable waste management methods, posing significant challenges for organizations and stakeholders.

14.6.4 COST CONSIDERATIONS AND FINANCIAL CONSTRAINTS

Financial constraints, such as lack of government subsidies, inadequate funding for training and awareness programs, high conversion costs, and high disposal costs, make sustainable waste management projects financially unsustainable, making it challenging to effectively manage waste.

14.7 POLICY IMPLICATIONS AND RECOMMENDATIONS

Sorting waste at the point of origin becomes crucial for effective waste management and urban sustainability. Depending primarily on public awareness campaigns for implementation produces a less impactful consequence. Preventive measures, including policy regulations, administrative guidance, and incentive programs, are critical for more effective waste management.

14.7.1 REGULATORY FRAMEWORK TO SUPPORT DATA-DRIVEN WASTE MANAGEMENT

The proposed framework focuses on a single residence rather than a whole district or city because residential communities manage waste independently. The analytical framework, which makes use of IoT and data mining technologies, is divided into five major components: (1) supervised waste disposal, (2) a database hub, (3) incentive schemes, (4) statistical data mining, and (5) participatory decision making. To begin the analytical framework, a waste management system trial run is required. During this trial period, flexible time windows are established, typically spanning an entire day. The management cycle has four steps, which are detailed below:

14.7.1.1 STEP 1: SOFT REGULATIONS

Rather than enforcing harsh regulations that demand precise and fixed time slots, the government implements flexible regulations that govern time slots, allowing the great majority of citizens to select the most convenient time according to themselves.

14.7.1.2 STEP 2: SUPERVISORY GUIDANCE

During the initial phase of implementing the regulations, residents receive guidance from supervisors to assist them in properly sorting their waste. This involves employing sanitation workers or volunteers to demonstrate the correct sorting procedures.

14.7.1.3 STEP 3: STIMULATING INCENTIVES

Use incentives to encourage residents to organize their objectives. Residents would get incentives for properly sorting and putting trash in designated collection areas during appropriate time frames.

14.7.1.4 STEP 4: DEMOCRATIC DECISION-MAKING

Making decisions together using gathered data, a variety of statistical techniques can be used to evaluate how residents dispose of their waste. The best times to visit are chosen by the locals themselves through data-driven behavior analysis.

14.7.2 INCENTIVES FOR BUSINESSES AND MUNICIPALITIES TO ADOPT INNOVATIVE PRACTICES

Governments should consider implementing taxation, fees, and levies related to waste collection to recover costs and reduce waste generation. Incorporating these fees into property taxes can improve operational effectiveness and promote sustainable waste management. Programs like "pay-as-you-throw" or weight-based treatment fees can reduce household and business waste output. However, challenges include limited fees and enforcement difficulties, requiring careful planning, data collection, and robust monitoring systems.

14.7.3 EDUCATION AND OUTREACH PROGRAMS TO PROMOTE AWARENESS OF DATA-DRIVEN SOLUTIONS

The subcommittee on education and awareness aims to enhance public involvement in recycling and trash reduction initiatives through educational and community engagement initiatives. They plan to establish a protocol for waste management reforms, improve accessibility to educational resources, and promote confidence and openness in waste management efforts through branding and organizational messaging.

14.7.4 COLLABORATION BETWEEN GOVERNMENT, INDUSTRY, AND ACADEMIA TO DRIVE INNOVATION

"Academic institutions" that receive funding from "governmental bodies" and "organizations" can improve waste management laboratories and advanced research programs. "Governments" and "organizations" can work together with "academic institutions" to develop and implement technical improvements for the collection and final disposal of garbage. Additionally, "academic institutions" and "organizations" can collaborate together to collect data to support waste management plans created by "governmental bodies." These collective endeavors, among others, hold promise for enhancing waste management practices.

- Generate software that makes it easier to promote recyclable materials. Establish a service that allows municipalities to track household waste using trash bags that have QR codes on them.

- Develop novel applications for satellite tracking devices to improve garbage collection vehicles' routes. Offer advisory services to increase awareness and educate staff members about waste separation.
- Raise public knowledge of the environment by outlining the appropriate disposal classification for urban solid waste. As an illustration, consider marketing campaigns, classroom instruction, and volunteer orientation at waste collection locations.

14.8 FUTURE DIRECTIONS AND OPPORTUNITIES

14.8.1 EMERGING TRENDS IN DATA-DRIVEN WASTE MANAGEMENT

Emerging trends in data-driven waste management are revolutionizing the approach to handling and reducing waste. Utilizing advanced technologies like IoT sensors, machine learning algorithms, and big data analytics, waste management systems are becoming more efficient and sustainable. These innovations enable real-time monitoring of waste generation, collection, and disposal processes, allowing for optimized routing of collection vehicles, predictive maintenance of waste infrastructure, and dynamic adjustments to waste management strategies. Moreover, data-driven insights facilitate the implementation of targeted interventions to reduce waste generation, promote recycling and composting, and minimize landfill usage, ultimately leading to more cost-effective and environmentally friendly waste management practices.

14.8.2 POTENTIAL APPLICATIONS OF ADVANCED TECHNOLOGIES

The potential applications of advanced technologies are vast and diverse, encompassing virtually every aspect of human life and industry. From healthcare to agriculture, transportation to manufacturing, and beyond, cutting-edge technologies such as AI, blockchain, augmented reality, and quantum computing hold promise for transforming processes, enhancing efficiency, and unlocking new possibilities. In healthcare, for instance, AI-powered diagnostics and personalized medicine offer the potential for more accurate diagnoses and tailored treatments. In agriculture, IoT sensors and drones enable precision farming techniques, optimizing resource usage and crop yields. Advanced technologies also play a crucial role in smart cities, facilitating efficient energy management, traffic control, and waste management systems. Moreover, in fields like finance and cybersecurity, blockchain technology offers decentralized and secure solutions. The potential applications of advanced technologies are continually expanding, shaping the future of how we live, work, and interact with the world around us.

14.8.3 RESEARCH PRIORITIES FOR FURTHER ENHANCING SUSTAINABILITY IN WASTE MANAGEMENT

Research priorities for enhancing sustainability in waste management include developing innovative technologies, interdisciplinary research to understand environmental, social, and economic impacts, community engagement and education for responsible consumption, policy frameworks and governance structures, and exploring circular economy models and digital technologies for better data management and decision-making. These efforts aim to create a more resource-efficient and environmentally friendly future.

14.9 CONCLUSION

Embracing data-driven solutions is crucial for sustainable waste management. Data-driven solutions enhance efficiency and cost-effectiveness, improve recycling rates and resource recovery, reduce emissions, and increase waste diversion through innovations and technological advancements. Utilizing data analytics, waste management systems can streamline collection routes, schedules, and resource distribution. This approach cuts down on fuel usage, lowers operational expenses, and improves the overall efficiency of waste collection operations. Sophisticated data systems enhance the precision in identifying the types and amounts of waste, leading to improved sorting and increased recycling rates. AI and machine learning technologies can automate the process of recognizing and separating recyclable materials from waste streams. Similarly, in addition to that data systems assists waste management companies in maintaining regulatory compliance by offering precise and prompt reporting. This guarantees that all operations adhere to local, national, and international environmental standards. Utilizing data-driven solutions enhances transparency and accountability in waste management processes, fostering trust among stakeholders such as communities, regulators, and investors.

The government authority should facilitate in the innovative waste management practices by creating the enabling policy condition. Governments should create rules and policies to promote innovation in waste management. This could involve setting recycling goals, implementing landfill taxes, and offering incentives for technologies that reduce waste. Governments should also offer financial backing for research and development in waste management innovations through grants, subsidies, and low-interest loans.

KEYWORDS

- **waste management**
- **data-driven solutions**
- **big data analytics**
- **resource allocation**

REFERENCES

1. Abbasi, M., & El Hanandeh, A. (2016). Forecasting municipal solid waste generation using artificial intelligence modeling approaches. *Waste Management*, *56*, 13–22.
2. Abdallah, M., et al. (2019). Simulation and optimization of dynamic waste collection routes. *Waste Management & Research*. 37(8), 793–802
3. Agagu, O. (2009). *Threats to the Nigerian Environment: A Call for Positive Action*. Nigerian Conservation Foundation
4. Agarwal, A., Singhmar, A., Kulshrestha, M., & Mittal, A. (2005). Municipal solid waste recycling and associated markets in Delhi, India. *Resources, Conservation and Recycling*, *44*(1), 73–90
5. Ahmad, R. W., Salah, K., Jayaraman, R., Yaqoob, I. & Omar, M. 2021. Blockchain for waste management in smart cities: A survey. *IEEE Access*, *9*, 131520–131541
6. Akhtar, M., et al. (2017). Backtracking search algorithm in CVRP models for efficient solid waste collection and route optimization. *Waste Management*, *61*, 117–128
7. Ambulkar, A., & Shekdar, A. (2004). Prospects of biomethanation technology in the Indian context: A pragmatic approach. *Resources, Conservation and Recycling*, *40*(2), 111–128

8. Attah, T. & Alausa, O. 2023. Geospatial analysis for effective waste management strategies. In *Proceedings of the 6th National Conference of School of Engineering, Federal Polytechnic Ilaro, 2023 Ogun State, Nigeria*
9. Bashir, M., Aziz, H., Amr, S., Sethupathi, S., Ng, C., & Lim, J. (2015). The competency of various applied strategies in treating tropical municipal landfill leachate. *Desalination and Water Treatment*, *54*(9), 2382–2395
10. Bigirimana, S. Managing Waste through the Internet of Things (IoT). In *Proceedings of the EAI International Conference on Research, Innovation and Development for Africa*, 2018. pp. 314–318
11. Bringezu, S. (2017). Visions of a sustainable resource use. In *Sustainable Resource Management*, pp. 155–215
12. da Costa, T. P., et al. (2022). A systematic review of real-time monitoring technologies and its potential application to reduce food loss and waste: Key elements of food supply chains and IoT technologies. *Sustainability*, *15*(1), 614.
13. Elshaboury, N., et al. (2021). Predictive analysis of municipal solid waste generation using an optimized neural network model. *Processes*, *9*(11), 2045
14. Fang, B., Yu, J., Chen, Z., Osman, A. I., Farghali, M., Ihara, I., Hamza, E. H., Rooney, D. W. & Yap, P.-S. 2023. Artificial intelligence for waste management in smart cities: A review. *Environmental Chemistry Letters, 21*, 1959–1989
15. Giusti, L. (2009). A review of waste management practices and their impact on human health. *Waste Management*, *29*(8), 2227–2239
16. Goddard, H. (1995). The benefits and costs of alternative solid waste management policies. *Resources, Conservation and Recycling*, *13*(3–4), 183–213
17. Guerrero, L. A., Maas, G., & Hogland, W. (2013). Solid waste management challenges for cities in developing countries. *Waste Management*, *33*(1), 220–232
18. Gundupalli, S. P., Hait, S., & Thakur, A. (2017). A review on automated sorting of source-separated municipal solid waste for recycling. *Waste Management*, *60*, 56–74
19. Gupta, P. K., Shree, V., Hiremath, L. & Rajendran, S. 2019. The use of modern technology in smart waste management and recycling: Artificial intelligence and machine learning. In *Recent Advances in Computational Intelligence*, pp. 173–188.
20. Islam, M. T., & Huda, N. (2018). Reverse logistics and closed-loop supply chain of Waste Electrical and Electronic Equipment (WEEE)/E-waste: A comprehensive literature review. *Resources, Conservation and Recycling*, *137*, 48–75
21. Jiang, P., Van Fan, Y., Zhou, J., Zheng, M., Liu, X., & Klemeš, J. J. (2020). Data-driven analytical framework for waste-dumping behaviour analysis to facilitate policy regulations. *Waste Management*, *103*, 285–295
22. Karadimas, N. V., Papatzelou, K., & Loumos, V. G. (2007). Optimal solid waste collection routes identified by the ant colony system algorithm. *Waste Management & Research*, *25*(2), 139–147
23. Khan, S. A. (2016). E-products, E-waste and the basel convention: Regulatory challenges and impossibilities of international environmental law. *Review of European, Comparative & International Environmental Law*, *25*(2), 248–260
24. Komane, B. L. & Mathonsi, T. E. 2023. Design of a smart waste management system for the city of Johannesburg. arXiv preprint arXiv:2303.14436
25. Kulisz, M., & Kujawska, J. (2020). Prediction of municipal waste generation in Poland using neural network modeling. *Sustainability*, *12*(23), 10088

26. Likotiko, E., et al. (2018). Real time solid waste monitoring using cloud and sensors technologies. *The Online Journal of Science and Technology—January*, *8*(1).
27. Lino, F., & Ismail, K. (2012). Analysis of the potential of municipal solid waste in Brazil. *Environmental Development*, *4*, 105–113.
28. Liu, W., Cui, P., Nurminen, J. K., & Wang, J. (2017). Special issue on intelligent urban computing with big data. *Machine Vision and Applications*, *28*, 675–677.
29. Liu, Z., Guan, D., Wei, W., Davis, S. J., Ciais, P., Bai, J., Peng, S., Zhang, Q., Hubacek, K., Marland, G., Andres, R. J., Crawford-Brown, D., Lin, J., Zhao, H., Hong, C., Boden, T. A., Feng, K., Peters, G. P., Xi, F., Liu, J., & He, K. (2015). Reduced carbon emission estimates from fossil fuel combustion and cement production in China. *Nature*, *524*(7565), 335–338
30. Luo, S., Xiao, B., Hu, Z., Liu, S., Guan, Y., & Cai, L. (2010). Influence of particle size on pyrolysis and gasification performance of municipal solid waste in a fixed bed reactor. *Bioresource Technology*, *101*(16), 6517–6520
31. Malek, J., & Desai, T. N. (2019). Prioritization of sustainable manufacturing barriers using Best Worst Method. *Journal of Cleaner Production*, *226*, 589–600
32. Makarenko, N., & Budak, O. (2017). Waste management in Ukraine: Municipal solid waste landfills and their impact on rural areas. *Annals of Agrarian Science*, *15*(1), 80–87
33. Marchettini, N., Ridolfi, R., & Rustici, M. (2007). An environmental analysis for comparing waste management options and strategies. *Waste Management*, *27*(4), 562–571
34. Matter, A., Ahsan, M., Marbach, M., & Zurbrügg, C. (2015). Impacts of policy and market incentives for solid waste recycling in Dhaka, Bangladesh. *Waste Management*, *39*, 321–328
35. Minghua, Z., Xiumin, F., Rovetta, A., Qichang, H., Vicentini, F., Bingkai, L., Giusti, A., & Yi, L. (2009). Municipal solid waste management in Pudong new area, China. *Waste Management*, *29*(3), 1227–1233
36. Niska, H. & Serkkola, A. 2018. Data analytics approach to create waste generation profiles for waste management and collection. *Waste Management*, *77*, 477–485
37. Olusina, J., & Shyllon, D. (2014). Suitability analysis in determining optimal landfill location using multi-criteria evaluation (MCE), GIS & remote sensing. *International Journal of Computer Engineering Research*, *04*(6), 7–20
38. Oppelt, E. (1987). Incineration of hazardous waste. 558–586
39. Orlins, S., & Guan, D. (2016). China's toxic informal e-waste recycling: local approaches to a global environmental problem. *Journal of Cleaner Production*, *114*, 71–80
40. Prechthai, T., Parkpian, P., & Visvanathan, C. (2008). Assessment of heavy metal contamination and its mobilization from municipal solid waste open dumping site. *Journal of Hazardous Materials*, *156*(1–3), 86–94.
41. Quina, M., Bordado, J., & Quinta-Ferreira, R. (2011). Air pollution control in municipal solid waste incinerators. In *The Impact of Air Pollution on Health, Economy, Environment and Agricultural Sources*, pp. 331–358
42. Rowan, N. J., & Casey, O. (2021). Empower Eco multiactor HUB: A triple helix 'academia-industry-authority' approach to creating and sharing potentially disruptive tools for addressing novel and emerging new Green Deal opportunities under a

United Nations Sustainable Development Goals framework. *Current Opinion in Environmental Science & Health, 21*, 100254
43. Shyam, G. K., Manvi, S. S. & Bharti, P. (2017). Smart waste management using Internet-of-Things (IoT). In *2017 2nd International Conference on Computing and Communications Technologies (ICCCT)*. IEEE, pp. 199–203.
44. Siskind, E., & Susskind, L. (1989). The incineration conflict: Addressing public concerns. *Environmental Impact Assessment Review*, *9*(3), 317–329
45. Tammemagi, H. (1999). *The Waste Crisis: Landfills, Incinerators, and the Search for a Sustainable Future*. Oxford University Press.
46. Vaverková, M. (2019). Landfill impacts on the environment. *Geosciences*, 431.
47. Viotti, P., Tatti, F., Rossi, A., Luciano, A., Marzeddu, S., Mancini, G., & Boni, M. (2020). An Eco-balanced and integrated approach for a more-sustainable MSW management. *Waste and Biomass Valorization, 11*, 5139–5150
48. Wigle, L. (2014). How big data will make us more energy efficient. In *World Economic Forum*. [Online]. Available: https://www. weforum. org/agenda/2014/05/big-data-will-make-us-energy-efficient
49. Mondal, T., Choudhury, M., Kundu, D., Dutta, D., & Samanta, P. (2023). Landfill: An eclectic review on structure, reactions and remediation approach. *Waste Management, 164*, 127–142. https://doi.org/10.1016/j.wasman.2023.03.034
50. Paul, S., Choudhury, M., Deb, U., Pegu, R., Das, S., & Bhattacharya, S. S. (2019). Assessing the ecological impacts of ageing on hazard potential of solid waste landfills: A green approach through vermitechnology. *Journal of Cleaner Production, 236*, 117643. https://doi.org/10.1016/j.jclepro.2019.117643
51. Choudhury, M., Jyethi, D. S., Dutta, J., Purkayastha, S. P., Deb, D., Das, R., Roy, G., Sen, T., & Bhattacharyya, K. G. (2022). Investigation of groundwater and soil quality near to a municipal waste disposal site in Silchar, Assam, India. *International Journal of Energy and Water Resources*, *6*(1), 37–47. https://doi.org/10.1007/s42108-021-00117-5
52. Choudhury, M., & Dutta, J. (2017). A comparative study of municipal solid waste management status for three major towns of Upper Assam-India. *International Journal of Waste Resources*, *7*(291), 2. https://doi.org/10.4172/2252-5211.1000291

CHAPTER 15

Smart Waste Management: Technologies for Monitoring and Effective Control

SHIPRA MOHAN TRIPATHI[1*] and RAVI KUMAR GANGWAR[2]

[1]*Department of Biotechnology, Utkarsh School of Management and Technology (Affiliated to MJP Rohilkhand University), Bareilly, UP, India*

[2]*Department of Soil Science, Institute of Environmental Sciences, Hungarian University of Agriculture and Life Sciences, Gödöllő, Hungary*

*Corresponding author

ABSTRACT

The primary waste management approach is mostly tedious and highly cost-effective. These methods lead to unnecessary carbon emissions, exceeding traffic congestion, overflowing bins, unsanitary environment, and so on. Traditional methods adopted by municipalities and waste management companies mostly act by their fixed schedules and procedures, that is, they trail unsustainable methods of solid waste management. Smart waste management employs waste fill levels, bin movements, locations, and other innovative technologies to transform waste management. Innovative technologies and IoT data analytics aid in deducing the challenges and improving them. These creative methods consist of smart garbage boxes, garbage volume detectors, automated recycling gadgets, garbage van loading systems, busty garbage runs, solar power junk composite, e-scrap booth, and recycling platforms. The smart city ecosystem abided by the

Data Analytics and Smart Technologies in Solid Waste Management: Advancing Sustainability.
Moharana Choudhury, Sushobhan Majumdar, and Avnish Chauhan (Eds.)

DOI: 10.1201/9781779640833-19

smart waste management system as an essential part, with the advancement of IoT-enabled smart waste sensors and waste collection in cities. The employment of IoT in waste management can potentially minimize the unnecessary expenses aroused through operational insufficiency during trash collection processes. This chapter discusses route optimization, smart recycling, and data analysis in smart waste management.

15.1 INTRODUCTION

Smart waste management is in-build with advance technologies that achieves waste management task professionally through the sensors implemented on the garbage bins. This waste management system is dependent on the various algorithms and machine-based objectives that enhance the waste disposal through sharp routes. The equipment's employed for the size determination of waste are especially designed by the professionals and are linked through internet having detailed information about waste management [1]. Smart waste management bids for a waste segmentation mechanism having a high-quality camera, which examine the waste pile into grille segments. The camera captures the detailed image and further helps to sort out particular waste item for recycling and thus help in maintaining a clean and green environment [2].

15.1.1 CHALLENGE IN TRADITIONAL WASTE MANAGEMENT

Waste management is the most consequential task encompassing the whole Earth with the excessive growth of population and resource utilization. Traditional waste management programs can have unhealthy effect on environment, atmospheric conditions, and the well-being of a human. The adverse effect on the environment includes landfills through which the soil and water gets contaminated by the waste loaded with chemicals contaminants and for wild animals littering can have drastic consequences, when microplastics are ingested accidently [46]. Additionally, if there is no periodic recycling or recovery of waste is performed, there are possibilities that the raw materials to be extracted and modified into a product are disoriented. The Earth's atmosphere and climatic systems are also affected by the old waste management practices, that is, methane is released (greenhouse gas) through landfills [3, 48]. Various studies have been carried in the recent years where concerns were shown toward the pollutants emerged in the water bodies, leads to hazards and negative impacts on both human and aquatic life. The emerging pollutants can be categorized as polyfluoroalkyl substances, flame retardants, pharmaceutical compounds, illicit drugs, pesticides, and so on [4].

15.1.2 ENVIRONMENTAL EFFECT

Inefficient waste management leads to significant environmental problems. Improper waste disposal results in soil and water pollution, depletion of natural resources, and loss of biodiversity, adversely affecting the environment, humans, and animals. Dumping waste on the outskirts of cities and towns, especially when uncompact, leads to putrefaction and a substantial increase in density. Soluble salts and metals from solid waste contaminate the soil, disrupting soil structure. Compostable portions of municipal solid waste can be separated for use as valuable manure. Traditional waste deposition sites lack impermeable linings, causing contamination of both surface and groundwater. Natural water sources become polluted from surface runoff and leachates from dump sites (Figure 15.1). Various factors such as waste composition, time, temperature, moisture, and oxygen presence influence the formation of leachates [5].

FIGURE 15.1 Effects of waste management.

15.1.3 ECONOMIC EFFECT

When waste is not sorted, recycled, or disposed of properly, it escalates costs for collection, transportation, and disposal. These expenses strain local governments and waste management companies, impeding their ability to allocate resources effectively and reducing productivity in various industries. For instance, inadequate waste management leads to waste accumulation in work areas, creating unhygienic environments and decreasing efficiency. Such practices pose health and safety risks, leading to workplace accidents or illnesses. These factors disrupt productivity, lower land values, decrease tourism, exploit resources, incur costly cleaning efforts, and ultimately hinder economic growth [6].

15.2 SMART WASTE MANAGEMENT: TRANSFORMATIVE SOLUTION

Earlier the waste management sector was considered as traditional static industry. Invention of Internet of Things (IoT) technology and other innovations have become commercially available to meet the budget cuts and ambitious sustainability targets as desired by waste authorities for smart solutions. Smart waste management prioritizes on resolving solid waste problems by implementing sensors, intelligent monitoring systems, and mobile applications. These devices and applications help to optimize the finest route for packed containers and create smart schedules for drivers. They inform the trash or municipal organizations regarding any severe accidents like high temperature or dislocation of garbage cans through GPS system [7, 8].

15.3 CURRENT ISSUES IN TRADITIONAL WASTE MANAGEMENT

Amount of waste produced varies from one place to another with significant impact on both the environment and public health leading to cause diseases. Improper waste disposal give rise to littering, unlawful dumping, and environmental pollution, affecting the ecosystem. If exposed for long time it puts impact on the quality of soil, water, and air.

The waste management problems can be resolved if it overcomes multiple challenges such as scarcity of finances, architecture, and general acknowledgement. Hence, the waste management systems must be acceptable, creative and reconciling [9].

15.4 DRAWBACKS OF CONVENTIONAL WASTE MANAGEMENT APPROACHES

Conventional waste management has hazardous and nonhazardous systems. Nonhazardous waste is disposed of through thermal treatment or landfilling, which can lead to soil contamination and air pollution, whereas hazardous waste poses risks to health and the environment [10].

15.4.1 CARBON EMISSIONS

Carbon dioxide (CO_2) emissions from cement production, fossil fuel burning, transportation, and industrial activities are a major contributor to global climate change [11].

15.4.2 TRAFFIC CONGESTION

Traffic congestion increases fuel consumption, leading to higher CO_2 emissions, air pollution, and noise pollution [12].

15.4.3 OVERFLOWING BINS

Overflowing waste bins contaminate water resources, impacting humans and wildlife. For example, food waste in landfills releases harmful methane gas [13].

15.4.4 UNSANITARY ENVIRONMENTS

Safe, healthy, and supportive environment is good for health. Human life style gets effected by unhealthy surrounding caused by the inappropriate sanitary and environmental practice which pollutes the life resources such as air, drinking water, food, and so on [14].

15.4.5 INEFFICIENCIES IN FIXED SCHEDULES AND PROCEDURES FOLLOWED BY MUNICIPALITIES AND WASTE MANAGEMENT COMPANIES

Municipal Solid Waste Management of various developing countries faces typical problems, such as

1. Inadequate service coverage
2. Operational inefficiencies of services
3. Limited utilization of recycling activities
4. Inadequate management of nonindustrial hazardous waste
5. Inadequate landfill disposal

The main reasons are lack of financial resources to handle the excess of waste produced by the rapid growing cities. Inadequate fee charges and insufficient funds from a central municipal budget do not provide the good finance for quality service. The available resources are mostly allocated to the posh areas having high tax yields with residents having political influence, which leads to peri-urban areas unserved [15].

Sometimes the technical issues hinder an efficient service and higher population coverage. Conventional approaches for waste collection applied in the industrialized countries are also introduced to developing countries. Vehicles allotted for waste management programs are sophisticated, expensive, and difficult to operate and maintain, proves to be inadequate for the conditions in developing countries [15].

15.5 SMART WASTE MANAGEMENT TECHNOLOGIES

15.5.1 SMART GARBAGE BOX

A smart garbage box is equipped with sensors and smart technologies like Wi-Fi/Bluetooth for data collection on waste volume and types. It includes GPS tracking and data analysis software to optimize waste collection routes, enhancing efficiency. These systems aid in improving waste collection and management, offering insights for effective waste reduction and recycling strategies.

Features and benefits of smart garbage boxes include wireless ultrasonic fill-level sensors and solar-powered trash compactors. Data gathered is further utilized to enhance services:

- Locations prone to garbage overflow.

- Necessary bin installations in densely populated areas to prevent overflow.
- Frequency of collection services to prevent overcrowding.
- Opportunities to reduce fuel consumption and expenses [16].

15.5.2 GARBAGE VOLUME DETECTOR

Garbage volume detector proves to have an important role in analysis of garbage collection processes and the reduction of operational costs. Introduction of wireless technologies, like IoT, helps to propel the market growth. IoT-enabled garbage volume detector provides timely accessed information on trash volume, that is, proper and precise waste collection with time-encoded programs. These innovations help in reduction of unnecessary collection trips, minimize fuel consumption, and decreases carbon emissions [17].

15.5.3 AUTOMATED RECYCLING GADGETS

An unparallel insight regarding the composition and potential recyclability of waste materials can be achieved through the automated systems as they have the ability to process in detail and optimize huge data. Automated gadgets along with robotics can easily create smarter, more efficient, and more flexible waste management systems [18].

An automated waste sorting machine is installed with the cameras having smart sensors. Along with conveyor belt, waste items move and are photographed simultaneously, and the captured images are then uploaded to a cloud database. In the final step machine learning models briefly examine the images with full descriptions and data on the physical, chemical, thermal, and biological properties of the waste [19].

15.5.4 GARBAGE VAN LOADING SYSTEM

In the waste management and recycling industry, lorry calibration system is widely implemented, approved, and resourceful weighing solution. In order to prevent unsafe overloading truck scales assist the operators to ensure their vehicles are loaded with precise capacity. Other than accurate weighing, these are heavy-duty scales and can overcome drastic climatic conditions and rough tours. For better tracking and monitor truck operations, the scales are integrated with radio-frequency identification (RFID) technology [20].

15.5.5 BUSTY GARBAGE RUNS

The waste management solutions become a necessity in the urban areas with denser population growth in order to lodge the high volume of trash. Various developed/developing cities have implemented busty garbage disposal bins that are connected through interlinked underground pipes. Through the pipes, trash travels directly to the waste collection hub, where it can be sorted or isolated for proper disposal or recycling. Food wastes are sorted through one channel and destined for a facility to convert it into biofuel. Wastes that are destined for recycle then sorted to another channel. In the end, mixed wastes are wind up in an incinerator. Trash collector truck pick up all the garbage in one trip from the collection point to the dump or recycling facility rather than making number of rounds and halting at every home and business in the neighborhood [21].

15.5.6 SOLAR POWER JUNK COMPOSITE

To enhance garbage collection efficiency and reduce trips to the dumping area, solar-powered junk compactors have been introduced. These compactors are three times more efficient than traditional trash bins, compressing trash to increase capacity. They collect and analyze data on load and collection times to optimize the gathering process. These innovative waste disposal units use solar energy for automatic compression. Equipped with solar panels, they convert solar energy into electrical power stored in a battery for the compaction mechanism. Sensors inside trigger compaction when trash reaches a certain level, reducing volume by up to 80%. Hydraulic or electric actuators compress the waste into a more condensed form, increasing bin capacity and minimizing odors while deterring pests [22].

15.5.7 E-SCRAP BOOTHS AND RECYCLING PLATFORMS

Nowadays, numerous companies and organizations have launched e-scrap recycling platforms where old electronic devices are accepted, and contributors receive reimbursement. These recycling units play a vital role in reducing e-scrap and recovering valuable materials from discarded electronics. They encompass systems, processes, and technologies for e-scrap:

- Collection
- Transportation
- Rectification
- Disposal

One challenge for recycling centers is sorting contaminated waste. To simplify recycling, specific apps for waste sorting have been developed, limiting unrecyclable materials entering recycling centers. These apps provide detailed data on recycling practices, location centers, and extensive lists of materials to assist users in identifying recyclable items [23].

15.6 INTEGRATION OF IOT AND DATA ANALYTICS

With IoT-enabled smart waste sensors, waste management is now a vital source in mod town ecosystem, authorizing cities/towns in brief analysis of garbage collection, reduction overflow of in garbage bins, and in proper resource utilization.

IoT technology and boosted optimization programs adapted for trash management had shown a great improvement in waste collection process, cost-effective, and sustainable for environment. A remarkable cost reduction can be attained by reducing overfilled trash cans and the prevalence of garbage collections after the detailed optimization through actual time analyzer using IoT sensors. Number of repetitive trips within the distance can be obtained by factual route optimization resulting in fuel consumption followed by low carbon emissions. This information-based scheme can also be recorded from recycling hubs in order to increase the amount of materials to be recycled [24].

15.7 SMART WASTE MANAGEMENT IN THE SMART CITY ECOSYSTEM

The smart city concept is gaining popularity in developing countries as a solution to urbanization challenges [25, 26]. Smart waste management is crucial for sustainable smart city ecosystems. Advanced technologies like IoT, AI, and data analytics are used in waste collection, recycling, and disposal [27]. Sensors in waste bins monitor fill levels for optimized collection routes. Additionally, AI algorithms analyze data to predict waste generation patterns for proactive management.

15.7.1 IOT SENSORS

IoT sensors in waste bins monitor fill levels in real-time, alerting authorities for timely collections, optimizing routes, reducing fuel consumption, and emissions [28].

15.7.2 ROUTE OPTIMIZATION

Advanced algorithms optimize waste collection routes considering traffic, closures, and demand, reducing costs, congestion, and pollution [29].

15.7.3 WASTE SEGREGATION AND RECYCLING

Encourages segregation at source, educates the public, offers incentives, and uses IoT for efficient recycling [30].

15.7.4 ENERGY RECOVERY

Explores energy recovery from waste through technologies like anaerobic digestion and waste-to-energy plants, reducing reliance on fossil fuels [31].

15.7.5 DATA ANALYTICS AND REPORTING

Platforms collect data for insights on waste, recycling, and environmental metrics, enabling continuous improvement and citizen accountability [32].

Smart waste management contributes significantly to urban sustainability by conserving resources through efficient waste collection, processing, and recycling. This helps minimize environmental impact by reducing fuel consumption, landfill usage, and greenhouse gas emissions [27, 33]. Additionally, by diverting waste from landfills and promoting recycling, smart waste management supports the circular economy by reintroducing valuable materials back into production cycles [34, 47].

15.8 CASE STUDIES AND SUCCESS STORIES

Smart cities face challenges with waste disposal due to population growth from rural migration, leading to health and environmental issues. Overflowing garbage bins pose health risks, highlighting the need for improved waste management. Authorities and researchers must address this issue by implementing IoT-based solutions for effective waste monitoring and control [35].

Several case studies worldwide demonstrate the effectiveness of smart waste management systems in improving sustainability, efficiency, and urban life quality.

15.8.1 SMART CITY OF SINGAPORE

Singapore, renowned for its cleanliness and efficiency, has implemented a comprehensive smart waste management system overseen by the National Environmental Agency (NEA). The city-state utilizes IoT sensors installed in waste bins to monitor fill levels in real-time, enabling dynamic route optimization for waste collection vehicles. Additionally, the NEA employs data analytics to forecast waste generation patterns and allocate resources effectively. This proactive approach has resulted in reduced collection times, minimized traffic congestion, and optimized resource utilization, contributing to Singapore's reputation as a model of urban sustainability [36].

15.8.2 BARCELONA'S WASTE MANAGEMENT SYSTEM

Barcelona has embraced smart waste management as part of its broader commitment to sustainable urban development. The city's waste management system incorporates IoT-enabled bins equipped with fill-level sensors and GPS tracking devices. These sensors transmit data to a centralized platform, allowing authorities to monitor waste accumulation and plan collection routes efficiently. Barcelona's smart waste management initiatives have led to significant improvements in waste collection efficiency, reduced operational costs, and enhanced environmental sustainability. Moreover, the city's emphasis on public engagement and education campaigns has fostered a culture of responsible waste disposal among residents [37].

15.8.3 WASTE-TO-ENERGY PLANTS IN SWEDEN

Sweden stands out for its innovative approach to waste management, with a focus on waste-to-energy conversion. Several municipalities in Sweden have invested in advanced waste-to-energy plants that utilize cutting-edge technologies to incinerate municipal solid waste and generate electricity and heat. Sweden imports approximately 2.7 million tons of waste per year (2014) to supply waste to energy facilities, and 50% of its own waste is burnt in the same facilities [38]. These facilities not only reduce the volume of waste sent to landfills but also provide a sustainable source of energy for local communities. By coupling smart waste management with energy recovery solutions, Sweden has achieved remarkable success in minimizing waste while generating renewable energy, positioning itself as a global leader in sustainable waste management practices [39].

15.8.4 SAN FRANCISCO'S ZERO WASTE INITIATIVE

In 2002, San Francisco set a target of achieving a 75% diversion rate by 2010 and aimed for zero waste in the long term. Surpassing the initial goal ahead of schedule, the city achieved a diversion rate of over 80% and reduced its waste disposal by half. In 2018, San Francisco revised its zero waste goals to decrease solid waste production by 15% and further reduce landfill or incineration disposal by 50% by 2030 [40]. Other localities on a national and international level are taking notice of the city's advancements in landfill diversion, which are seen as a model for sustainable waste management. San Francisco has made significant progress toward its zero-waste goal. Moreover, the city's emphasis on community engagement and partnerships with local businesses and organizations has fostered widespread support for sustainable waste management practices.

15.8.5 SURAT'S INTEGRATED SOLID WASTE MANAGEMENT SYSTEM (ISWMS)

In Surat, Gujarat, the Surat Municipal Corporation faces a challenge managing increasing municipal solid waste. They have implemented an ISWMS with decentralized waste processing units, composting facilities, and a waste collection network. RFID technology in waste bins monitors collection and waste movement. The ISWMS has improved waste collection efficiency, reduced landfill dependence, and enhanced city cleanliness [41].

15.8.6 INDORE'S WASTE MANAGEMENT SYSTEM

Indore in Madhya Pradesh is one of India's cleanest cities, with a population of over 3.2 million generating 1200 metric tons of waste daily and projected to increase to 2000 tons in next 10 years [42]. Previously facing waste disposal challenges, the city has transformed its cleanliness through initiatives like door-to-door waste collection, public awareness campaigns, and technology use. Indore has been named India's cleanest city for 6 years in a row, thanks to its innovative waste management strategies, including GPS-enabled waste collection vehicles and CCTV monitoring at dumping sites [43, 44].

15.8.7 DUBAI'S WASTE-TO-ENERGY PLANT

Dubai has invested in a waste-to-energy plant as part of its sustainable waste management strategy. The plant aims to divert 75% of municipal solid waste away from landfills by 2021 and reduce methane gas emissions. It converts city waste into electricity, aligning with Dubai's vision of sustainability and green technology leadership [44].

15.9 FUTURE TRENDS AND EMERGING TECHNOLOGIES

Future trends in smart waste management are poised to transform traditional practices, offering more efficient and eco-friendly solutions. Key upcoming technologies and their impact on waste management practices include:

15.9.1 ARTIFICIAL INTELLIGENCE (AI) AND MACHINE LEARNING

AI and machine learning algorithms analyze data to predict waste generation patterns, optimize collection routes, and identify recycling opportunities. These systems enable autonomous decision-making for dynamic adaptation.

15.9.2 IOT SENSORS AND SMART BINS

IoT sensors and smart bins detect fill levels, monitor waste properties, and provide actionable data to authorities, facilitating proactive decision-making and reducing costs.

15.9.3 ROBOTICS AND AUTOMATION

Advanced robotic systems with AI capabilities autonomously sort recyclable materials from mixed waste streams, enhancing efficiency, reducing labor costs, and improving material quality.

15.9.4 BLOCKCHAIN TECHNOLOGY FOR WASTE TRACEABILITY

Blockchain enhances transparency and traceability in waste management supply chains, verifying recycled product authenticity and ensuring compliance with regulations.

15.9.5 ADVANCED WASTE-TO-ENERGY TECHNOLOGIES

Technologies like pyrolysis and gasification convert waste to energy sustainably, recovering resources and reducing reliance on fossil fuels.

15.9.6 ENHANCED WASTE COLLECTION AND TRACKING

RFID tags on waste bins and items enable real-time monitoring of collection routes and schedules, enhancing efficiency and reducing collection times.

Embracing these innovations represents a pivotal step toward achieving waste reduction, promoting recycling, and preserving the environment in the future.

15.10 CHALLENGES AND LIMITATIONS

Smart Waste Management plans and techniques in cities are quite expensive as it employs both garbage collection and transportation for proper disposal. This detailed procedure sometimes seems to be easy but adapting in daily routine is quite problematic and challenging. Few challenges which arose are as follows.

15.10.1 MISCONCEPTIONS OF SMART SENSORS OPERATION

Most of the population consider it as complicated and expensive method for waste disposal, that is, a proper knowledge and handling procedures are required.

15.10.2 INSTALLATION OF THE SMART SENSOR

This includes detailed steps and procedures after installation in order to ensure its effectiveness. It collects information regarding garbage bin and alerts for any incidence.

15.10.3 UNORGANIZED TRUCK ROUTES

Improper route distribution and guidance leads to repetitive trips, longer routes, and excess fuel consumption.

15.10.4 RECYCLING

Collected waste from a bin is sorted on the basis of recycling and degradation by the waste management company. This method is manual and takes time to complete. When compared to manual method for waste collection the automated systems are loaded with the technical information about the product, that is, material made up of, the appropriate method of recycling and so on to improve the recycling within less cost.

15.10.5 UNEVEN WASTE LEVELING IN WASTE BINS

Sometimes the sensors only sense or read the fill levels of the waste bins, that is, waste has unevenly filled the half of bin than the other and the sensors read that the bin is full, when in reality it is only half full [45].

15.11 CONCLUSION

Emerging technologies and innovative strategies are shaping the future of waste management, addressing challenges like waste generation, environmental sustainability, and resource management. Technologies such as AI, IoT, robotics, blockchain, waste-to-energy, and RFID promise to optimize waste processes and promote sustainability. Collaborative efforts are needed to realize the potential of these technologies and create a more sustainable waste management ecosystem. Advanced technologies, strategic policies, and public engagement will be crucial in shaping the future of waste management for cleaner and greener environments.

KEYWORDS

- **Internet of Things**
- **busty garbage runs**
- **garbage volume detectors**
- **waste management**
- **e-scrap**

REFERENCES

1. Umer, H., Ahmed, A., Ali, F., Ali, S. S., & Khan, M. A. (2022 October). A systematic literature review on smart waste management using machine learning. In *2022 Mohammad Ali Jinnah University International Conference on Computing (MAJICC)*. IEEE, pp. 1–9.
2. Cheema, S. M., Hannan, A., & Pires, I. M. (2022). Smart waste management and classification systems using cutting edge approach. *Sustainability*, *14*(16), 10226. https://doi.org/10.3390/su141610226.
3. Fei, T. P., Kasim, S., Hassan, R., Ismail, M. N., Salikon, M. Z. M., Ruslai, H., Jahidin, K. & Arshad, M. S. (2017). SWM: Smart waste management for green environment.

In *2017 6th ICT International Student Project Conference (ICT-ISPC)*. IEEE, pp. 1–5.

4. Munyaneza, J., Jia, Q., Qaraah, F. A., Hossain, M. F., Wu, C., Zhen, H., & Xiu, G. (2022). A review of atmospheric microplastics pollution: In-depth sighting of sources, analytical methods, physiognomies, transport and risks. *Science of the Total Environment.*, *822*, 153339
5. Hamer, G. (2003). Solid waste treatment and disposal: Effects on public health and environmental safety. *Biotechnology Advances*, *22*(1–2), 71–79.
6. Effect of Waste on Population in Umbada Locality, Khartoum State, Fangama, S. I. M., & Hamed, W. B. (2018). *International Journal of Current Microbiology and Applied Sciences (Excellent Publishers)*, 7(11), 1327–1333
7. Jajoo, P., Mishra, A., Mehta, S., & Solvande, V. (2018). Smart garbage management system. *2018 International Conference on Smart City and Emerging Technology (ICSCET)*, Mumbai, India, pp. 1–6.
8. Kannan, D., Khademolqorani, S., Janatyan, N., & Alavi, S. (2024). Smart waste management 4.0: The transition from a systematic review to an integrated framework. *Waste Management*, *174*, 1–14. https://doi.org/10.1016/j.wasman.2023.08.041. Epub 2023 Sep 22. PMID: 37742441.
9. Moghadam, M. A., Mokhtarani, N., & Mokhtarani, B. (2009). Municipal solid waste management in Rasht City, Iran. *Waste Management*, *29*(1), 485–489.
10. Glouche, Y., & Couderc, P. (Jun 2013). A smart waste management with self-describing objects. *The Second International Conference on Smart Systems, Devices and Technologies (SMART'13)*, IARIA, Rome, Italy.
11. IPCC. (2022). Climate change 2022: Mitigation of climate change. In Shukla, P. R., Skea, J., Slade, R., Al Khourdajie, A., van Diemen, R., McCollum, D., Pathak, M., Some, S., Vyas, P., Fradera, R., Belkacemi, M., Hasija, A., Lisboa, G., Luz, S., & Malley, J. (Eds.), *Contribution of Working Group III to the Sixth Assessment Report of the Intergovernmental Panel on Climate Change*, Cambridge University Press, Cambridge: UK and New York, NY, USA. https://doi.org/10.1017/9781009157926
12. Sulaiman, H., & Wahyuni, W. (2023). Traffic congestion effect on socio-economic of road users in Palembang city. https://doi.org/10.2991/978-2-38476-072-5_9.
13. Arthur, M. P., Shoba, S. & Pandey, A. (2024). A survey of smart dustbin systems using the IoT and deep learning. *Artificial Intelligence Review 57*, 56. https://doi.org/10.1007/s10462-023-10646-6
14. Schultz, J. A. M. (2002). A study of changing traditional urban solid waste management approaches to sustainable practices. In *Managing Intermediate Size Cities: Sustainable Development in a Growth Region of Thailand.* Dordrecht: Springer Netherlands, pp. 209–222.
15. Pardini, K., Rodrigues, J. J. P. C., Diallo, O., Das, A. K., de Albuquerque, V. H. C., & Kozlov, S. A. (2020). A smart waste management solution geared towards citizens. *Sensors* , *20*, 2380.
16. Folianto, F., Low, Y. S., & Yeow, W. L. (2015). Smartbin: Smart waste management system, *IEEE Tenth International Conference on Intelligent Sensors, Sensor Networks and Information Processing (ISSNIP)*, Singapore, pp. 1–2.
17. Wahab, A., M. H., Kadir, A. A., Tomari, M. R., & Jabbar, M. H. (2014). Smart recycle bin: A conceptual approach of smart waste management with integrated web based

system. *International Conference on IT Convergence and Security (ICITCS)*, Beijing, China, pp. 1–4.
18. Fang, B., Yu, J., Chen, Z. *et al.* 2023. Artificial intelligence for waste management in smart cities: A review. *Environmental Chemistry Letters 21*, 1959–1989. https://doi.org/10.1007/s10311-023-01604–3
19. Bakhshi, T., & Ahmed, M. (2018). IoT-enabled smart city waste management using machine learning analytics. In *2nd International Conference on Energy Conservation and Efficiency (ICECE)*, Lahore, Pakistan, pp. 66–71.
20. Bharadwaj, A. S., Rego, R., & Chowdhury, A. (2016). IoT based solid waste management system: A conceptual approach with an architectural solution as a smart city application. In *IEEE Annual India Conference (INDICON)*, Bangalore, India, pp. 1–6.
21. Miller, B., Spertus, J., & Kamga, C. (2014). Costs and benefits of pneumatic collection in three specific New York City cases. *Waste Management*, *34*(11), 1957–1966.
22. Lakhotia, R., Fazal, A., Yadav, A., & Bhattacharjee, A. (2021). Design and development of an economical and reliable solar-powered trash compactor. In Bose, M., & Modi, A. (Eds.), *Proceedings of the 7th International Conference on Advances in Energy Research*. Springer Proceedings in Energy, Springer, Singapore.
23. Snehil, S., Vindhyachalam, O., & Singh, K. K. (2023). Application of the internet of things in E-waste management. In *Applications of Optimization and Machine Learning in Image Processing and IoT*. Chapman and Hall/CRC, pp. 67–78.
24. Abdullah, N., Alwesabi, O. A., & Abdullah, R. (2019). IoT-based smart waste management system in a smart city. In Saeed, F., Gazem, N., Mohammed, F., & Busalim, A. (Eds.), *Recent Trends in Data Science and Soft Computing. IRICT 2018. Advances in Intelligent Systems and Computing*. Springer: Cham, p. 843.
25. Silva, B. N., Khan, M., & Han, K. (2018) Towards sustainable smart cities: A review of trends, architectures, components, and open challenges in smart cities. *Sustainable Cities and Society*, *38*, 697–713.
26. Kim, J. (2022). Smart city trends: A focus on 5 countries and 15 companies. *Cities*, *123*, 103551.
27. Shukla, S., & Hait, S. (2022). Smart waste management practices in smart cities: Current trends and future perspectives. In *Advanced Organic Waste Management*. Elsevier, pp. 407–424.
28. Wang, C., Qin, J., Qu, C., Ran, X., Liu, C., & Chen, B. (2021). A smart municipal waste management system based on deep-learning and Internet of Things. *Waste Management*, *135*, 20–29.
29. Shyam, G. K., Manvi, S. S., & Bharti, P. (February 2017). Smart waste management using Internet-of-Things (IoT). In *2nd International Conference on Computing and Communications Technologies (ICCCT)*). IEEE, pp. 199–203.
30. Varudandi, S., Mehta, R.,Mahetalia, J., Parmar, H., & Samdani, K. (2021). A smart waste management and segregation system that uses internet of things, machine learning and android application. In *2021 6th International Conference for Convergence in Technology (I2CT)*. IEEE, pp. 1–6.
31. Nižetić, S., Djilali, N., Papadopoulos, A., & Rodrigues, J. J. (2019) Smart technologies for promotion of energy efficiency, utilization of sustainable resources and waste management. *Journal of Cleaner Production*, *231*, 565–591.

32. Niska, H., & Serkkola, A. (2018). Data analytics approach to create waste generation profiles for waste management and collection. *Waste Management*, *77*, 477–485.
33. Laurieri, N., Lucchese, A., Marino, A., & Digiesi, S. (2020) A door-to-door waste collection system case study: a survey on its sustainability and effectiveness. *Sustainability*, *12*(14), 5520.
34. Mandpe, A., Paliya, S., Gedam, V. V., Patel, S., Tyagi, L., & Kumar, S. (2023). Circular economy approach for sustainable solid waste management: A developing economy perspective. *Waste Management & Research*, *41*(3), 499–511.
35. Saji, R. M., Gopakumar, D., Kumar, S. H., Sayed, K. M., & Lakshmi, S. (2016). A survey on smart garbage management in cities using IoT. *International Journal of Engineering Computational Science*, *5*(11), 18749–18754.
36. Wong, C., Wood, J., & Paturi, S. (2021). Managing waste in the smart city of Singapore. In *Tropical Constrained Environments and Sustainable Adaptations: Businesses and Communities*, pp. 225–241.
37. Atanasova, A., & Naydenov, K. (2020). The innovative approaches for the development of smart cities. In *Smart Geography: 100 Years of the Bulgarian Geographical Society*, pp. 237–245.
38. Ram, C., Kumar, A., & Rani, P. (2021). Municipal solid waste management: A review of waste to energy (WtE) approaches. *Bioresources*, *16*(2), 4275.
39. Avest, A., & Scherman, J. (2022). Waste to energy in Sweden-A study of Sweden's and Renova's operation.
40. US-EPA (United States Environment Protection Agency). Zero Waste Case Study: San Francisco. Updated on 22 November 2023. (accessed: 5 April, 2024).
41. Bhakkad, M. (2019). Solid waste management: A case study of Surat city. *International Journal of Management and Applied Science*, *5*, 34–36.
42. Singh, R. (2021). Municipal solid waste management in the city of Indore—A case study. *Journal of Civil Engineering and Environmental Sciences*, *7*, 8–17.
43. SwachhSurvekshan. (2022). *Ranking Report*. Ministry of Housing and Urban Affairs, Government of India, p. 11.
44. Shareefdeen, Z., Youssef, N., Taha, A., & Masoud, C. (2020). Comments on waste to energy technologies in the United Arab Emirates (UAE). *Environmental Engineering Research*, *25*(1), 129–134.
45. Anagnostopoulos, T., Zaslavsky, A., Kolomvatsos, K., Medvedev, A., Amirian, P., Morley, J., and Hadjieftymiades, S. (2017). Challenges and opportunities of waste management in IoT-enabled smart cities: A survey. *IEEE Transactions on Sustainable Computing*, *2*(3), 275–289.
46. Choudhury, M., Jyethi, D. S., Dutta, J., Purkayastha, S. P., Deb, D., Das, R., Roy, G., Sen, T., & Bhattacharyya, K. G. (2022). Investigation of groundwater and soil quality near to a municipal waste disposal site in Silchar, Assam, India. *International Journal of Energy and Water Resources*, *6*(1), 37–47. https://doi.org/10.1007/s42108-021-00117-5
47. Paul, S., Choudhury, M., Deb, U., Pegu, R., Das, S., & Bhattacharya, S. S. (2019). Assessing the ecological impacts of ageing on hazard potential of solid waste landfills: A green approach through vermitechnology. *Journal of Cleaner Production*, *236*, 117643. https://doi.org/10.1016/j.jclepro.2019.117643

48. Mondal, T., Choudhury, M., Kundu, D., Dutta, D., & Samanta, P. (2023). Landfill: An eclectic review on structure, reactions and remediation approach. *Waste Management*, *164*, 127–142. https://doi.org/10.1016/j.wasman.2023.03.034

CHAPTER 16

Emerging Waste Contaminants Such as Macroplastics/Microplastics: Strategies to Combat Plastic Pollution

KIRAN BALA[1*], ARUSHI DOGRA[2], FARHANA RAHMAN[3], SAKSHI VERMA[4], ANSHUL[5], PUSHP LATA[6], and SANGEETA JAISWAL[7]

[1]*Department of Zoology, Deshbandhu College, University of Delhi, New Delhi, Delhi, India*

[2]*Faculty of Life Sciences and Biotechnology, South Asian University, New Delhi, Delhi, India*

[3]*University of Technology Sydney (UTS), Ultimo, NSW, Australia*

[4]*Department of Nutrition Biology, Central University of Haryana, Mahendergarh, Haryana, India*

[5]*Institute of Environmental Studies, Kurukshetra University, Kurukshetra, Haryana, India*

[6]*Department of Zoology, University of Delhi, New Delhi, Delhi, India*

[7]*Department of Internal Medicine, Division of Gastroenterology, University of Michigan, Ann Arbor, MI, USA*

*Corresponding author

ABSTRACT

Plastics are widespread contaminants that affect almost all life forms in different habitats. This chapter discusses two types of plastic pollution: macroplastics and microplastics (MPs). It provides information regarding

Data Analytics and Smart Technologies in Solid Waste Management: Advancing Sustainability. Moharana Choudhury, Sushobhan Majumdar, and Avnish Chauhan (Eds.)

DOI: 10.1201/9781779640833-20

the frequency of contact between flora, wildlife, and contaminants. Macroplastics, while being more visible, represent substantial threats not only to wildlife but also to human societies. Issues like clogged drainage systems, increasing waste, and overflowing landfills highlight the crucial need for fast and effective remediation technology. MPs, while less visible, have significantly more damaging effects on environmental and biological processes. These particles tend to collect in living organisms, affecting the nutrient cycle, soil productivity, and water quality. The chapter investigates both preventive methods and advanced technology solutions for tackling the widespread problem of plastic pollution, with a focus on new techniques to mitigate its negative environmental impacts.

16.1 INTRODUCTION

Plastics have gained immense popularity during the last few decades because of their marvelous combination of features like durability, almost weightlessness, resistance to corrosion, and cost-friendliness. This leads to 348 million tons of global production in 67 years and is projected to reach 33 billion tons by 2025. Microplastics (MPs), which are 1–5000 μm particles found in terrestrial and aquatic ecosystems, can potentially harm plants and animals [1]. There are two primary and secondary types of MPs. Primary MPs are mostly found in the textile, footwear, and cosmetic industries; they can also be found in sewage disposal sites that contain microfibers and plastic resin spills. Secondary MPs are primarily obtained through plastics, such as single-use plastic, which end up in the ocean through surface runoff. Research indicates that soil erosion eventually washes these MPs away, releasing the pollutants into water bodies [2].

16.1.1 MACROPLASTICS

Plastic polymer with a size greater than 5 mm is usually considered as macroplastics. However, this is not a standard internationally. Over half (51%) of the world's 359 million tons of plastic produced in 2018 came from Asia (30% from China, 4% from Japan, and 17% from the rest of Asia). A total of 61.8 million tons, or 17.2% of the world's production, was produced in Europe [3]. Macroplastics are an emerging threat to rivers, seas, and oceans. The quantification of macroplastics is another crucial method for developing strategies. It has been demonstrated that plastic debris settling on the seafloor modifies the environments of marine communities, mainly because it can create anoxic conditions in sediment, as seen in Ireland's intertidal shores [62]. Plastic debris left on a riverbed can create artificial complex grounds [4]. It is widely acknowledged that in freshwater systems, the surface river flow velocity is the primary environmental factor determining the velocity, distribution, and distance travelled by floating plastic. Plastic bottles, fishing nets, food packets, straws, and poly bags significantly contribute to plastics in oceans and other water bodies. In a study, it has been presented that the top three items responsible for plastic pollution are from single-use plastics [5].

16.1.2 MICROPLASTICS

MPs are smaller than 5 mm in size, may seem harmless, but have the potential to affect every ecosystem. Their long-term effects can be seen in the body tissues of fish, other animals, and flora [6]. MPs are mainly categorized into two types, namely primary and secondary MPs, based on their origin. Primary sources include MPs from cosmetic microbeads and microfibers released from drying clothes in washing machines. In contrast, secondary sources of MPs include large pieces of plastic chunks disposed of in rivers and later degraded because of physiological breakdown [7]. A total of 11% of the 19–23 million metric tons of plastic waste produced in 2016 has already found its way into the ocean. The data also showed the startling possibility that by 2030, the scale could increase to 53 million metric tons if action is not taken [8]. The concentration of these MPs is rising at an alarming rate because of ineffective management, which impacts marine life and the marine environment. Recent studies have revealed that the marine environment near urban areas contains higher concentrations of MPs, and the tissues of aquatic animals living in these areas have high microplastic accumulation [9]. MPs are found in different shapes (e.g., beads, fragments, and strips), so a broad strategy is needed to tackle them. MPs' small size gives them an edge to escape wastewater treatment plants, making their presence felt from deep down in the ocean to unexplored lands. Because of their durability and weightlessness, MPs can be easily carried by wind and water currents. Introducing MPs into the human body and accumulation in fish can result from unknowingly inhaling and ingesting these minute particles [10].

16.2 SOURCES AND IMPACT OF PLASTIC POLLUTION

Sources from land include plastics generated from textile industries, microbeads from the cosmetic industry and personal care, rubber from vehicle tires, paints, and single-use plastics, a few of the top-rated plastic sources [11]. According to the report of UNEP, it is estimated that 5 trillion plastic bags are consumed annually, and it is shocking to reveal that around 13 million tons of plastic waste end up in water bodies each year. India is one of the world's top consumers of plastic, with an average of 5.6 million tons produced annually, according to a 2014 report by the Central Pollution Control Board in New Delhi [12, 13].

Water bodies situated near landfills were found to have traces of MPs in water, fish, and plants, indicating landfills as one of the significant sources of MPs [14, 60]. Improper storage, silage, and bale nets, and previously mentioned equipment are examples of direct methods of plastic dumping because they release MP fragments or pellets into the soil and contaminate it. The release of untreated sewage sludge from residential areas, compost containing single-use plastics, polythenes, leaky drainage systems, and numerous other instances are examples of indirect plastic dumping [15, 61]. MPs can be ingested and inhaled unknowingly, affecting the food chain. The concentration, size, type, and oxygen-containing functional groups in MPs may alter the growth rate of plants and cause physical injuries. MPs can also serve as pollutant carriers, contributing to the deterioration of the environment [16]. Figure 16.1 summarizes different sources and how they affect organisms in various ecosystems.

16.3 INNOVATIVE SOLUTION TO COMBAT PLASTIC POLLUTION

This chapter divides strategies to deal with plastic pollution into two ways:

1. Preventive strategy
2. Technological strategy

16.3.1 PREVENTIVES STRATEGIES

This strategy works on the principle of not generating more plastics, knowingly or unknowingly, by identifying the sources of plastic waste generation. The preventive strategy includes a few practices in day-to-day life that can make a big difference. Some of those measures will be discussed below.

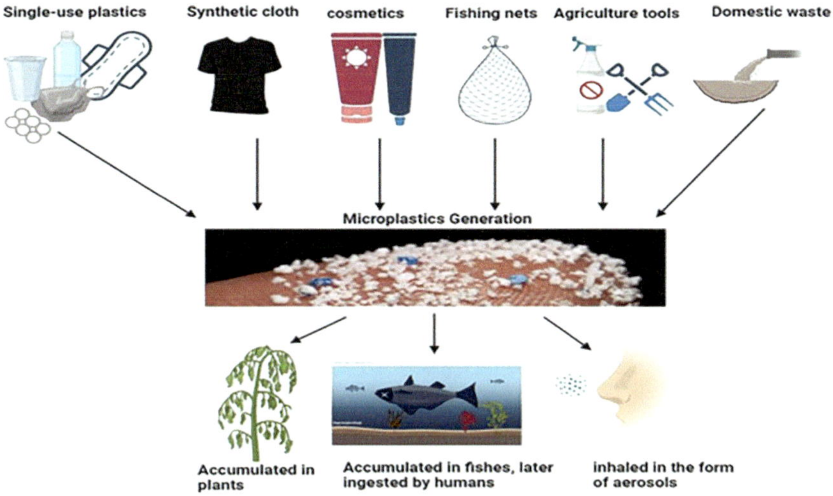

FIGURE 16.1 Different sources of plastics and their impact on flora and fauna.

16.3.2 FINDING ALTERNATIVES

16.3.2.1 BEESWAX

Beeswax, extracted from honeycombs, has hydrophobic properties and can be used to make beeswax wrap. This wrap has antibacterial properties and can provide a good alternative to plastic wraps. This wrap will protect the environment and prevent humans from ingesting MPs via food wraps [17].

3.2.2 WOODEN COMBS AND TOOTHBRUSHES

Another option to prevent plastic from entering the environment is to use bamboo toothbrushes. Bamboo brushes require 97% less plastic to produce, package, and transport than standard plastic brushes [18].

16.3.2.3 ECO-FRIENDLY CLOTHING

Cotton clothing prevents the release of microfibers into the environment, while other synthetic clothing may threaten the ecosystem. Moreover, air drying should be adopted instead of using a washing machine and dryer to prevent the release of microfibers into the drainage system [19].

16.3.2.4 DISPOSABLE PLATES

Disposable plates have gained much popularity in the past decade, resulting in huge piles of cutlery enough to pollute the environment. Hence, eco-friendly cutlery can prevent more plastic dumping into the soils and oceans [20]. Some alternatives are represented in Figure 16.2.

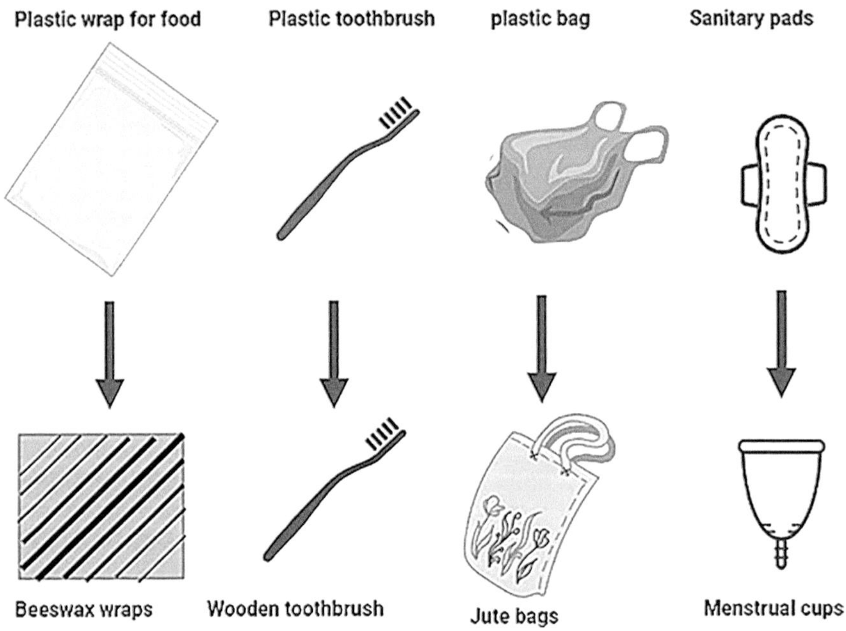

FIGURE 16.2 Alternatives for plastic consumption in day today life.

16.3.2.5 ECO-FRIENDLY AGRICULTURE PRACTICES

Because of a lack of knowledge, farmers have been using many agricultural practices that can harm the soil's nature, and surface runoff will soon result in dumping those MPs into water bodies. Agriculture should be practiced to reduce plastic usage as much as possible [15]. The leaching of degraded plastic caused by UV radiation and other natural forces is one of the main concerns with plastic films. Despite the introduction of green plastics into the world, numerous recent studies have demonstrated that their "greenness" is still debatable [21]. Besides these innovative agricultural techniques, afforestation ought to be promoted. This technique will help in gaining time for soil bioremediation while simultaneously reducing soil erosion. Because roots anchor the soil, surface runoff is inhibited, which keeps MPs out of the ocean and other aquatic ecosystems. These preventive strategies hence prevent the addition of more plastic to the ecosystem. However, there are several other ways also to prevent the contribution of more plastics, like choosing a zero-waste lifestyle, opting for wooden toys, using eco-friendly kitchen utensils, and eco-friendly personal care products (choosing a menstrual cup over pads and cotton nappies instead of diapers) [22].

16.3.3 TECHNOLOGICAL STRATEGY

16.3.3.1 MEMBRANE BIOREACTORS

Wastewater treatment plants employ the principle of membrane bioreactors. When appropriate aeration is provided, the membranes in these bioreactors typically sink in the water tank. Suction pumps remove filtered matter from the membrane, and air blowers are essential for maintaining the uniformity of the suspended solids. Additionally, the airflow helps prevent membrane fouling [23]. MPs can be removed from wastewater at each stage of treatment with an overall efficiency of up to 90% or even 99.9%. However, very high concentrations of MPs in wastewater and unique forms of MPs may sometimes escape the mechanism [10].

Optimizing current treatments aims to identify the optimal flocculant materials that can agglomerate MPs, with a particular emphasis on coagulation and settling capabilities. Thus far, some substitutes for traditional coagulants have been studied. Biopolymers, such as extracellular polymeric substances (EPS) produced from microalgae, provide an intriguing solution. Since EPS exhibits strong flocculants' action even at low concentrations, they offer a viable substitute for dangerous synthetic flocculants [24].

However, membrane bioreactors, still being one of the most used strategies for MPs filtration, still lags with some setbacks like low efficiency if the concentration of MPs is too high. Hence, more research for the optimization of the technology is still needed. Supplementary scaling and fouling of the membranes are issues for the membrane bioreactors [25]. A typical membrane bioreactor has been represented in Figure 16.3.

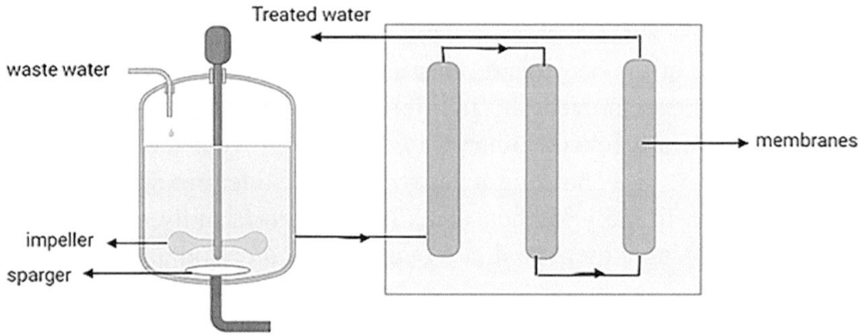

FIGURE 16.3 Membrane bioreactor.

16.3.3.2 ORGANOSILANE

Conventional strategies fail to purify MPs, and new technologies are still being modified because of their lower efficiency, technical support, and not-so-pocket-friendly nature. Organosilanes propose a promising approach to removing MPs from wastewater. This approach has been developed in the past few years and is still being tested for pilot-scale use [10]. The principle of organosilanes involves the physical aggregation of MPs, followed by chemical binding through a water-induced sol–gel process, which solidifies and facilitates the removal of the agglomerates. This method was proposed in 2016 and tested successfully in potable drinking water with high PE- and PP-based polymer concentrations. Organosilanes are composed of three reactive groups, an organic group, and a silica atom acting as the central atom. The organic group uses van der Waals forces to interact with the surface of the MPs. As a result, the organosilane binds to the surface of the MPs, aggregating them into clusters. The agglomerated MPs are then chemically bonded to form a three-dimensional hybrid silica network, as represented in Figure 16.4, which facilitates their easy removal. Organosilanes' high diversity and fast adaptability add to their advantages, making them suitable for pilot-scale use. However, it is difficult to comment on the disadvantages because of its naïve development [25].

16.3.3.3 ELECTROOXIDATION

It is another promising application for the degradation of MPs. Unlike other methods in this chapter, EO does not break down MPs but transforms them

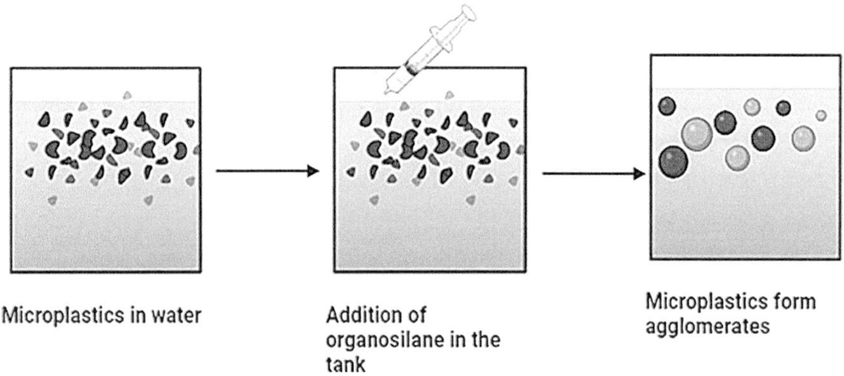

FIGURE 16.4 Functioning of organosilane.

into nontoxic gaseous compounds. In a case study done on polystyrene, the mineralization of MPs during the application of the EO method was further supported by scanning electron microscopy, total organic carbon, and Fourier transform infrared spectroscopy analyses, which showed no broken MPs, no organic carbon, and no oxidized functional groups following MP oxidation [26].

The method relies on producing oxidizing radicals *in situ* through direct and indirect electrochemical reactions. The radicals' typical reduction potential enables them to cleave and destroy MPs' polymeric bonds. However, electro-oxidation has only been tested at a laboratory scale and still demands wider research for pilot-scale usage. Nevertheless, generating nontoxic gases and leaving no residues can be advantageous for this technology. Hence, EO can be expected to flourish as a novel technology in the near future [10].

16.3.3.4 BIOREMEDIATION

The cosmopolitan presence of MPs makes it an alarming situation to deal with the growing plastic pollution. However, many technologies have been discussed above, but it is interesting to know that Mother Nature itself also has some creatures that can degrade plastics. Hence, using living organisms to degrade plastics and restore a healthy environment is referred to as bioremediation [27].

Bioremediation can be of two types based on the location of the treatment.

1. *In situ* bioremediation—when the contaminated soil or water is treated without being removed.
2. *Ex situ* bioremediation—when treatment is given to remove soil or water samples to treat contamination [28].

Some microbes form a biofilm around the plastic fragments and degrade them, as shown in Figure 16.5.

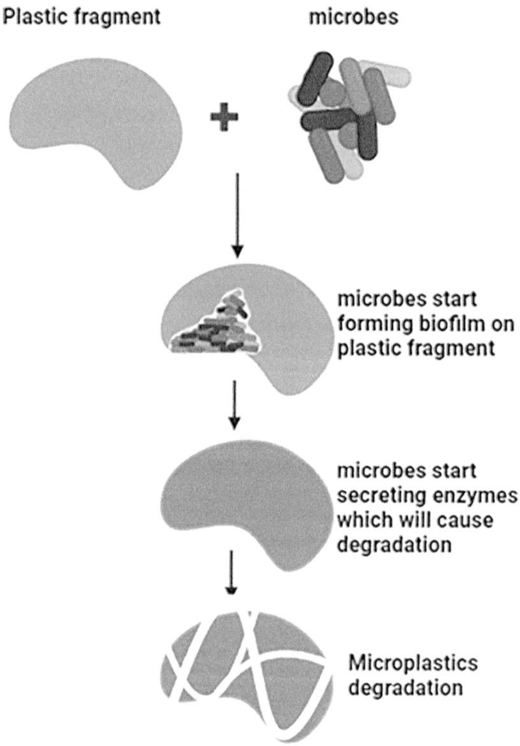

FIGURE 16.5 Biofilm formation by microbes.

The steps included in biofilm formation are as follows:

- The microbe adheres to the microplastic fragment's surface.
- The microbes on the MPs' surface release extracellular polymers.
- The microorganisms divide and multiply, covering the MPs fragment's surface.

This causes the biofilm to fully form and releases degrading enzymes to further break down MPs[29]. The biofilm formation has been described in Figure 16.5.

16.3.3.5 BIOREMEDIATION BY BACTERIA

A case study on the bacteria *Ideonella sakaiensis* shows how this bacterium degrades polyethylene terephthalate (PET). The bacteria "eat" PET and use its enzymes to break it down. The enzyme, PETase, breaks down PET into mono (2-hydroxyethyl) terephthalic acid, which then further breaks down into terephthalic acid and ethylene glycol (Sharma, 2018). According to another case study, *Pseudomonas alcaligenes* breaks down polyethylene bags more effectively than *Desulfotomaculum nigrificans*. Once the incubation period was extended, *Pseudomonas alcaligenes* was found to more successfully reduce the weight of the polythene bag. These two are soil bacteria that use enzymes and gene cloning to break down plastic polymers. These soil bacteria often contain lignin peroxidase enzymes and manganese peroxidases [30]. Another polymer broken down by bacteria such as *Clostridium botulinum* and *Alcaligenes faecalis* is polycaprolactone. The so-called biodegradable plastics, which are often shown to be "not so biodegradable," may be broken down by the thermophilic bacteria *Bacillus brevis*. The depolymerization of Polyhydroxyalkanoates has been observed by bacteria such as *Pseudomonas stutzeri* and *Streptomyces* sp. *[31]*.

16.3.3.6 BIOREMEDIATION BY FUNGI

A study found that *Trichoderma viride* and *Aspergillus nomius*, two species obtained from Indonesian landfills, are known to decrease the weight and tensile strength of LDPE. It was discovered that *Aspergillus* species, such as *Aspergillus clavatus*, *Aspergillus flavus*, and *Aspergillus terreus*, were effective against PE polymers [32]. Another study reported that *Pleurotus ostreatus* has been shown through experimentation to utilize plastic garbage as a substrate. It was discovered that the fungus immobilized the plastic remnants and secreted high concentrations of enzymes that broke down lignin molecules. This is projected to change the plastic material's molecular weight [33].

16.3.3.7 BIOREMEDIATION BY WORMS

Galleria mellonella and other *Lepidoptera* and *Coleoptera* are useful in degrading PE. Saliva, which is thought to be secreted by the anterior part of the gut, contains certain enzymes that can break down MPs without any prior preparation [34]. Three species of earthworms—*Lumbricus terrestris*, *Eisenia fetida*, and *Eisenia Andrei—that* were involved in their study, which provides insight for future research on topics including how earthworms mineralize plastic polymer monomers and their success rate. The study shows a reduction in weight and size of the plastic fragments [35]. Another study found that *Tenebrio molitor*, a mealworm species, can break down LDPE and a few other plastic polymers. Even though the plastic pieces fed their larvae, they still need food to thrive. Nonetheless, mealworms do affect the breakdown of polymers, with scattered extracellular depolymerases being the primary contributor [36]. Major case studies showing a positive response in bioremediation of plastics have been summarized in Table 16.1.

TABLE 16.1 Bioremediation by Bacteria, Fungi, and Worms

S. No.	Name of the Plastic	Name of the Organism	Reference
A. Bioremediation by bacteria			
1.	Polyethylene terephthalate	*Ideonella sakaiensis*	[37]
2.	Polyethylene	*Pseudomonas alcaligenes*	[30]
3.	Polycaprolactone	*Clostridium botulinum* and *Alcaligenes faecalis*	[31]
B. Bioremediation by fungus			
1.	Low-density polyethylene	*Trichoderma viride* and *Aspergillus nomius*	[32]
2.	Polyethylene	*Aspergillus clavatus*, *Aspergillus flavus*, and *Aspergillus terreus*	[32]
3.	Polylactic acid	*Fusarium moniliforme*, *Penicillium roqueforti*	[31]
C. Bioremediation by worms			
1.	Polyethylene	*Galleria mellonella*	[34]
2.	Low-density polyethylene	*Tenebrio molitor*	[36]

16.4 NANOMATERIAL-BASED SOLUTIONS

Plastic pollution presents serious environmental risks because it contains macroplastics and MPs. To address this, scientists are investigating novel approaches, such as nanomaterials like nanocomposites and nanofibers. By making plastics more resilient and biodegradable, these substances drastically cut down on waste [38, 63]. Furthermore, to successfully remove MPs from the environment, nanomaterials are being developed for air and water filtration systems [39]. New sensor technologies identifying and tracking MPs and providing crucial information for cleanup operations also show promise in pollution monitoring [38]. It is possible to significantly reduce the ubiquitous problem of plastic pollution by incorporating these cutting-edge materials into plastic production and environmental management strategies [40, 41].

16.5 STRATEGIES TO LIMIT PLASTIC POLLUTION IN DEVELOPING REGIONS

Plastic pollution is a global environmental issue that poses significant challenges, especially in developing countries. Lack of infrastructure, such as ineffective and inefficient waste management systems, recycling centers, and insufficient resources to properly handle and dispose of plastic waste, is a major issue in these areas [42–45]. This deficiency frequently results in the accumulation of plastic waste in rivers, oceans, and landfills, accelerating environmental degradation and endangering wildlife. Plastic pollution management strategies require strong infrastructure, making them harder to implement. Without adequate facilities, single-use plastics banning policies and regulations are difficult to enforce [46, 47]. Limited facilities and transportation networks make recycling systems difficult to establish and maintain [48]. Lack of awareness and education about plastic pollution's effects hinder plastic waste reduction [49–51].

A comprehensive approach that acknowledges barriers and develops new solutions is needed to address this global issue [52]. Lack of waste management systems and infrastructure, difficulty managing plastic waste, especially in rural areas, and limited recycling are major challenges [53, 54]. Education about the environmental impacts of plastic pollution, waste reduction, and community-based clean-up should be prioritized [55, 56]. Additionally, corporations are crucial to change. Business collaboration can improve regional waste management and invest in green packaging [57, 58]. In addition, the government must regulate plastic production and use and provide financial incentives to environmentally friendly businesses [59]. These strategies are the main way to reduce plastic pollution in developing nations through cooperative, multistakeholder methods.

16.6 FUTURE PERSPECTIVE

This chapter outlines the sources of different types of plastics to design technologies or find alternatives to combat the problem of plastic pollution. Using more alternatives will prevent us from contributing further to the issue and help restore a "green" planet. This will help us lead a sustainable life and experience a healthy environment.

This chapter focused on four major technologies: membrane-based wastewater treatment plants, organosilanes, electrooxidation, and bioremediation. Membrane bioreactor-based technology has been used for a long time and needs more advancement to deal with the enormous concentration of MPs. Bioremediation uses living organisms to degrade MPs, for which many case studies have been presented. However, modern-day gene-editing tools like CRISPR-Cas technology and gene cloning can be used to carry out these processes at a pilot scale. Organosilanes and electrooxidation look promising at laboratory or small scale, but a lot more work needs to be done before launching them on a large scale.

16.7 CONCLUSION

The adverse effects of plastic pollution were discussed to provide an idea of how fragile the situation is and how urgently the problem needs attention. Plastic pollution is a huge danger to biodiversity and requires urgent action on both practical and moral levels. Research on more possible sources will help to find the root causes of plastic generation and develop efficient technologies. Everyone should practice the preventive strategy while scientists work on developing more efficient techniques to combat plastic pollution. This will help combat the issue of plastic pollution.

KEYWORDS

- **plastic pollution**
- **microplastics**
- **macroplastics**
- **environmental remediation**
- **waste management**
- **ecosystem impact**
- **pollution control technologies**
- **soil productivity**
- **water quality**
- **sustainable solutions**

REFERENCES

1. Hale, R. C., Seeley, M. E., La Guardia, M. J., Mai, L., & Zeng, E. Y. (2020). A global perspective on microplastics. *Journal of Geophysical Research: Oceans*, 125(1). e2018JC014719. https://doi.org/10.1029/2018JC014719.
2. Xu, C., Zhang, B., Gu, C., Shen, C., Yin, S., Aamir, M., & Li, F. (2020). Are we underestimating the sources of microplastic pollution in terrestrial environment? *Journal of Hazardous Materials*, *400*, 123228. https://doi.org/10.1016/j.jhazmat.2020.123228.
3. Lechthaler, S., Waldschläger, K., Stauch, G., & Schüttrumpf, H. (2020). The way of macroplastic through the environment. *Environments*, *7*(10), 1–30. https://doi.org/10.3390/environments7100073.
4. Al-Zawaidah, H., Ravazzolo, D., & Friedrich, H. (2021). Macroplastics in rivers: Present knowledge, issues and challenges. *Environmental Science: Processes & Impacts*, *23*(4), 535–552. https://doi.org/10.1039/d0em00517g.
5. Winton, D. J., Anderson, L. G., Rocliffe, S., & Loiselle, S. (2020). Macroplastic pollution in freshwater environments: Focusing public and policy action. *Science of the Total Environment*, *704*, 135242. https://doi.org/10.1016/j.scitotenv.2019.135242.
6. Wang, C., Zhao, J., & Xing, B. (2021). Environmental source, fate, and toxicity of microplastics. *Journal of Hazardous Materials*, *407*, 124357. https://doi.org/10.1016/j.jhazmat.2020.124357.
7. Bui, X. T., Vo, T. D. H., Nguyen, P. T., Nguyen, V. T., Dao, T. S., & Nguyen, P. D. (2020). Microplastics pollution in wastewater: Characteristics, occurrence and removal technologies. *Environmental Technology and Innovation*, *19*, 101013. https://doi.org/10.1016/j.eti.2020.101013.

8. Borrelle, S. B., et al. (2020). Mitigate plastic pollution. *Science*, *369*(6510), 1515–1518. https://science.sciencemag.org/content/369/6510/1515.
9. Gola, D., et al. (2021). The impact of microplastics on marine environment: A review. *Environmental Nanotechnology, Monitoring & Management*, *16*, 100552. https://doi.org/10.1016/j.enmm.2021.100552.
10. Fiore, M., Fraterrigo Garofalo, S., Migliavacca, A., Mansutti, A., Fino, D., & Tommasi, T. (2022). Tackling marine microplastics pollution: An overview of existing solutions. *Water, Air, & Soil Pollution*, *233*(7), 276. https://doi.org/10.1007/s11270-022-05715-5.
11. Xu, S., Ma, J., Ji, R., Pan, K., & Miao, A. J. (2020). Microplastics in aquatic environments: Occurrence, accumulation, and biological effects. *Science of the Total Environment*, *703*, 134699. https://doi.org/10.1016/j.scitotenv.2019.134699.
12. Yang, L., Zhang, Y., Kang, S., Wang, Z., & Wu, C. (2021). Microplastics in soil: A review on methods, occurrence, sources, and potential risk. *Science of the Total Environment*, *780*, 146546. https://doi.org/10.1016/j.scitotenv.2021.146546.
13. Laskar, N., & Kumar, U. (2019). Plastics and microplastics: A threat to environment. *Environmental Technology and Innovation*, *14*, 100352. https://doi.org/10.1016/j.eti.2019.100352.
14. Golwala, H., Zhang, X., Iskander, S. M., & Smith, A. L. (2021). Solid waste: An overlooked source of microplastics to the environment. *Science of the Total Environment*, *769*, 144581. https://doi.org/10.1016/j.scitotenv.2020.144581.
15. Lwanga, E. H., et al. (2022). Review of microplastic sources, transport pathways and correlations with other soil stressors: a journey from agricultural sites into the environment. *Chemical and Biological Technologies in Agriculture*, *9*(1), 1–20. https://doi.org/10.1186/s40538-021-00278-9.
16. Ge, J., Li, H., Liu, P., Zhang, Z., Ouyang, Z., & Guo, X. (2021). Review of the toxic effect of microplastics on terrestrial and aquatic plants. *Science of the Total Environment*, *791*, 148333. https://doi.org/10.1016/j.scitotenv.2021.148333.
17. Saini, Y., Gandhi, H., Garg, S., Joshi, K., & Singhal, D. (2022). EPRA International Journal of Research and Development: Consumer behaviour analysis: Sustainable food wrapping with beeswax. *EPRA International Journal of Research and Development (IJRD)*, 7838(September), 65–71.
18. Lyne, A., Ashley, P., Saget, S., Porto Costa, M., Underwood, B., & Duane, B. (2020). Combining evidence-based healthcare with environmental sustainability: using the toothbrush as a model. *British Dental Journal*, *229*(5), 303–309. https://doi.org/10.1038/s41415-020-1981-0.
19. Hasan, M. M., Cai, L., Ji, X., & Ocran, F. M. (2022). Eco-friendly clothing market: A study of willingness to purchase organic cotton clothing in Bangladesh. *Sustainability*, *14*(8), 1–13. https://doi.org/10.3390/su14084827.
20. Dordevic, D., Necasova, L., Antonic, B., Jancikova, S., & Tremlová, B. (2021). Plastic cutlery alternative: Case study with biodegradable spoons. *Foods*, *10*(7), 1612. https://doi.org/10.3390/foods10071612.
21. Hofmann, T., et al. (2023). Plastics can be used more sustainably in agriculture. *Communications Earth & Environment*, *4*(1), 1–11. https://doi.org/10.1038/s43247-023-00982-4.

22. Velasco Perez, M., Sotelo Navarro, P. X., Vazquez Morillas, A., Espinosa Valdemar, R. M., & Hermoso Lopez Araiza, J. P. (2021). Waste management and environmental impact of absorbent hygiene products: A review. *Waste Management and Research*, *39*(6), 767–783. https://doi.org/10.1177/0734242X20954271.
23. Naga Vignesh, S. (2020). Working principle of typical bioreactors. In Bioreactors: Sustainable Design and Industrial Applications in Mitigation of GHG Emissions. Retrive from: https://doi.org/10.1016/B978-0-12-821264-6.00010-3.
24. Lapointe, M., Farner, J. M., Hernandez, L. M., & Tufenkji, N. (2020). Understanding and improving microplastic removal during water treatment: Impact of coagulation and flocculation. *Environmental Science & Technology*, *54*(14), 8719–8727. https://doi.org/10.1021/acs.est.0c00712.
25. Sturm, M. (2022). Microplastics Removal from Water with Organosilanes.
26. Kiendrebeogo, M., Karimi Estahbanati, M. R., Khosravanipour Mostafazadeh, A., Drogui, P., & Tyagi, R. D. (2021). Treatment of microplastics in water by anodic oxidation: A case study for polystyrene. *Environmental Pollution*, *269*, 116168. https://doi.org/10.1016/j.envpol.2020.116168.
27. Ramezani, M., Ghahari, S., & Ghahari, S. (2021). Microbial Rejuvenation of Polluted Environment.
28. Mondal, S., & Palit, D. (2019). Effective role of microorganism in waste management and environmental sustainability. In: Jhariya, M., Banerjee, A., Meena, R., Yadav, D. (Eds.), *Sustainable Agriculture, Forest and Environmental Management*. Springer: Singapore, pp. 485–515. https://doi.org/10.1007/978-981-13-6830-1_14.
29. Sun, X. L., Xiang, H., Xiong, H. Q., Fang, Y. C., & Wang, Y. (2023). Bioremediation of microplastics in freshwater environments: A systematic review of biofilm culture, degradation mechanisms, and analytical methods. *Science of the Total Environment*, *863*, 160953. https://doi.org/10.1016/j.scitotenv.2022.160953.
30. Begum, M. A., Varalakshmi, B., & Umamagheswari, K. (2015). Biodegradation of polythene bag using bacteria isolated from soil. *International Journal of Current Microbiology and Applied Sciences*, *4*(11), 674–680. http://www.ijcmas.com/vol-4-11/M.%0AAriba%0ABegum,%0Aet%0Aal.pdf.
31. Caruso, G. (2015). Plastic degrading microorganisms as a tool for bioremediation of plastic contamination in aquatic environments. *Journal of Pollution Effects & Control*, 3, 3. https://doi.org/10.4172/2375-4397.1000e112.
32. Vaksmaa, A., Guerrero-Cruz, S., Ghosh, P., Zeghal, E., Hernando-Morales, V., & Niemann, H. (2023). Role of fungi in bioremediation of emerging pollutants. *Frontiers in Marine Science*, 10(March), 1–21. https://doi.org/10.3389/fmars.2023.1070905.
33. Ratuchne, A., et al. (2023). Pleurotus ostreatus and a novel fungal composite: Development and bioremediation of plastic wastes. *Resources, Conservation and Recycling Advances*, *19*, 200167. https://doi.org/10.1016/j.rcradv.2023.200167.
34. Sanluis-Verdes, A., et al. (2022). Wax worm saliva and the enzymes therein are the key to polyethylene degradation by Galleria mellonella. *Nature Communications*, 13(1), 5568. https://doi.org/10.1038/s41467-022-33127-w.
35. Sanchez-Hernandez, J. C., Capowiez, Y., & Ro, K. S. (2020). Potential use of earthworms to enhance decaying of biodegradable plastics. *ACS Sustainable Chemistry and Engineering*, *8*(11), 4292–4316. https://doi.org/10.1021/acssuschemeng.9b05450.

36. Zhong, Z., Nong, W., Xie, Y., Hui, J. H. L., & Chu, L. M. (2022). Long-term effect of plastic feeding on growth and transcriptomic response of mealworms (Tenebrio molitor L.). *Chemosphere*, *287*(P1), 132063. https://doi.org/10.1016/j.chemosphere.2021.132063.
37. Sharma, S. R. (2018). Bioremediation of polythenes and plastics: A microbial approach. In *Nanotechnology in the Life Sciences*, Springer: Cham, pp. 97–114. https://doi.org/10.1007/978-3-030-02369-0_6.
38. Mandal, G., & Mishra, S. (2023). A review on emerging micro and nanoplastic pollutants, heavy metals and their remediation techniques. *Nanofabrication*, *8*(1), 302. https://doi.org/10.37819/nanofab.008.302.
39. Kuswandi, B., & Asrofi, M. (2020). Nano-technologies and reinforcements. In *Processing and Development of Polysaccharide-Based Biopolymers for Packaging Applications*,Elsevier, pp. 185–205.
40. Ogunola, O. S. (2017). Current advances in strategies to mitigate the impacts of micro/nano plastics: A review. *Journal of Environmental Analysis Toxicology*, *7*(447), 2161–0525.
41. Pathakoti, K., Manubolu, M., & Hwang, H. M. (2018). Nanotechnology applications for environmental industry. In *Handbook of Nanomaterials for Industrial Applications*. Elsevier, pp. 894–907.
42. Browning, S., Beymer-Farris, B., & Seay, J. R. (2021). Addressing the challenges associated with plastic waste disposal and management in developing countries. *Current Opinion in Chemical Engineering*, *32*, 100682.
43. Rustagi, N., Pradhan, S. K., & Singh, R. (2011). Public health impact of plastics: An overview. *Indian Journal of Occupational and Environmental Medicine*, *15*(3), 100–103.
44. Angnunavuri, P. N., Attiogbe, F., & Mensah, B. (2020). Consideration of emerging environmental contaminants in Africa: Review of occurrence, formation, fate, and toxicity of plastic particles. *Scientific African*, *9*, e00546.
45. Alpízar, F., Carlsson, F., Lanza, G., Carney, B., Daniels, R. C., Jaime, M., Ho, T., Nie, Z., Salazar, C., Tibesigwa, B., & Wahdera, S. (2020). A framework for selecting and designing policies to reduce marine plastic pollution in developing countries. *Environmental Science & Policy*, *109*, 25–35.
46. Pilapitiya, P. N. T., & Ratnayake, A. S. (2024). The World of Plastic Waste: A Review. Cleaner Materials, 100220.
47. Diana, Z. T., Virdin, J., Valiyaveettil, S., Li, H. X., & Rittschof, D. (2023). Emerging challenges and solutions for plastic pollution. *Frontiers in Marine Science*, *10*, 1162680.
48. Joshi, C., Seay, J., & Banadda, N. (2019). A perspective on a locally managed decentralized circular economy for waste plastic in developing countries. *Environmental Progress & Sustainable Energy*, *38*(1), 3–11.
49. Tibbetts, J. H. (2015). Managing marine plastic pollution: Policy initiatives to address wayward waste. *Environmental Health Perspectives*, *123*(4), A90–A93.
50. Sogbanmu, T. O. (2022). Plastic pollution in Nigeria is poorly studied but enough is known to urge action. *The Conversation*.
51. Diggle, A., & Walker, T. R. (2022). Environmental and economic impacts of mismanaged plastics and measures for mitigation. *Environments*, *9*(2), 15.

52. Horton, A. A. (2022). Plastic pollution: When do we know enough? *Journal of Hazardous Materials*, *422*, 126885.
53. Singh, J., Laurenti, R., Sinha, R., & Frostell, B. (2014). Progress and challenges to the global waste management system. *Waste Management & Research*, *32*(9), 800–812.
54. Thacker, S., Adshead, D., Fay, M., Hallegatte, S., Harvey, M., Meller, H., O'Regan, N., Rozenberg, J., Watkins, G., & Hall, J. W. (2019). Infrastructure for sustainable development. *Nature Sustainability*, *2*(4), 324–331.
55. Li, P., Wang, X., Su, M., Zou, X., Duan, L., & Zhang, H. (2021). Characteristics of plastic pollution in the environment: A review. *Bulletin of Environmental Contamination and Toxicology*, *107*, 577–584.
56. Williams, A. T., & Rangel-Buitrago, N. (2022). The past, present, and future of plastic pollution. *Marine Pollution Bulletin*, *176*, 113429.
57. Martinho, G., Pires, A., Portela, G., & Fonseca, M. (2015). Factors affecting consumers' choices concerning sustainable packaging during product purchase and recycling. *Resources, Conservation and Recycling*, *103*, 58–68.
58. Svanes, E., Vold, M., Møller, H., Pettersen, M. K., Larsen, H., & Hanssen, O. J. (2010). Sustainable packaging design: A holistic methodology for packaging design. *Packaging Technology and Science: An International Journal*, *23*(3), 161–175.
59. Wiesinger, H., Wang, Z., & Hellweg, S. (2021). Deep dive into plastic monomers, additives, and processing aids. *Environmental Science & Technology*, *55*(13), 9339–9351.
60. Paul, S., Choudhury, M., Deb, U., Pegu, R., Das, S., & Bhattacharya, S. S. (2019). Assessing the ecological impacts of ageing on hazard potential of solid waste landfills: A green approach through vermitechnology. *Journal of Cleaner Production*, *236*, 117643. https://doi.org/10.1016/j.jclepro.2019.117643
61. Mondal, T., Choudhury, M., Kundu, D., Dutta, D., & Samanta, P. (2023). Landfill: An eclectic review on structure, reactions and remediation approach. *Waste Management*, *164*, 127–142. https://doi.org/10.1016/j.wasman.2023.03.034
62. Choudhury, M., Sharma, A., Pervez, A., Upadhyay, P., & Dutta, J. (2022). Growing menace of microplastics in and around the coastal ecosystem. In *Coastal Ecosystems: Environmental Importance, Current Challenges and Conservation Measures*, pp. 117–137. https://doi.org/10.1007/978-3-030-84255-0_6
63. Roy, P., Choudhury, M. (2022). Bioeconomy and green plastic production. In Baskar, C., Ramakrishna, S., & Daniela La Rosa, A. (Eds.), *Encyclopedia of Green Materials*. Springer: Singapore. https://doi.org/10.1007/978-981-16-4921-9_189-1

CHAPTER 17

Waste Management in a Changing Climate: Building Resilience for the Future

PUJYASMITA NAYAK[1*], SUMAN KUMAR DAS[2], SAMIKSHYA MISHRA[3], and AMIT KUMAR PANDIT[4]

[1]*India Meteorological Department, Bhubaneswar, Odisha, India*

[2]*University of Brescia, Brescia, Italy*

[3]*Department of Environmental Sciences, Sambalpur University, Sambalpur, Odisha, India*

[4]*National Institute of Aerospace, Hampton, VA, USA*

*Corresponding author

ABSTRACT

Waste management and climate change emphasize the imperative of enhancing resilience amidst increasing challenges. It underscores the escalating impacts of climate change on waste management systems, such as extreme weather events, rising temperatures, and shifting precipitation patterns, which exacerbate existing challenges like inadequate infrastructure for waste management and escalating poor waste treatment. This chapter emphasizes the critical link between waste management and climate change (waste management–climate change nexus). It highlights the need for a deeper understanding of this interconnectedness and the importance of building resilience within waste management systems. Such resilience requires technological advancements, robust policy frameworks, community engagement, and cross-sectoral collaboration.

Data Analytics and Smart Technologies in Solid Waste Management: Advancing Sustainability.
Moharana Choudhury, Sushobhan Majumdar, and Avnish Chauhan (Eds.)

DOI: 10.1201/9781779640833-21

Furthermore, it emphasizes the intricate relationship between waste management and climate change, including feedback loops, impacts on agriculture, including their significance in water and soil management, and the significance of Industry 4.0 technologies for monitoring and understanding climate change.

17.1 INTRODUCTION

The increase in world population has led to a growing demand for consumer goods, driven by rapid and dynamic technological changes. However, this increasing need is seriously affecting the environment through resource depletion and waste generation (solid and liquid waste generation). This human and industrial activity, as illustrated in Figure 17.1, has complicated the issue of climate change by generating different significant emissions levels that impact the environment, sociopolitical systems, ecological balance, and socioeconomic spheres [14, 19]. Despite aiming to limit global warming below 1.5°C, as outlined in the Paris Agreement of 2015, anticipated changes in precipitation patterns and rising temperatures threaten infrastructure, communities, and agriculture, posing a significant risk to global economics and societies.

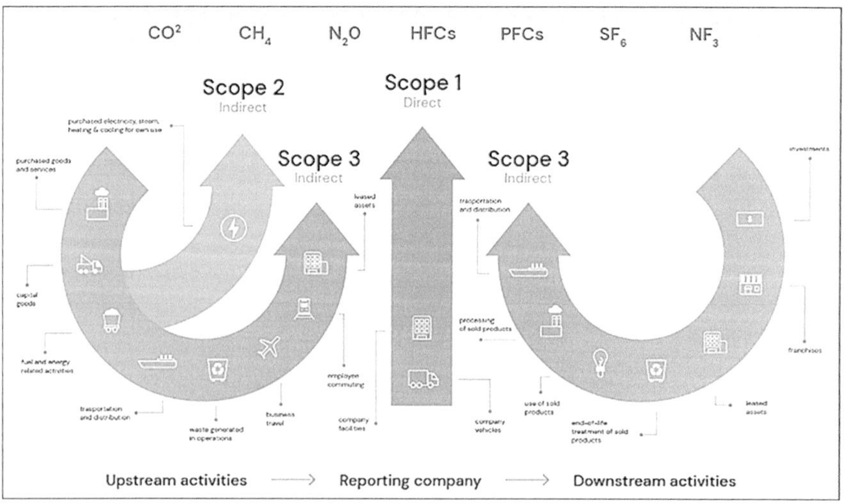

FIGURE 17.1 Greenhouse gas emission due to different industrial activities.

Source: Ref. [19] discussed scope 1, 2, and 3 levels of emission in the industrial value chain.

Overgeneration of solid waste contributes significantly to climate change, accounting for an estimated 80% of global greenhouse gas (GHG) emissions [25]. Hence, urban areas are consequently playing an essential role in the global effort to address climate change. Inefficient waste management practices directly contribute to greenhouse gas emissions. To address this challenge and achieve a more sustainable waste management system, prioritizing decarbonization and implementing smart waste management solutions are crucial steps. Hence, efficient and sustainable waste management operations are essential at a micro level to achieve this goal.

Furthermore, waste management and climate change are closely linked concepts, as numerous studies have emphasized the interdependence between them. Changing weather patterns significantly impact waste generation. Droughts can lead to crop failure and food spoilage, while excessive rainfall disrupts agricultural practices, both resulting in increased food waste [2, 10]. Extreme and frequent weather events, such as floods, hurricanes, and wildfires, also contribute to waste generation and the waste management system. These disasters have a negative impact on waste management systems by causing damage to collection and disposal facilities. Another limitation is that rising temperatures create operational difficulties for waste management. High temperatures rise can speed up the breakdown of organic waste, necessitating changes in how waste is collected and treated. Additionally, climate change can affect water levels, potentially limiting water availability for essential processes like water treatment and waste-to-energy conversion.

This chapter has explored the interconnectedness of waste management and climate change. It has also discussed various actions, policy frameworks, and the need for future research to address the critical uncertainties arising from these interrelated challenges. To effectively address these complex issues, a deeper understanding of their interplay is essential. Recent research suggests that advanced technologies have been identified as the most promising solution for understanding the complex interplay between waste management and climate change [25]. Additionally, by enabling the exploration of diverse waste reduction and management approaches, they can contribute to building a more robust system for resource circulation. This comprehensive approach ensures that sustainability is addressed across all three dimensions: environmental, social, and economic.

17.2 THE INTERPLAY OR RELATIONSHIP BETWEEN WASTE MANAGEMENT AND CLIMATE CHANGE

17.2.1 OVERVIEW OF THE IMPACTS OF CLIMATE CHANGE ON WASTE MANAGEMENT

17.2.1.1 INCREASED FREQUENCY AND INTENSITY OF EXTREME WEATHER EVENTS

The interplay between waste management and climate change is intricate and multifaceted. Climate change can affect waste management through extreme weather events, changes in waste composition, and implications for waste collection and disposal [19]. These events can significantly impact waste management infrastructure by causing damage to facilities, disrupting waste collection services, and creating challenges in waste disposal and treatment [Protocol]. Conversely, poor waste management contributes to climate change through greenhouse gas emissions, but it also offers opportunities to mitigate climate change through sustainable practices like recycling, composting, and waste-to-energy technologies.

17.2.1.2 CHANGES IN WASTE COMPOSITION AND GENERATION PATTERNS

Changes in waste composition and generation patterns can occur as a result of various factors, including climate change. Shifts in environmental conditions can influence the types and amounts of waste produced by communities. For example, changes in consumption patterns, population growth, and economic activities can impact the composition of waste generation [10]. Climate change may also affect the decomposition rates of organic waste and the overall volume of waste produced. Adapting waste management strategies to address these changing patterns is crucial for effectively managing waste and minimizing environmental impacts.

17.2.2 IMPLICATIONS FOR WASTE COLLECTION, TREATMENT, AND DISPOSAL

Climate change can have significant implications for waste collection, treatment, and disposal processes. Increased frequency and intensity of extreme weather events can disrupt waste collection services, leading to delays or interruptions in waste pickup. Changes in waste composition may require adjustments in treatment methods to address new types of waste materials or contaminants. Additionally, climate change impacts can affect waste disposal facilities, such as landfills and incinerators, potentially leading to operational challenges or increased risks of environmental contamination. Adapting smart waste management practices to address these implications is essential for building resilience in waste systems and reducing the environmental footprint of waste management activities.

17.2.3 FEEDBACK LOOPS BETWEEN WASTE MANAGEMENT PRACTICES AND CLIMATE CHANGE

Feedback loops exist between waste management practices and climate change, influencing each other in various ways:

17.2.3.1 CONTRIBUTION OF WASTE MANAGEMENT TO GREENHOUSE GAS EMISSIONS

Waste management processes such as landfill decomposition and incineration release greenhouse gases like methane and carbon dioxide, contributing to climate change [3]. Improper waste disposal practices can exacerbate these emissions, creating a feedback loop that amplifies the impact of waste on climate change. Landfill methane emissions can be reduced by diverting organic waste through composting or anaerobic digestion [5]. Resource extraction and production can be reduced through recycling and reuse programs. Landfill gas can capture methane for energy production, reducing emissions.

17.2.3.2 POTENTIAL FOR WASTE MANAGEMENT TO MITIGATE CLIMATE CHANGE

Waste management is crucial in mitigating climate change by reducing greenhouse gas emissions, promoting resource efficiency, and fostering a circular economy. Implementing sustainable waste management practices can create a positive feedback loop by reducing the environmental impact of waste and contributing to climate change mitigation efforts. Cities can implement comprehensive recycling programs, mandate organic waste composting, invest in advanced waste-to-energy technologies, and implement landfill gas capture systems. These measures reduce reliance on fossil fuels and produce renewable energy [4]. By implementing effective waste management strategies, such as extended producer responsibility programs and policies, communities and governments can contribute significantly to global efforts to combat climate change [23].

17.3 MANAGING WASTE WITH CONTEXT TO CLIMATE CHANGE

Waste management, however, is a major yet silent source of greenhouse gas emissions. A total of 42% of all greenhouse gas emissions in the US are thought to be related to trash management, according to a recent analysis from the US Environmental Protection Agency. Global problems are caused by improper handling of solid waste, such as industrial solid pollution, municipal solid waste, and marine litter. Inadequate disposal of heavy metals also affects water, soil, and vegetation, hindering sustainable growth. Treatment techniques, incineration, and landfills are known aspects of waste management [9–26]. The most common source of gaseous emissions into the atmosphere that contribute to global warming is fossil fuel combustion. In an attempt to slow down global warming, there have been recent concentrated efforts undertaken, particularly in developed countries, to lessen reliance on fossil fuels [12]. Diverse perspectives, best practices, and teamwork are essential. Organizations like UNEP (United Nations Environment Programme) and SBC (Sustainable Business Council) can perform tasks at national, international, or subnational levels.

Figure 17.2 shows the contributions of these gases to global warming, and a simple analysis shows that 26% of CH_4 emissions are generated by energy production, whereas 57% are related to garbage and waste management activities. A significant amount of greenhouse gases is released by waste management techniques, such as composting, mass fires, incineration, anaerobic digestion, and biogas plants. Similarly, Biogas, which is mostly composed of CO_2 and CH_4 with traces of other gases, is produced through anaerobic degradation of biodegradable carbon. Over time, bacteria in landfills break down organic matter, generating around 50% CO_2 and 50% CH_4, as well as trace amounts of other gaseous chemicals [10].

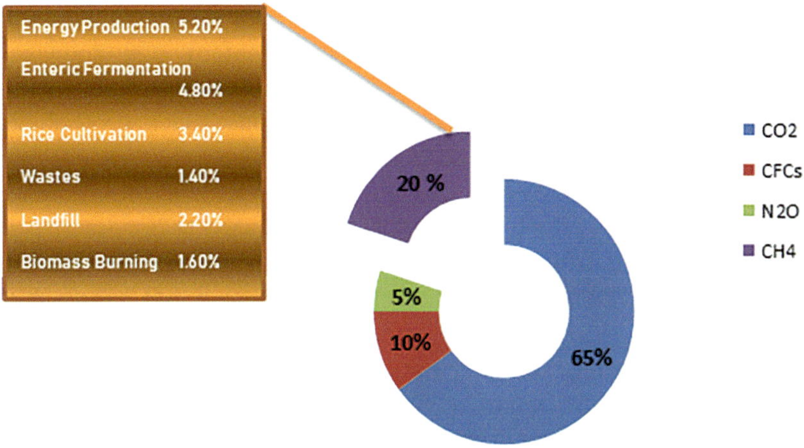

FIGURE 17.2 Contribution of different gases to global warming.

Source: Ref. [10] discussed the pictorial illustration of percentage of GHG gases.

17.4 TECHNOLOGIES FOR MANAGEMENT AND UNDERSTANDING OF CLIMATE CHANGE

It is essential to comprehend the effects of climate change to take proactive and anticipatory waste management measures, and vice versa. Monitoring and assessing climatic conditions and its effects on air, water, and soil quality is necessary. To create a resilient smart city and a smart waste management system need for technological solutions that can observe changes in air, water, and soil quality. These solutions are often based on Industry 4.0 technologies, which support the transformation of data into knowledge at different levels, as shown in Figure 17.3. This framework should be followed by sources that collect raw data from observations and require contextualization to provide direct value [6]. In this framework, the information collection used various techniques, RFID tags and reader's cameras, GPS sensors, two-dimensional code labels and readers' terminals, sensor networks and wireless networks [6, 8]. These methods collect information, such as pressure, temperature, gas concentrations, and humidity. The data is then aggregated and sorted, and information is extracted to answer fundamental questions such as who, what, where, and when regarding climatic changes. However, smart enabling technologies, such as Internet of Things (IoT), RFID, NFC sensors, LoRaWAN sensor, and so on, offer a compelling solution for waste collection systems. By facilitating real-time data collection, these technologies enable smart bin systems that monitor container fullness levels [6, 25]. Moving beyond information, knowledge transforms into practical guidance, expertise, and valuable insights. Further, this wisdom combines actionable knowledge with independent decisions and behaviors through interactive procedures and adaptive judgment [6]. These technological solutions support data analytics for utilizing data collected from different sources. For example, "Descriptive analytics" summarize and contextualize raw data from IoT devices to illustrate what has occurred [1]. The emergence of technologies like artificial intelligence (AI), and machine/deep learning has significantly impacted waste recovery. By gathering and processing information, these technologies support waste sorting and enable the development of functional applications based on advanced decision-support systems [7]. In contrast, "Discovery analytics" manage large data volumes to extract nontrivial knowledge, identifying trends and novel findings for comprehensive understanding. For example, remote sensing allows us to observe changes in land cover and water levels, as well as predict potential climatic impacts resulting from changes in temperature and humidity. This early warning system helps identify deviations from normal weather conditions that could disrupt waste disposal operations [1, 7, 15]. Combining remote sensing data with other local data can lead to the generation of more information on resource management, such as water and air pollution. The use of digital data visualization and cyber-physical systems aids in simulating environmental data for analysis purposes [15, 17].

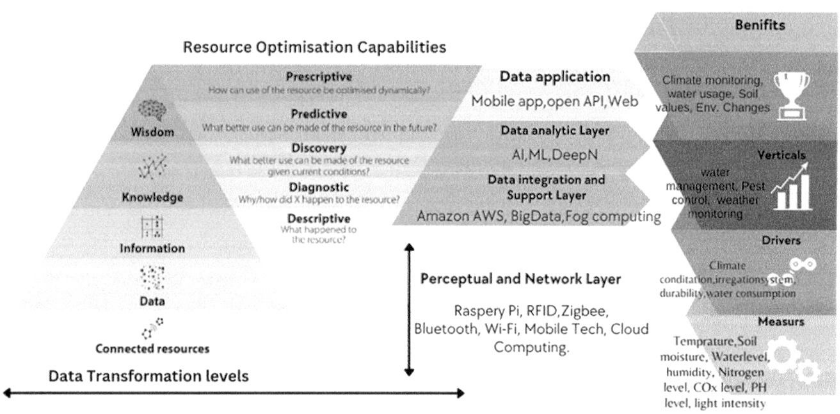

FIGURE 17.3 An intelligent framework for managing and measuring climate change.

Source: Adapted from Ref. [6] discussed the schematic illustration of digital technology based on the ability to support decision making.

"Predictive analytics" use historical data to forecast future trends and probabilities, providing insights into likely future events. Technological solutions, such as IoT-based smart agricultural devices, have significantly contributed to monitoring and tracking changes in soil quality. The accuracy of the proposed device is satisfactory for farmers and shows improvement in farming production [15, 17]. Finally, "prescriptive analytics" take these predictions a step further by recommending actions based on these forecasts, considering potential uncertainties, and employing advanced optimization techniques to suggest the best courses of action [16]. Additionally, by informing users about proper practices through digital technologies, manufacturers can empower them to extend product lifespans, further reducing waste. Consequently, this approach fosters stakeholder engagement and unlocks synergies between waste management and climate change through digitalization. This technology is recognized as a facilitator in promoting environmental sustainability, which enhances visibility of energy consumption and greenhouse gas emissions across various processes. By leveraging this technology, organizations and cities can effectively monitor, analyze, and manage their energy, resource usage and targeted interventions to reduce environmental impact and GHG emissions. For example, big data helps in analyzing the significant amount of data collected from various sources for wastewater management [8, 17].

17.6 POLICY IMPLICATIONS AND RECOMMENDATIONS

17.6.1 POLICY INTERVENTIONS TO ENHANCE RESILIENCE IN WASTE MANAGEMENT

17.6.1.1 STRENGTHENING REGULATORY FRAMEWORKS

Governments should establish robust regulations that promote sustainable waste management practices. This includes enforcing waste reduction, recycling, and disposal guidelines. Governments can strengthen waste management by establishing comprehensive legislation, enforcing regulations through inspections and penalties, setting performance standards, and incentivizing sustainable practices. These frameworks prioritize pollution prevention, habitat conservation, and ecosystem preservation, minimizing adverse environmental impacts. Regulatory frameworks also adapt to climate change challenges by incorporating resilience-building measures into waste management strategies. By implementing these frameworks, governments can create an environment conducive to sustainable waste management practices, mitigate environmental risks, and promote a circular economy.

17.6.1.2 INVESTING IN RESEARCH AND INNOVATION

Policymakers should allocate resources for research and development in waste management technologies. Innovations can lead to more efficient waste processing, reduced environmental impact, and improved resilience. The government allocates funds for research on advanced waste-to-energy technologies or biodegradable materials. Governments invest in research and innovation to improve waste management technologies, processes, and systems. This includes funding for research initiatives, developing advanced technologies, supporting pilot projects, collaborating with industry partners, offering incentives for innovation adoption, promoting knowledge sharing and capacity building, integrating data analytics and digital technologies into waste management processes, and supporting interdisciplinary collaboration. These efforts aim to enhance resilience in waste management by driving technological advancements, fostering collaboration, and promoting sustainable practices.

17.6.1.3 PROMOTING PUBLIC–PRIVATE PARTNERSHIPS

Collaboration between government agencies, private companies, and civil society organizations is crucial. Partnerships can enhance waste collection, recycling infrastructure, and public awareness. The government collaborates with private waste management companies to improve collection infrastructure and public education campaigns. Governments provide grants, subsidies, tax breaks, or preferential procurement for goods and services that use sustainable practices as a means of encouraging the adoption of new ideas. Exchange of knowledge and ability efforts for technical help, training courses, and workshops all support building. Research and innovation expenditures make it possible to incorporate digital technology and data analytics into waste management procedures, improving resource allocation and fortifying the system's resistance to environmental threats.

In order to solve complicated waste management concerns, interdisciplinary collaboration is encouraged.

17.6.2 INTEGRATION OF WASTE MANAGEMENT INTO BROADER CLIMATE CHANGE ADAPTATION STRATEGIES

17.6.2.1 RISK ASSESSMENT

Evaluate waste-related vulnerabilities to climate change impacts (e.g., extreme weather events, rising sea levels). Develop strategies to mitigate these risks. Local government conducts a vulnerability assessment of waste management infrastructure to identify climate-related risks.

17.6.2.2 INFRASTRUCTURE RESILIENCE

Strengthen waste treatment facilities, landfills, and recycling centers to withstand climate-related disruptions. Retrofitting waste treatment facilities with flood barriers or elevating landfill sites in coastal areas to mitigate sea-level rise impacts.

17.6.2.3 COMMUNITY ENGAGEMENT

Involve local communities in waste management planning. Educate them about climate risks and encourage responsible waste disposal. Hosting community workshops to educate residents on proper waste disposal practices and the importance of waste reduction in mitigating climate change impacts.

17.6.2.4 CIRCULAR ECONOMY

Promote circular economy principles, emphasizing resource efficiency, reuse, and recycling. Implementing policies to incentivize businesses to adopt circular economy practices, such as product redesign for recyclability, or facilitating material recovery and reuse.

17.7 RECOMMENDATIONS FOR FUTURE RESEARCH AND ACTION

17.7.1 SMART WASTE MANAGEMENT SYSTEM

Gather comprehensive data on waste generation, disposal patterns, and climate-related risks. Establishing a national database to monitor waste generation trends, disposal rates, and climate-related impacts on waste management infrastructure.

17.7.2 BEHAVIORAL INSIGHTS

Understand public behavior regarding waste disposal and design targeted interventions. Conducting surveys or behavioral studies to understand public attitudes and perceptions toward waste disposal and recycling.

17.7.3 CAPACITY BUILDING

Train waste management professionals and raise awareness among policy-makers. Offering training programs and workshops for waste management professionals on climate-resilient practices and innovative technologies.

17.7.4 INNOVATION

Explore emerging technologies for waste reduction and resource recovery. Investing in research on emerging technologies, such as AI for waste sorting or bio-based materials for packaging.

17.8 CONCLUSION

Socio-agricultural, socio-economic, and physical systems that are essential to psychological health are being significantly impacted by climate change. The Nexus method considers water regimes in soils and landscapes, dynamic soil processes, and the complete waste creation cycle, which is essential for understanding the transformation of waste. Research should prioritize understanding these processes, as many studies highlight the negative effects of waste dumping on soil, such as water contamination and soil deterioration. To mitigate these problems, future strategies include the implementation of public awareness programs like Swachh Bharat, data modeling using machine learning and AI, and the application of LoRaWAN-connected Low-Power Wide-Area Networks for precision agriculture and water and soil management. Sustainable development practices and research information are crucial for effective pollution control. Effective pollution management requires international collaboration in administration, politics, research and development, policy, and monitoring. It provides valuable insights into the current state of knowledge on waste management resilience in a changing climate. Moving forward, further research and practical interventions are needed to enhance waste management resilience and build sustainable futures for communities worldwide.

AUTHOR CONTRIBUTION STATEMENT

PN and SKD contribute to conceptualization, theoretical framework, and technical details; SM contributes to data collection and initial draft preparation; AKP revised the document.

DECLARATION OF COMPETING INTEREST

Regarding the study discussed in this chapter, the authors indicated that they had no competing interests.

KEYWORDS

- **climate change**
- **industry 4.0**
- **policy frameworks**
- **resilience**
- **sustainability**
- **waste management**

REFERENCES

1. Raneesha Madushanki, A. A., Halgamuge, M. N., Surangi Wirasagoda, W. A. H., & Syed, A. (2019). Adoption of the Internet of Things (IoT) in agriculture and smart farming towards Urban greening: A review. *International Journal of Advanced Computer Science and Applications (IJACSA)*, *10*(4), 11–28. http://dx.doi.org/10.14569/IJACSA.2019.0100402
2. Chikodzi, D. (2016). Crop yield sensitivity to climatic variability as the basis for creating climate resilient agriculture. *American Journal of Climate Change*, *5*(1), 69–76.
3. Friedrich, J., Stahl, J., Hoogendoorn, G., & Fitchett, J. M. (2020). Exploring climate change threats to beach tourism destinations: Application of the hazard–activity pairs methodology to South Africa. *Weather, Climate, and Society*, *12*(3), 529–544.
4. Gómez-Sanabria, A., Kiesewetter, G., Klimont, Z., Schoepp, W., & Haberl, H. (2022). Potential for future reductions of global GHG and air pollutants from circular waste management systems. *Nature Communications*, *13*(1), 106.
5. Hajam, Y. A., Kumar, R., & Kumar, A. (2023). Environmental waste management strategies and vermi transformation for sustainable development. *Environmental Challenges*, *13*, 100747.
6. Kristoforsen, https://doi.org/10.1016/j.jbusres.2020.07.044
7. Kurniawan, T. A., Meidiana, C., Goh, H. H., Zhang, D., Othman, M. H. D., Aziz, F., Anouzla, A., Sarangi, P. K., Pasaribu, B., & Ali, I. (2024). Unlocking synergies between waste management and climate change mitigation to accelerate decarbonization through circular-economy digitalization in Indonesia. *Sustainable Production and Consumption*, *46*, 522–542.
8. Maity, T., Paul, S., Samanta, J. et al. (2024). Design and development of IoT-Based SmartTech-Agri devices for smart agriculture crop field. *Journal of The Institution of Engineers (India): Series B*, *105*, 753–762. https://doi.org/10.1007/s40031-024-01002-5
9. Mondal, T., Choudhury, M., Kundu, D., Dutta, D., & Samanta, P. (2023). Landfill: An eclectic review on structure, reactions and remediation approach. *Waste Management*, *164*, 127–142. https://doi.org/10.1016/j.wasman.2023.03.034
10. Nnaji, C. C., & Utsev, J. T. (2011). Climate change and waste management: A balanced assessment. *Journal of Sustainable Development in Africa*, *13*(7), 17–34.
11. Panel, I. (2003). IPCC. *Global Environmental Challenges of the Twenty-first Century: Resources, Consumption, and Sustainable Solutions*, (3), 113.
12. Paul, S., Choudhury, M., Deb, U., Pegu, R., Das, S., & Bhattacharya, S. S. (2019). Assessing the ecological impacts of ageing on hazard potential of solid waste landfills: A green approach through vermitechnology. *Journal of Cleaner Production*, *236*, 117643. https://doi.org/10.1016/j.jclepro.2019.117643
13. Poulton, P., Johnston, J., Macdonald, A., White, R., & Powlson, D. (2018). Major limitations to achieving "4 per 1000" increases in soil organic carbon stock in temperate regions: Evidence from long-term experiments at Rothamsted Research, United Kingdom. *Global Change Biology*, *24*(6), 2563–2584.

14. Protocol, G. G. (2018). About the GHG protocol. Recuperado de https://ghgprotocol.org/about-us.
15. Rangaswamy, M., & Arabia, S. (2012). *Internet of Things (IOT) and Cloud Computing for Agriculture: An Overview*. Agro-Informatics and Precision Agriculture 2012 (AIPA 2012).
16. Rumana, H. S., Sharma, R. C., Beniwal, V., & Sharma, A. K. (2014). A retrospective approach to assess human health risks associated with growing air pollution in urbanized area of Thar Desert, western Rajasthan, India. *Journal of Environmental Health Science and Engineering*, *12*, 1–9.
17. S. Liu, Guo, L., Webb, H., Ya, X., & Chang, X. (2019). Internet of Things monitoring system of modern eco-agriculture based on cloud computing. *IEEE Access*, 7, 37050–37058. https://doi.org/10.1109/ACCESS.2019.2903720.
18. Saxena, A., & Raj, S. (2021). Impact of lockdown during COVID-19 pandemic on the air quality of North Indian cities. *Urban Climate*, *35*, 100754.
19. Teske, S., & Nagrath, K. (2022). Global sector-specific Scope 1, 2, and 3 analyses for setting net-zero targets: Agriculture, forestry, and processing harvested products. *Discover Applied Sciences*, *4*, 221. https://doi.org/10.1007/s42452-022-05111-y
20. Tshuma, M. C. (2009). *A Socio-economic Impact Assessment (SEIA) of the Best Management Practices (BMP) Project of the Zanyokwe Irrigation Scheme at Farm Level* (Doctoral dissertation, University of Fort Hare).
21. Tubiello, F. N., Salvatore, M., Cóndor Golec, R. D., Ferrara, A., Rossi, S., Biancalani, R., Federici, S., Jacobs, H., & Flammini, A. (2014). *Agriculture, Forestry and other Land use Emissions by Sources and Removals by Sinks. Rome, Italy*
22. Tubiello, F. N., Salvatore, M., Ferrara, A. F., House, J., Federici, S., Rossi, S., Biancalani, R., Condor Golec, R. D., Jacobs, H., Flammini, A., Prosperi, P., Cardenas-Galindo, P., Schmidhuber, J., Sanz Sanchez, M. J., Srivastava, N., & Smith, P. (2015). The contribution of agriculture, forestry and other land use activities to global warming, 1990–2012. *Global Change Biology*, *21*(7), 2655–2660.
23. Un, C. (2023). A Sustainable approach to the conversion of waste into energy: Landfill gas-to-fuel technology. *Sustainability*, *15*(20), 14782.
24. Wu, F., Fang, X., Yang, Y., Dupont-Nivet, G., Nie, J., Fluteau, F., Zhang, T., & Han, W. (2022). Reorganization of Asian climate in relation to Tibetan Plateau uplift. *Nature Reviews Earth & Environment*, *3*(10), 684–700.
25. Zhang, A., Venkatesh, V. G., Liu, Y., Wan, M., Qu, T., & Huisingh, D. (2019). Barriers to smart waste management for a circular economy in China. *Journal of Cleaner Production*, *240*, 118198.
26. Choudhury, M., Jyethi, D. S., Dutta, J., Purkayastha, S. P., Deb, D., Das, R., Roy, G., Sen, T., & Bhattacharyya, K. G. (2022). Investigation of groundwater and soil quality near to a municipal waste disposal site in Silchar, Assam, India. *International Journal of Energy and Water Resources*, *6*(1), 37–47. https://doi.org/10.1007/s42108-021-00117-5

CHAPTER 18

Current Trends in Waste Management: Case Studies from the African Continent

RAKESH PANT,[1,3] KHUSHI SAH,[1] ANUSHKA DABRAL,[4] HARJEET TOMER,[1] SHEETAL,[1] SHIVANGI SINGH,[1] AMIT GUPTA,[2*] and DINAOL ZEWUDE[5]

[1]*Department of Microbiology, Graphic Era (Deemed to Be) University, Dehradun, Uttarakhand, India*

[2]*Department of Zoology, University of Jammu, Baba Saheb Ambedkar Road, Jammu Tawi, Jammu and Kashmir, India*

[3]*Department of Quality, Eureka Forbes Ltd Lal Tappar, Dehradun, Uttarakhand, India*

[4]*UCL School of Pharmacy, University College London, Gower Street, London, UK*

[5]*Department of Biology, School of Applied Natural Sciences, Adama Science and Technology University, Adama, Oromia, Ethiopia*

*Corresponding author

ABSTRACT

Waste is any substance or objects the holder removes or plans to discard. Every object, dangerous or not, falls within this category. The waste products generated by human activity are typically highly tenacious. Hazardous waste has been shown to have the potential to endanger both human health and the environment, necessitating special, separate processing. Chemical and physical qualities dictate exactly how materials

Data Analytics and Smart Technologies in Solid Waste Management: Advancing Sustainability.
Moharana Choudhury, Sushobhan Majumdar, and Avnish Chauhan (Eds.)

DOI: 10.1201/9781779640833-22

are collected and recycled. The main characteristics of hazardous waste are its explosiveness, poisonous consequences, corrosion, and ecological toxicity. The African continent's municipal solid waste collection rate is only 55%, with further reductions to 44% in sub-Saharan Africa. Over 90% of the waste is dumped in open dumpsites and unregulated landfills, with the population of African cities growing faster than other urban agglomerations. The current situation is primarily the result of organizational flaws and governance breakdowns, which put private sector investments at risk and resulted in subpar service performance. To safeguard the public's health, legislation requiring gathering solid trash in cities and restricting the released form of human waste was first proposed in the 19th century. Most developing nations cannot offer their whole populace adequate facilities for the assembly and removal of solid waste produced in the community.

18.1 INTRODUCTION

Waste is any substance or object that the holder gets rid of or plans to discard. Each and every object, dangerous or not, falls within this category. It may encompass a broad variety of things that the generator no longer needs or desires, including solid, liquid, and gaseous components. Waste may have a variety of detrimental impacts on human well-being, environment, and natural systems. For instance, it can contaminate the air, water, and soil [1].

1. The burning of certain types of waste products releases poisons into the atmosphere, worsening air pollution. In addition to causing global warming, these pollutants can aggravate respiratory ailments.
2. Improper disposal of garbage, especially hazardous waste, can harm water supplies. Such pollution not only damages ecosystems and jeopardizes aquatic life, but it may also make humans sick by drinking contaminated water.
3. Erroneous dumping and landfills can lead to soil damage. The metals and toxins included in waste products have the ability to permeate the earth's pores and alter its composition and utility [2].
4. Exposure to toxic materials or contaminated water can cause respiratory issues, skin diseases, and other problems.

5. Unauthorized dumping and other waste disposal practices have the potential to upset natural ecosystems. This might have long-term negative effects on biological diversity and the ecosystem.
6. The continuous reliance on virgin resources that results from the nonrecycling and reuse of materials exacerbates the degradation of resources and the environmental harm caused by resource misuse. Methane, a potent greenhouse gas, is generated in landfills as decomposing materials break down. Methane increases temperature and contributes to climate change by trapping heat in the atmosphere [3].

The waste products generated by human activity are typically highly tenacious and cannot degrade, in contrast to the natural environment where waste products like carbon dioxide, oxygen, and decomposed organic matter are utilized as fuels and reactants. Legislators and other authorities must thus define and categorize garbage according to the risks it poses to the environment and public health in order to guarantee appropriate and efficient waste management. Establishing whether a substance qualifies as trash is essential to figuring out if the owner or maker needs to follow waste legislation. Terminology is crucial for both local and international reporting, as well as for gathering and analyzing waste data [4]. The majority of countries have established the term waste, which is typically associated with the idea of disposal. There are two main categories of garbage that may be determined, hazardous waste, and nonhazardous or solid rubbish, based on the numerous laws and policy instruments that are usually in existence. This categorization is also used by the Basel Convention [5].

Hazardous waste is a waste that has been shown to possess the potential to endanger both human health and the environment, necessitating special, separate processing. Chemical and physical qualities dictate exactly how materials are collected and recycled. The main characteristics of hazardous waste are its explosiveness, poisonous consequences, corrosion, and ecological toxicity. Waste that is liquid, powdered, or gaseous needs to be treated differently at all times to stop it from spreading [6]. Any garbage that is not labeled as hazardous is considered nonhazardous waste. Examples of this type of waste include glass, metal, paper products, plastics, organic materials, and beverage cans. Solid waste is not hazardous, but improper collection and disposal can have negative effects on the surroundings and public health. Targeted garbage collection is considered to be one of the most challenging aspects of waste management [37] (Figure 18.1). It entails sorting and collecting waste prior to reuse or recycling a substantial percentage of solid waste [7].

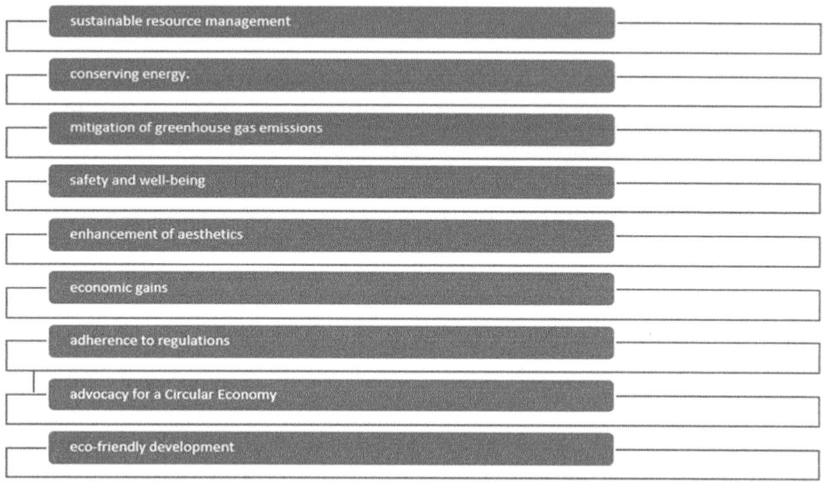

FIGURE 18.1 Goals of waste management [5–7].

Source: Adapted from Ref. [5-7]

Now coming to atmospheric pollution, a major outcome of poor disposal of waste is the leakage of hazardous gases via garbage dumps which has an adverse effect on the natural world and human health. As organic waste breaks down in landfills, gases including methane, also known as CH_4, and volatile organic compounds are produced [8–36] (Figure 18.2).

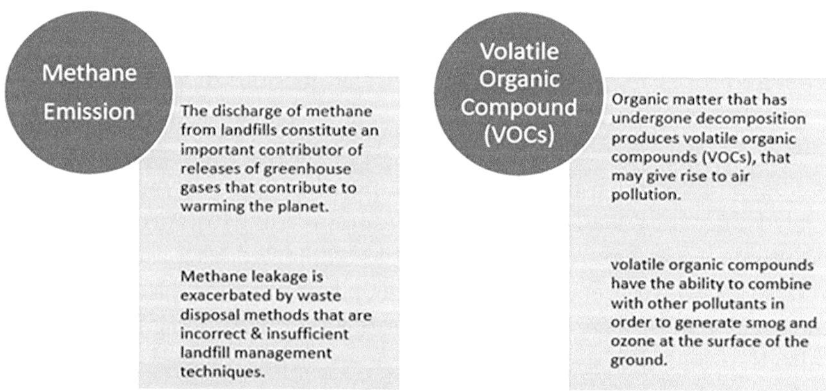

FIGURE 18.2 Categories in which organic waste breakdown in landfills.

Source: Adapted from Ref. [8, 9]

18.2 SPREAD OF DISEASES

Inappropriate waste management practices carry possible health and environmental problems. Workers in this industry are primarily concerned about direct health concerns, and it is important to keep them as safe from waste contact as possible. Additionally, processing hospital and clinic trash carries some dangers [10]. Certain solid wastes disintegrate quickly, emit an unpleasant stench, and unintentionally draw numerous disease vectors, particularly insect vectors, which in turn causes direct pollution of the environment [11]. Many pest insects, also referred to as synanthropic or pestiferous insects are dipteran insects that breed on trash or other surfaces and pose a threat to human health in various ways. An increasing number of vector-borne diseases, including chikungunya, dengue, zika, malaria, filariasis, and zoonotic reservoirs, are being nourished in plastic garbage, which poses a threat to public health. Chikungunya, dengue, and zika are commonly spread by the arbovirus vector *Aedes* species. Both the *Albopictus* species and *Aedes aegypti* favor artificial breeding habitats. Polyvinyl chloride bags, teacup, containers, car tires, drums, coconut shells, and other *Anunobi* are among the materials found mostly in these breeding sites made of plastic. Numerous parasite infections can be transmitted by microplastics. Since there are many different wastes (both organic and inorganic) in the contemporary urbanized environment, landfill debris changes the density of insects. Over the course of four months, 26,434 flies belonging to the Calliphoridae and Psychodidae genera were found in trash dumps in the United States, indicating that calliphorid flies were extremely prevalent across the population.

Hence, there is a need of a proper waste disposal system to reduce the risks mentioned above. In addition to these benefits, effective waste management also reduces the release of greenhouse gases and improves air and water quality. By managing your waste appropriately, you may help preserve natural resources such as wood, water, and minerals. Thus, this is how recycling, reusing, and reducing work [10].

18.3 IMPACT ON BIODIVERSITY

Biodiversity and human health are closely related to ecosystem services. Threats to biodiversity and ecosystems as a result include habitat loss, contamination of the environment, and climate change. Plants suffer greatly from pollutants in both the air and water since most plants will die in contaminated soil. Rainwater carries toxins that can damage plants, even in cases where the soil is clean and uncontaminated. Changes in soil ecology have an impact on the diversity, composition, and efficiency of the soil microbiota. Contamination gradients were highly connected with microbial community abundance and diversity, which likely has an impact on the functions of soil ecosystems. Environmental pollution has the potential to significantly affect soil microbial diversity, community's composition, and ecological clusters. In this way, improper waste management causes irreversible harm to ecosystems worldwide by directly contributing to the extinction of many species each day [12]. The health of all species on the planet is seriously hampered by our waste problem in terms of biodiversity. Therefore, always make sure you are contributing to the appropriate disposal of waste instead of hastening the devastation of the ecosystem.

18.4 CASE STUDY

Sub-Saharan Africa (SSA) accounted for 65% of the 125 million tons of municipal solid waste (MSW) produced in Africa in 2012. This volume is expected to increase to 244 million tons by 2025. Having an average garbage collection rate of 55%, cities and municipalities dispose of more than half of the waste generated. Although coverage varies widely between cities, SSA has a lower average of 44%. The average MSW collection rate is predicted to reach 69% by 2025.

Over 90% of the garbage is dumped on land, despite 80–90% being recyclable. Only 4% of the garbage is recycled, often by marginalized informal reclaimers. Waste plastic, making up 13% of the MSW, is causing concern for the continent [13] (Figure 18.3).

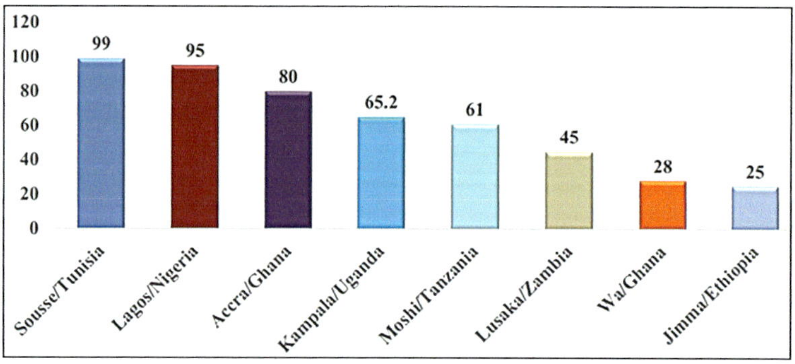

FIGURE 18.3 MSW collection coverage for cities in Africa.

Source: Adapted from Ref. [13, 14]

The African continent's MSW collection rate is only 55%, with further reductions to 44% in SSA. Over 90% of the waste is dumped in open dumpsites and unregulated landfills, with the population of African cities growing at a faster rate than other urban agglomerations. This leads to an increased prevalence of diseases like malaria, typhoid, cholera, dengue, yellow fever, gastroenteritis, and hepatitis in areas close to garbage dumps. Inadequate MSW management systems in Africa pollute air, soil, and water bodies, exacerbated by natural disasters like floods and inundation. The degradation of 58% of organic waste contributes significantly to global warming [14].

Improper solid-waste management in Africa leads to soil fertility issues, plant development issues, and contamination of surface and groundwater bodies. This has negative impacts on aquatic life and human health. The shattered window hypothesis suggests that a messy environment encourages crime and antisocial behavior. National governments are recognizing the need for action and foreign funding in waste management. Kenya has increased garbage recycling by 90% through a partnership with Taka, making waste collection, sorting, composting, plastic recycling, and waste purchase more affordable in poorer regions. South Africa uses advanced technology to simplify rubbish collection services.

In 2017, 24 African countries, the Japan International Cooperation Agency, the Ministry of Environment in Japan, Yokohama City, the UN Environment Programme, and UN-Habitat formed the African Clean Cities Platform. The platform seeks to fulfill the Sustainable Development Goals for waste management in addition to fostering clean, safe cities.

SSA faces challenges such as lack of political will, inadequate waste management policies, insufficient funding, environmental education, weak infrastructure, and limited collection coverage. Trash collection is crucial for public health protection in poor nations [15].

1. According to a study conduct in 2019 by Zamriet et al., waste management initiatives in developing nations present a number of difficulties for society, the economy, and the environment. Waste that is not adequately handled poses a hazard to human health and causes other serious environmental impacts by contaminating the soil, water, and air. For example, this leads to the proliferation of zoonotic diseases in rodents, insects, mosquitoes, and various other animals, in addition to air- and water-borne illnesses and potential cancer risks, in contaminated habitats. This suggests that waste management affects social, economic, and environmental

well-being in addition to being a technological or infrastructure issue an estimated 9 million individuals worldwide pass away from illnesses linked to various forms of contamination and severe poverty every year.

2. Another study conducted in Ghana, 12,710 t of household garbage are produced per day, or 0.47 kg per person. But only 10% of solid garbage is disposed of correctly, with the rest finding their way into sewers, open spaces, and streams. Because of this, managing solid waste in many Ghanaian cities and towns is a challenge [16]. It is imperative that methods for guaranteeing appropriate waste management be developed, as the general population of Ghana is growing [5]. As a result, numerous studies looking into Ghana's solid-waste management challenges have been conducted. But even though financial inclusion—which is defined as providing people and businesses with sustainable, cost-effective access to financial services and products—is recognized as a key tactic for enhancing trash disposal, none of the research mentioned above focused on it. Therefore, it has been discovered that financial inclusion, particularly for the impoverished, improves economic growth, decreases income disparity, and reduces poverty [13–17].
3. Africa, the second-largest continent with the highest population growth, faces challenges in waste management, education, infrastructure, agriculture, urbanization, and electricity generation. Poverty and limited resources contribute to poor waste management, particularly in impoverished areas. An estimated 67,740 health institutions and 282,447 t of medical waste are produced annually in Africa. Most African nations lack medical waste management regulations, leading to increased use of incinerators and poor disposal practices. The Health Professional Council of South Africa created a comprehensive waste management guideline booklet, while Ethiopia and Kenya lack a national strategy for regulating the safe elimination of unwanted medications. The World Health Organization introduced a global guide for disposing of undesired medication waste in 1999, which has helped countries dispose of their trash efficiently. In developing countries, there are no long-term trends regarding the disposal of classified healthcare waste (HCW), and many healthcare workers receive inadequate training on waste disposal. Nigeria has made

changes to improve waste management, with public sector healthcare workers receiving more training on proper disposal [18–20].

4. The reengineering of primary health care in South Africa aims to increase community-based care (CBC) availability, primarily for impoverished and marginalized communities. However, this approach also leads to an increase in HCW produced in homes, highlighting the need for more focus on HCW management in these settings. This study investigated HCW management in CBC from the perspective of stakeholders and policymakers in 29 resource-poor communities outside Durban, KwaZulu-Natal. The study included interviews with nine policymakers and 21 stakeholders from 29 neighborhoods in Durban. The findings revealed that peri-urban communities, established by the apartheid government, are characterized by poverty, food insecurity, and unemployment. Rural communities, which are often isolated from major cities, lack affordable transportation options and high rates of poverty and unemployment. The KwaZulu-Natal municipality of eThekwini provides basic and free municipal services to these communities [20–22].
5. Zimbabwe produces a variety of solid-waste types, including hazardous and nonhazardous, from various sources such as streets, homes, supermarkets, manufacturing companies, and organizations. These waste types require attention and disposal methods, such as burning, garbage dumps, open-air dumps, and open pits. However, these methods are often at the bottom of the waste management hierarchy, potentially causing negative environmental effects. Zimbabwe's poor solid-waste management poses a threat to human life and is linked to pollution of land, water, and air. The country's "Not in My Backyard" syndrome and urbanization contribute significantly to its garbage output. About 90% of solid trash produced in Zimbabwe's cities is disposed of, with the remaining 10% being recycled or converted. This linear waste management approach is linked to environmental health issues. The study aimed to review the life cycle assessment model's relevance to Zimbabwe's solid-waste management, comparing it with literature on solid waste from other nations. The study found that Zimbabwe is experiencing a significant increase in solid waste from food, electronics, and wood, indicating the need for attention to detail and adherence to waste separation [22–24].

18.5 FINANCIAL

In the upcoming years, it is anticipated that environmental clean-up regulations will result in billions of dollars in corporate expenses. In some cases, litigation may be required due to the possible loss of revenue. Taxpayers can be quite certain that a significant amount of their environmental cleanup expenses are deductible under current legislation. Businesses that have purchased stock or assets from companies that have environmental clean-up obligations are in a riskier situation, and no business should buy anything without carefully considering how environmental clean-up liabilities will be handled [25–28].

Globally, 2.01 billion tons of municipal solid trash is produced each year; by 2050, this amount is predicted to rise to 3.40 billion tons. By 2050, solid-waste growth in SSA is predicted to more than triple. However, at least 33% of trash worldwide is not managed in an environmentally sound way. For example, almost 90% of garbage in low-income nations is burned outdoors or dumped in uncontrolled landfills. These methods of disposing of trash have detrimental influences on the atmosphere, public safety, and human health since they can cause climate change by producing methane and act as places of breeding for vectors of disease [28–30].

18.6 TECHNOLOGICAL REMEDIES FOR THE NEGATIVE EFFECTS OF POOR WASTE MANAGEMENT SYSTEM MANAGEMENT ON THE ENVIRONMENT AND HUMAN HEALTH

The primary generators of waste from healthcare include healthcare facilities (the HCFs). Waste from healthcare facilities (HCW) is the expression most frequently utilized to describe the garbage produced by healthcare facilities (Figure 18.4). Some additional terms that can be utilized are waste from hospitals, waste from biomedical procedures, waste from hospitals, and trash from healthcare facilities. All the waste types produced from HCFs are referred to as deadly building materials (HCW), irrespective of whether the substance being disposed of is unsafe or not, contagious, or not, or biochemical in character [31]. According to figures, healthcare workers account for between 1% and 2% of all urban trash generated [32] (Figure 18.5).

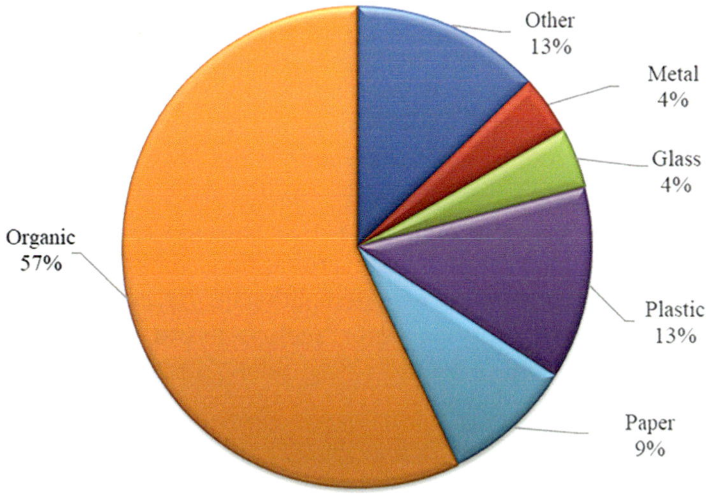

FIGURE 18.4 MSW composition in sub-Saharan Africa.

Source: Adapted from Ref. [14]

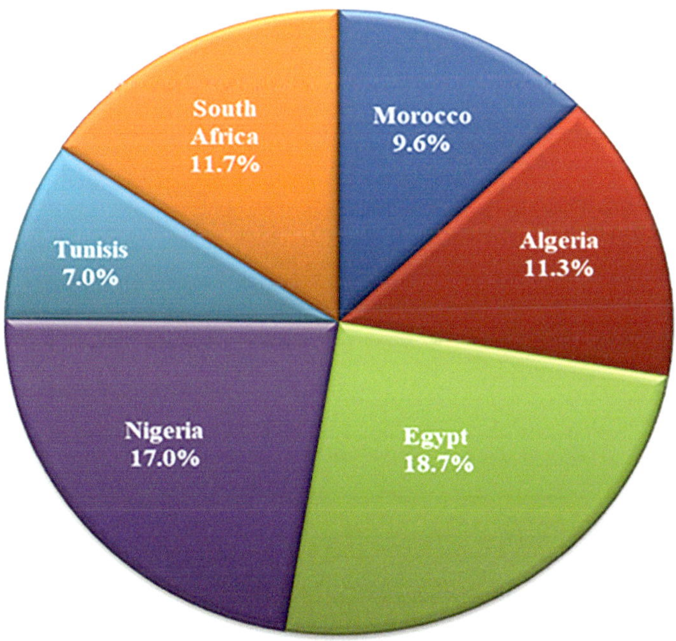

FIGURE 18.5 Use of plastic in African countries.

Source: Adapted from Ref. [15]

Africa has the capacity to enforce environmental laws, making it difficult to execute programs aimed at addressing waste and pollution. This has reduced the efficacy of trash management and fostered a culture of impunity. Consequently, some African nations have imposed product restrictions on hazardous items, especially single-use plastics. Twenty-nine African nations have previously enacted laws governing plastic, which range from prohibitions on all plastic bags to single-use bag laws. Governments are considering more prohibitions on single-use plastic products, including polyethylene terephthalate beverage bottles and items used in the food service sector, such as straws, cups, containers, and utensils, in response to the rising public uproar against plastic [33–35] (Figure 18.6).

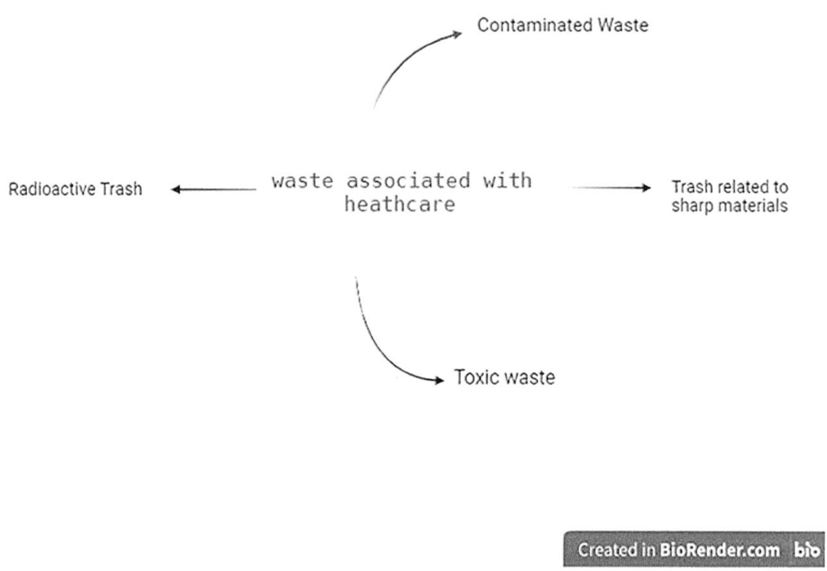

FIGURE 18.6 Types of waste associated with healthcare [32].

18.7 CONCLUSION

The ineffective handling of garbage in Africa today may be attributed to several factors, such as weak organizational structures, insufficient funding, limited expertise, casual legislation, inadequate enforcement, minimal general understanding, political instability, violence, bribery, and an absence of political will. This inefficiency impacts the atmosphere and human health. The current situation is primarily the result of organizational flaws and governance breakdowns, which put private sector investments at risk and resulted in subpar service performance. Action is required to realize the 2030 Agenda for Sustainable Development. Building capacity is essential for ecologically sound waste management, and good governance is vital to establishing an atmosphere for sustainable materials management. Africa has made solid-waste management a political priority. To solve these problems, regional and subregional strategies are required. During most of the development of humanity, the environment was referred to as the "global commons" emissions and wastes from the extraction, production, distribution, and use of materials, agriculture, manufacturing processes, and resource extraction, as well as human wastes, could be freely released into the air, water, or land. To safeguard the public's health, legislation requiring gathering solid trash in cities and restricting the released form of human waste was first proposed in the 19th century. But until the final half of the 20th century, the disposal of hazardous, industrial, and MSWs remained unregulated and largely "out of sight, out of mind." Uncontrolled air pollutants were more "visible;" the London haze led to the UK's 1956 Clean Air Act, which outlawed the burning of solid fuels in the home. Similar laws were passed in other affluent nations as well. Establishing an atmosphere that is conducive to sustainable materials management requires effective governance. Strengthening the ability to promote ecologically appropriate waste management is therefore necessary.

KEYWORDS

- **waste management**
- **environmental pollution**
- **solid waste**
- **health**
- **African countries**

REFERENCES

1. Proshad, R., Kormoker, T., Islam, M. S., Haque, M. A., Rahman, M. M., & Mithu, M. M. R. (2018). Toxic effects of plastic on human health and environment: A consequences of health risk assessment in Bangladesh. *International Journal of Health*, *6*, 1–5.
2. Adeniran, A. A., & Shakantu, W. (2022). The health and environmental impact of plastic waste disposal in South African townships: A review. *IJERPH*, *19*(2), 779.
3. Dikgang, J., Leiman, A., & Visser, M. (2012). Analysis of the plastic-bag levy in South Africa. *Resources, Conservation and Recycling*, *66*, 59–65.
4. Silva, A. L., Patrício, J. C., Prata, T. R., Walker, D., Campos, A. C., Duarte, A. M. V. M., Soares, D. B., & Santos, R. T. (2020). Rethinking and optimising plastic waste management under COVID-19 pandemic: Policy solutions based on redesign and reduction of single-use plastics and personal protective equipment. *Science of the Total Environment*, *742*, 140565.
5. Revuelta, B. M. (2021). Plastics. In *Construction Materials*. Springer Cham: London, UK. pp. 477–502.
6. Jambeck, J. R., Geyer, R., Wilcox, C., Siegler, T. R., Perryman, M., Andrady, A., Narayan, R., & Law, K. L. (2015). Plastic waste inputs from land into the ocean. *Science*, *347*, 768–771.
7. Punjab Pollution Control Board. (2010). *Status Report on Municipal Solid Waste in Punjab*. Punjab Pollution Control Board: Patiala, pp. 1–21.
8. Narayana, T. (2009). Municipal solid waste management in India: From waste disposal to recovery of resources. *Waste Management, 29*(3), 1163–1166.
9. Amasuomo, E., & Jim, B. (2016). The concept of waste and waste management. *Journal of Management and Sustainability*, *88*, 10–55.
10. Rasi, S., Lantela, J., Veijanen, A., & Rintala, J. (2008). Landfill gas upgrading with countercurrent water wash. *Waste Manage*, *28*, 1528–1534.
11. Marco, D., Paolo, C., Pierre, G., & Ioannis, A. (2018). Best environmental management practice for the waste management sector. JRC Science for Policy Report; http://dx.doi.org/10.2760/50247: 1-658.
12. Marco, D., Paolo, C., Pierre, G., & Ioannis, A. (2018). Best environmental management practice for the waste management sector. JRC Science for policy report, http://dx.doi.org/10.2760/50247: 1-658.
13. Godfrey, L., Ahmed, M. T., Gebremedhin, K. G., Katima, J. H. Y., Oelofse, S., Osibanjo, O., Richter, U. H., & Yonli, A. H. (2019). Solid waste management in Africa: Governance failure or development opportunity? Regional *Development in Africa*, 1–15. http://dx.doi.org/10.5772/intechopen.86974.
14. Babayemi, J. O., Nnorom, I. C., Osibanjo, O., & Weber, R. (2019). Ensuring sustainability in plastics use in Africa: Consumption, waste generation, and projections. *Environmental Sciences Europe*, *31*(60), 1–21.
15. Kang, Y. O., Yabar, H., Mizunoya, T., & Higano, Y. (2023). Environmental and economic performances of municipal solid waste management strategies based on LCA method: A case study of Kinshasa. *Heliyon*, *9*, 1–15.

16. Pichtel, J. (2005). *Waste Management Practices: Municipal, Hazardous, and Industrial*, CRC Press: Boca Raton. http://dx.doi.org/10.1201/9781420037517.
17. Bell, M. L., & Samet, M. J. (2005). Air pollution in environmental health. In Frumkin, H. (Ed.), *From Global to Local*. Jossey-Bass: San Francisco, pp. 387–415.
18. Lal, R. (2015). Restoring soil quality to mitigate soil degradation. *Sustainability*, *7*(5), 5875–5895.
19. Montgomery, D. R. (2007). Soil erosion and agricultural sustainability. *Proceedings of the National Academy of Sciences of the United States of America*, *104*(33), 13268–13272.
20. Bell, M. L., Dominican, F., & Samet, J. M. A. (2005). Meta-analysis of time-series studies of ozone and mortality with comparison to the national morbidity, mortality, and air pollution study. *Epidemiology*, *16*(4), 436–445.
21. Patz, J. A., Lendrum, D. C., Holloway, T., & Foley, J. A. (2005). Impact of regional climate change on human health. *Nature*, *438*(7066), 310–317.
22. Ustün, P. A., Bartram, J., Clasen, T., Colford, J. J. M., Cumming, O., Curtis, V., & Wolf, J. (2014). Burden of disease from inadequate water, sanitation and hygiene for selected adverse health outcomes: An updated analysis with a focus on low- and middle-income countries. *International Journal of Hygiene and Environmental Health*, *221*(5), 284–297.
23. Jones, K. E., Patel, N. G., Levy, M. A., Storeygard, A., Balk, D., Gittleman, J. L., & Daszak, P. (2008). Global trends in emerging infectious diseases. *Nature*, *451*(7181), 990–993.
24. Fobil, J. N., Armah, N. A., Hogarh, J. N., & Carboo, D. (2008). The influence of institutions and organizations on urban waste collection systems: An analysis of waste collection system in Accra, Ghana (1985–2000). *Journal of Environmental Management*, *86*(1), 262–271.
25. Medina, M. (1997). Informal recycling and collection of solid wastes in developing countries: Issues and opportunities.
26. Daszak, P., Cunningham, A. A., & Hyatt, A. D. (2000). Emerging infectious diseases of wildlife—Threats to biodiversity and human health. *Science (New York, N.Y.)*, *287*(5452), 443–449.
27. WHO. (2016). *Preventing Disease Through Healthy Environments: A Global Assessment of the Burden of Disease from Environmental Risks*. World Health Organization.
28. Kühn, S., Ree, V. D., Müller, M., Smits, T. H., & Höfle, M. G. (2011). Shifted dynamics of soil-borne microorganisms at elevated atmospheric CO2 and their impact on carbon cycling and related soil processes. *Applied and Environmental Microbiology*, *77*(24), 8710–8718.
29. Andrade, M. F., Kumar, P., Freitas, E. D., Ynoue, R. Y., & Martins, J. (2017). Air quality in the megacity of São Paulo: Evolution over the last 30 years and future perspectives. *Atmospheric Environment*, *159*, 66–82.
30. Gupta, S., & Pandey, A. (2017). Solid waste management in Delhi: A social cost-benefit analysis. *International Journal of Environmental Studies*, *74*(1), 35–53.
31. Sharma, R. K., Agrawal, M., & Marshall, F. (2008). Heavy metal contamination of soil and vegetables in suburban areas of Varanasi, India. *Ecotoxicology and Environmental Safety*, *66*(2), 258–266.

32. Biswas, A. K., Kumar, S., Babu, S. S., Bhattacharyya, J. K., & Chakrabarti, T. (2010). Studies on environmental quality in and around municipal solid waste dumpsite. *Resources, Conservation and Recycling*, *55*(2), 129–134.
33. Abubakar, I. R., & Aina, Y. A. (2016). Achieving sustainable cities in Saudi Arabia: Juggling the competing urbanization challenges. In *Population Growth and Rapid Urbanization in the Developing World*. IGI Global: Hershey, PA, USA, pp. 234–255.
34. Abul, S. (2010). Environmental and health impact of solid waste disposal at Mangwaneni dumpsite in Manzini: Swaziland. *Journal of Sustainable Development in Africa*, *12*, 64–78.
35. Pant, R., Gupta, A., Sing, A., Srivastava, S., & Patrick, N. (2023). A Summary of the role of Microorganisms in waste management. In *Microbial Technology for Sustainable E-Waste Management*, pp. 337–352.
36. Mondal, T., Choudhury, M., Kundu, D., Dutta, D., & Samanta, P. (2023). Landfill: An eclectic review on structure, reactions and remediation approach. *Waste Management*, *164*, 127–142. https://doi.org/10.1016/j.wasman.2023.03.034
37. Paul, S., Choudhury, M., Deb, U., Pegu, R., Das, S., & Bhattacharya, S. S. (2019). Assessing the ecological impacts of ageing on hazard potential of solid waste landfills: A green approach through vermitechnology. *Journal of Cleaner Production*, *236*, 117643. https://doi.org/10.1016/j.jclepro.2019.117643

CHAPTER 19

Current Trends in Waste Management: Case Studies from the Oceania Subcontinent

TSENBENI N. LOTHA[1*], LATONGLILA JAMIR[1*], PRABHAT KUMAR PATHAK[2], and TUANI LIDIAWATI SIMANGUNSONG[3,4]

[1]*Department of Environmental Science, Nagaland University, Lumami, Nagaland, India*

[2]*Department of Biochemistry and Molecular Biology, Oklahoma State University, Stillwater, OK, USA*

[3]*Department of Engineering Department, University of Surabaya, Surabaya, Indonesia*

[4]*Center for Environmental and Renewable Energy Studies, University of Surabaya, Surabaya, Indonesia*

*Corresponding author

ABSTRACT

Waste management is a critical environmental issue worldwide, with varying approaches and challenges across different regions. In the Oceania subcontinent, comprising Australia, New Zealand, and the Pacific Islands, waste management practices are influenced by unique geographical, cultural, and socioeconomic factors. Thus, this chapter delved into the ever-changing landscape of waste management in the Oceania subcontinent's countries and investigates contemporary trends impacting waste

Data Analytics and Smart Technologies in Solid Waste Management: Advancing Sustainability. Moharana Choudhury, Sushobhan Majumdar, and Avnish Chauhan (Eds.)

DOI: 10.1201/9781779640833-23

management techniques in nations across Oceania. It also explores the diverse methods to address the challenges of waste generation, disposal, and environmental effects through a case study approach. It examines governmental regulations, community-led initiatives, and technological breakthroughs shaping waste management practices. The effectiveness of technology-driven solutions is evaluated by assessing both shared successes and distinctive issues countries face. Ultimately, this chapter provides insights into sustainable waste management strategies tailored to the region's unique characteristics and catalyzes informed decision-making and the creation of focused strategies to enhance sustainability in waste management practices across the Oceania subcontinent.

19.1 INTRODUCTION

Solid waste is a byproduct of human activity, and the introduction of new materials, changing consumption habits, and increased rubbish output, particularly in metropolitan areas, have all contributed to waste management problems [1]. Many solid wastes are not only solid but also liquid, semisolid, or gaseous and the problem is caused mostly by increased daily solid waste output as an outcome of population growth, rising production, and consumption rates, particularly in urban areas [2–35]. The Oceania subcontinent, which includes Australia, New Zealand, and the Pacific Islands, presents substantial waste management difficulties due to its distinct geographical, environmental, and socioeconomic characteristics [3] According to the World Bank, the region produces more than 150 million tons of garbage every year, with Australia and New Zealand being the major producers [4]. The Pacific Islands, despite their modest populations, present distinct issues due to their remote locations and insufficient waste management infrastructure [5]. Oceania's waste management practices vary greatly, reflecting the region's different socioeconomic and environmental settings. Modern waste management systems exist in Australia and New Zealand, with facilities for collection, recycling, and disposal [6]. Oceania's unique biodiversity and fragile ecosystems render it especially vulnerable to the detrimental consequences of waste mismanagement and suffers serious social and economic costs as a result of inadequate waste management. Furthermore, the financial burden of waste management lies primarily on local governments and communities, diverting funding away from other critical services. Addressing these difficulties necessitates innovative and sustainable waste management strategies that are adapted to the region's individual demands and environment. Policy and regulatory frameworks Oceania's governments have acknowledged the importance of comprehensive waste management laws and regulations to meet the growing garbage challenge and have established waste management systems that prioritize trash reduction, recycling, and resource recovery. Furthermore, there is a growing interest in circular economy ideas, which promote resource efficiency and waste reduction. It looks at current trends, challenges, and opportunities in Oceania's waste management, highlighting the need for creative and sustainable solutions given that Australia produces over 67 million tons of garbage per year, with a recycling rate of about 58%. Approximately 37% of the 15.5 million tons of garbage produced in New Zealand each year is recycled. Due to a lack of resources and infrastructure, waste management is difficult in the Pacific Islands, which makes waste collection and disposal extremely difficult. This chapter evaluates the efficacy of technology-driven solutions in tackling these issues and investigates the ways in which cultural factors influence waste patterns and attempts to offer insights into sustainable waste management techniques suited to the particulars of the Oceania subcontinent by highlighting both common triumphs and specific challenges faced by nations in the region.

19.2 LITERATURE REVIEW

Solid waste management entails tracking (handling and storage), collection, transportation, processing, and disposal [7]. A literature survey on waste management in Oceania reveals a range of challenges and opportunities in the region [8]. Numerous studies have shown that trash generation rates are increasing in nations such as Australia and New Zealand, owing to population increase and urbanization [9, 10]. Some Pacific Island countries also suffer from a lack of proper waste management infrastructure, which has led to environmental contamination and public health problems. Oceania waste management research frequently highlights the need of environmentally friendly activities like recycling, composting, and waste-to-energy (WTE) technology. Marine litter, particularly plastic garbage, has become a major issue, endangering marine biodiversity and human health [36]. Researchers underline the critical need for immediate action to eliminate marine litter and encourage sustainable waste management procedures. Policy and regulation play a vital part in Oceania's waste management and emphasize the importance of comprehensive waste management policies that prioritize waste reduction and resource recovery. Some studies also stress the significance of community involvement and education in establishing sustainable waste management techniques. While obstacles exist, there is hope that with the proper policies, technologies, and community involvement, the region will make substantial progress toward a more sustainable waste management system. For example, Fiji has undertaken efforts to enhance waste collection and recycling through community engagement and awareness programs. Similarly, in Papua New Guinea, efforts focusing on sustainable waste management in urban areas have yielded good results but face difficulties in expanding to rural areas [5, 11].

19.3 OCEANIA SUBCONTINENTS: CASE STUDIES

Case studies from various Oceania countries are examined, focusing on waste management practices, difficulties, and innovative solutions.

19.3.1 STUDY SITE (FIGURE 19.1)

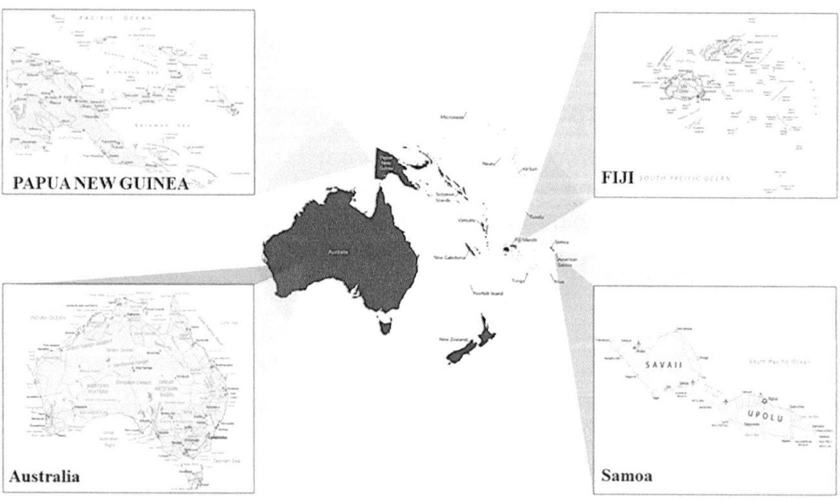

FIGURE 19.1 Map illustrating the study area's location.

19.3.1.1 CASE STUDY I: FIJI

Fiji, like many Pacific Island countries, confronts issues in waste management because of its remote location, limited resources, and expanding population [12]. Between 2013 and 2016, Fiji's gross domestic product increased by 11.5%, but waste quantity increased by 25.4% [13]. To address these challenges, Fiji has implemented innovative recycling initiatives [14]. One such initiative is the introduction of intelligent recycling bins in urban areas equipped with sensors to detect the level of waste and send alerts to waste management authorities when they need to be emptied [12]. Another brilliant recycling initiative is the establishment of community recycling centers, which provide facilities for residents to sort and recycle their waste, reducing the amount of waste sent to landfills.

19.3.1.2 CASE STUDY II: PAPUA NEW GUINEA

Due to the lack of infrastructure and resources, waste management is a major difficulty in this country. Open dumping, waste burning, and insufficient waste collection services all pose risks to the environment and public health [15]. To address these challenges, Papua New Guinea (PNG) has initiated several WTE initiatives. One notable project is the WTE plant in Port Moresby, the capital city where it converts municipal solid waste into electricity, by reducing the volume of waste sent to landfills and has been successful in providing a sustainable source of energy. It has helped alleviate the environmental burden of waste disposal and contributed to the country's energy needs [16].

19.3.1.3 CASE STUDY III: SAMOA

The Samoa Trash Authority (SWA) was established in 2009 to supervise trash management and recycling initiatives as part of the nation's efforts to enhance its waste management practices [17]. One of SWA's primary objectives is the development of an Integrated Waste Management Strategy, which aims to reduce waste generation while promoting recycling and composting. As part of this plan, SWA launched a recycling program in collaboration with local communities and businesses, diverting a substantial amount of waste from landfills while simultaneously providing options for cash production through the sale of recyclable products. Another critical component of Samoa's waste management policy is the promotion of composting to minimize organic waste [18]. Samoa has worked with local farmers to encourage composting and has offered training and support to help them set up composting facilities.

19.3.1.4 CASE STUDY IV: AUSTRALIA

Waste management practices in Australia have evolved dramatically over time, owing to environmental concerns and the necessity to manage the country's rising waste stream. With the intention of lowering garbage creation, raising recycling, and enhancing waste management techniques, Australia's waste management initiatives reached a significant turning point in 2009 with the introduction of the National Waste Policy [19, 20]. Several state and territory governments have endorsed this approach, having established their own waste management goals and policies. Additionally, great progress has been made in managing organic waste; many local governments have started recycling programs that collect organic waste (like leftover food and garden trash) to turn it into compost or soil conditioners. Another significant initiative is the Sydney Energy Recovery Facility, which turns nonrecyclable garbage into energy that has helped lessen the city's dependency on landfills and fossil fuels, resulting in environmental benefits such as lower greenhouse gas emissions [21].

19.4 OCEANIA WASTE MANAGEMENT TECHNOLOGICAL INNOVATIONS

In recent years, countries in the Oceania subcontinent have adopted more technological advances to improve garbage monitoring and management procedures which have been helpful in solving the myriad issues that these countries confront, including limited resources, remote regions, and expanding populations [22]. The systems employ sensors and Internet of Things technology to monitor waste generation, collection, and disposal in real-time. Smart dumpsters are another technical invention that has been widely implemented where bins are equipped with sensors that detect when they are full and immediately send a collection request [23, 24]. In Oceania, cutting-edge recycling techniques like pyrolysis and gasification are being researched to extract valuable materials from waste streams and convert nonrecyclable waste into syngas, bio-oil, and char, which can subsequently be used as raw materials for manufacturing or as renewable energy sources.

19.5 CASE STUDIES ON SMART WASTE SOLUTIONS AND WTE INITIATIVES

New Zealand has been aggressively implementing innovative waste solutions [25] and contributed to cost savings and less environmental effects during waste collecting operations, such as using intelligent bins with sensors that track their fill levels in real-time, allowing waste management businesses to optimize collection routes and timetables. It has also invested in WTE systems to generate renewable electricity with incineration and anaerobic digestion to transform trash into energy. One such plant is Western Australia's Kwinana Waste-to-Energy Plant, which started its operations in 2020 [20], with the potential to handle 400,000 t of residual garbage annually and produce sufficient electricity to power about 50,000 homes. Another WTE project is the Eastern Creek Energy from Waste facility in New South Wales [26]. Once operational; the facility will treat up to 320,000 t of garbage annually and create approximately 50 MW of electricity. Another innovative recycling solution adopted in Fiji is the use of mobile recycling containers [27], which are outfitted with compactors and sorting facilities, allowing them to handle recyclable materials on-site. This strategy lowers the need for transportation to recycling facilities and uses digital platforms and mobile applications to encourage people to recycle and reduce their waste [28]. The WTE plant in Port Moresby is intended to convert municipal solid waste to power [15] and involves burning garbage at elevated temperatures to produce steam, which then powers turbines to generate electricity. The initiative is intended to have various advantages, including minimizing environmental pollution and producing renewable energy. It also reflects PNG's commitment to sustainable development and lowering its carbon footprint [29].

19.6 CHALLENGES AND SUCCESSES IN IMPLEMENTING TECHNOLOGICAL ADVANCEMENTS

Despite the benefits of technical developments in waste management, there are various obstacles to their implementation in Oceania. The initial expenditure necessary to establish innovative waste solutions and WTE efforts might be significant, making it difficult for some Oceania countries to adopt these technologies. A lack of suitable infrastructure, such as collection systems and recycling facilities, can impede the efficient adoption of waste management technology improvements. There is a need for increased public knowledge and education where policies may be incompatible with adopting new technology, making their implementation difficult. Despite these limitations, Oceania has seen several triumphs in incorporating technical breakthroughs in garbage management, such as intelligent trash solutions, which have enhanced the efficiency of waste collection and handling, resulting in cost savings and cleaner environments.

19.7 ENVIRONMENTAL AND SOCIAL CHALLENGES IN OCEANIA

19.7.1 IMPROPER WASTE DISPOSAL IN OCEANIA

When heavy metals from leachate travel into groundwater, they can cause bioaccumulation and biomagnification, harming the ecosystem [30, 31]. Oceania's extensive coastlines make it susceptible to marine debris, primarily plastic waste affecting marine ecosystems [31]. When heavy metals are dumped into aquatic systems, they are typically attached to particle debris, which settles and becomes part of the sediment [17]. Hazardous gases, like SO_2 and CO, are among the most pressing, devastating impacts on marine life. Open burning of waste is another common practice in some parts of Oceania, particularly in rural and remote areas [26], which may result in the loss of biodiversity and ecosystem services that potentially produce chlorosis, slow plant development, and yield depression [32].

19.7.2 SOCIAL AND ECONOMIC IMPLICATIONS OF CURRENT WASTE MANAGEMENT PRACTICES

Improper trash disposal harms the environment, well-being, and livelihoods of communities throughout the region. When released into the air, land, and water by open dumping and waste burning, hazardous chemicals and pollutants increase the risk of respiratory illnesses, skin infections, and other health issues. Trash mounds and congested areas also contribute to unhealthy surroundings and social degradation, which hurt local economies, particularly those that depend on tourism and fishing. Landfilling organic waste that may be composted into energy subsidizes resource depletion and environmental degradation [33]. It is recommended to produce biogas from organic waste, such as food and agricultural waste because composting organic waste lowers methane emissions from landfills and improves soil quality. Programs for recycling plastics, glass, paper, and metals are essential for educating the public about the benefits of sustainable waste management practices and the adverse environmental effects of improper waste disposal.

19.7.3 POLICY-MAKING, SUSTAINABLE DEVELOPMENT, AND ENVIRONMENTAL CONSERVATION EFFORTS

The countries of the Oceania subcontinent have advanced significantly in the areas of environmental preservation, sustainable development, and policymaking. By 2024, Australia hopes to have banned the export of tires, plastic, paper, and glass for recycling. It has also implemented the National Waste Policy, which focuses on resource recovery, waste prevention, and safe waste disposal. The Waste Minimization Act of New Zealand seeks to lessen waste's adverse environmental effects by encouraging expanded producer responsibility and product stewardship. To address marine pollution and promote sustainable waste management practices, the Pacific Islands have also started regional initiatives like the Pacific Regional Waste and pollution management program. These initiatives also assess the efficacy of technology-driven solutions in resolving waste management issues.

19.8 OPPORTUNITIES FOR SUSTAINABLE WASTE MANAGEMENT IN OCEANIA

Oceania offers a unique opportunity to implement sustainable waste management strategies due to its diversified ecosystems, biodiversity, and cultural history, with several opportunities that bode well for the region's environmental and social well-being [10]. Furthermore, there is a growing trend toward using innovative waste solutions, such as sensors and data analytics, to optimize waste collection routes, resulting in more efficient waste management procedures and lowering costs and environmental implications [23]. Another possibility is to promote circular economy principles that emphasize resource efficiency and waste minimization and create new economic opportunities by encouraging reuse, recycling, and material recovery from waste streams. Additionally, governments, industry stakeholders, and communities must build alliances and collaborate to support sustainable waste management programs to exchange best practices, invest in infrastructure, and raise awareness about the significance of responsible waste management.

19.9 FUTURE PROSPECT

As the region continues to urbanize and grow, waste generation will likely increase, putting further strain on existing waste management facilities. However, this allows introducing innovative solutions and technologies that can enhance waste management practices while reducing environmental effects. One promising development is the growth of WTE technology, which can help minimize dependency on landfills and fossil fuels for electricity generation. Another interesting path is the creation of circular economy efforts that attempt to reduce waste generation while increasing material reuse, recycling, and recovery [34]. By adopting a more circular strategy, Oceania may minimize its reliance on finite resources and build a more sustainable and resilient economy. Furthermore, improved public awareness and involvement in waste management concerns might result in more environmentally friendly consumption and waste reduction methods. Education campaigns and community recycling initiatives can promote a sustainable culture in the region.

19.10 CONCLUSION

In conclusion, the importance of effective waste management strategies in the Oceania subcontinent emphasizes stakeholders' need to collaborate to achieve sustainable waste management practices. Waste management in Oceania is a multidimensional challenge that demands immediate attention and coordinated efforts from various stakeholders. Despite the region's specific environmental and social constraints, there are multiple chances to apply sustainable waste management strategies. Oceania can transition to a more sustainable waste management system by implementing circular economy principles, engaging local communities, leveraging innovation and technology, investing in green infrastructure, conserving natural resources, improving climate resilience, and strengthening policy and governance. Nevertheless, addressing the environmental and social consequences of existing waste management systems necessitates a comprehensive method for the interdependence of environmental, economic, and societal variables.

KEYWORDS

- **waste management**
- **community-led initiatives**
- **sustainability**
- **technological innovations**
- **smart waste solutions**

REFERENCES

1. Joseph, L. P., & Prasad, R. (2020). Assessing the sustainable Municipal Solid Waste (MSW) to electricity generation Potentials in Selected Pacific Small Island Developing States (PSIDS). *Journal of Cleaner Production*, *248*(3), 119222.
2. Das, S., Lee, S.-H., Kumar, P., Kim, K.-H., Lee, S. S., & Bhattacharya, S. S. (2019). Solid waste management: Scope and the challenge of sustainability. *Journal of Cleaner Production*, *228*(6), 658–678. https://doi.org/10.1016/j.jclepro.2019.04.323.
3. Oliveira, A. de L., & Turra, A. (2015). Solid waste management in coastal cities: Where are the gaps? Case study of the north coast of são paulo. *Brazil. J. Integr. Coast. Zo. Manag. 15*(4), 453–465. https://doi.org/10.5894/rgci544.
4. Gellert, P. K., & D'Onofrio, S. (2024). Flex commodities and intertwining world-ecologies: Indonesian palm waste as an environmental fix in the new zealand dairy industry. *Political Geography*, *108*(1), 103038.
5. Begg, S. S., De Ramon N'Yeurt, A., & Begg, S. (2023). Interweaving resource management with indigenous knowledge to build community resilience in the pacific Islands: Case of the Waimanu catchment in Viti Levu, Fiji. *Regional Environmental Change*, *23*(3), 1–18. https://doi.org/10.1007/s10113-023-02079-2.

6. Scott, J., Beydoun, D., Amal, R., Low, G., & Cattle, J. (2005). Landfill management, leachate generation, and leach testing of solid wastes in Australia and overseas. *Critical Reviews in Environmental Science and Technology*, *35*(3), 239–332.
7. Chandel, A. S., Weto, A. E., & Bekele, D. (2024). Geospatial technology for selecting suitable sites for solid waste disposal: A case study of shone town, Central Ethiopia. *Urban, Planning and Transport Research*, *12*(1). https://doi.org/10.1080/21650020.2024.2302531.
8. Shahabuddin, M., Uddin, M. N., Chowdhury, J. I., Ahmed, S. F., Uddin, M. N., Mofijur, M., & Uddin, M. A. (2023). A review of the recent development, challenges, and opportunities of electronic waste (e - Waste). *International Journal of Environmental Science and Technology*, *20*(4), 4513–4520. https://doi.org/10.1007/s13762-022-04274-w.
9. Blake, V., Farrelly, T., & Hannon, J. (2019). Is voluntary product stewardship for E-waste working in New Zealand? A whangarei case study. *Sustain*, *11*(11), 1–26. https://doi.org/10.3390/su11113063.
10. Ferdous, W., Manalo, A., Siddique, R., Mendis, P., Zhuge, Y., Wong, H. S., Lokuge, W., Aravinthan, T., & Schubel, P. (2021). Recycling of landfill wastes (Tyres, Plastics and Glass) in construction—A review on global waste generation, performance, application and future opportunities. *Resources, Conservation and Recycling*, *173*(10), 105745.
11. Fernandez, G., Asari, M., Uy, N., Veitata, S., Fayazi, M., Ling, Y., Xu, Q., Wang, H., Ramos, L. V., & Singh, S. (2023). Developing capacity for post-typhoon disaster waste management in Lautoka, Fiji and Makati, Philippines. *APN Science Bulletin*, *13*(1), 153–162. https://doi.org/10.30852/SB.2023.2226.
12. Sewak, A., Deshpande, S., Rundle-Thiele, S., Zhao, F., & Anibaldi, R. (2021). Community perspectives and engagement in sustainable Solid Waste Management (SWM) in Fiji: A socioecological thematic analysis. *Journal of Environmental Management*, *298*(11), 113455.
13. Guerrero, M. (2016). Statistical Yearbook for Asia and the Pacific 2016: SDG Baseline Report United.
14. Farrelly, T. A., Borrelle, S. B., & Fuller, S. (2021). The strengths and weaknesses of pacific Islands plastic pollution policy frameworks. *Sustainability*, *13*(3), 1–36.
15. Doaemo, W., Dhiman, S., Borovskis, A., & Zhang, W. (2021). Assessment of municipal solid waste management system in Lae City, Papua new guinea in the context of sustainable development. *Environment, Development and Sustainability*, *23*(12), 18509–18539. https://doi.org/10.1007/s10668-021-01465-2.
16. Mudd, G. M., Roche, C., Northey, S. A., Jowitt, S. M., & Gamato, G. (2020). Mining in Papua New Guinea: A complex story of trends, impacts and governance. *Science of the Total Environment*, *741*(11), 140375.
17. Diarra, I., & Prasad, S. (2021). The current state of heavy metal pollution in pacific Island countries: A review. *Applied Spectroscopy Reviews*, *56*(1), 27–51.
18. Mohee, R., Mauthoor, S., Bundhoo, Z. M. A., Somaroo, G., & Soobhany, N. (2015). Current status of solid waste management in small Island developing states: A review. *Waste Management*, *43*(6), 539–549. https://doi.org/10.1016/j.wasman.2015.06.012.

19. He, H., Reynolds, C. J., & Boland, J. (2018). Assessment of solid waste generation and treatment in the Australian economic system: A closed waste supply-use model. *Waste Management*, *78*(8), 346–355.
20. Shooshtarian, S., Maqsood, T., Wong, P. S. P., Khalfan, M., & Yang, R. J. (2019). Review of energy recovery from construction and demolition waste in Australia. *Journal of Construction Engineering, Management & Innovation*, *2*(3), 112–130. https://doi.org/10.31462/jcemi.2019.03112130.
21. Islam, T., & Huda, N. (2020). Assessing the recycling potential of "Unregulated" e-Waste in Australia. *Resources, Conservation and Recycling*, *152*(9), 104526.
22. Halog, A., & Anieke, S. (2021). A review of circular economy studies in developed countries and its potential adoption in developing countries. *Circular Economy and Sustainability*, *1*(6), 209–230. https://doi.org/10.1007/s43615-021-00017-0.
23. Yken, J. Van; Boxall, N. J., Cheng, K. Y., Nikoloski, A. N., Moheimani, N. R., & Kaksonen, A. H. (2021). E-waste recycling and resource recovery: A review on technologies, barriers and enablers with a focus on Oceania. *Metals (Basel)*, *11*(8), 1313.
24. Mihai, F.-C., Gnoni, M.-G., Meidiana, C., Ezeah, C., & Elia, V. (2019). Waste Electrical and Electronic Equipment (WEEE): Flows, Quantities and Management - a Global Scenario.
25. Hannon, J., & Zaman, A. U. (2018). Exploring the phenomenon of zero waste and future cities. *Urban Science*, *2*(3), 90.
26. Yang, D., Liu, J., Wang, Q., Hong, H., Zhao, W., Chen, S., Yan, C., & Lu, H. (2019). Chemosphere geochemical and probabilistic human health risk of chromium in mangrove sediments: A case study in Fujian, China. *Chemosphere*, *233*(10), 503–511. https://doi.org/10.1016/j.chemosphere.2019.05.245.
27. Tudor, D. T., & Williams, A. T. (2021). The effectiveness of legislative and voluntary strategies to prevent ocean plastic pollution: Lessons from the UK and South pacific. *Marine Pollution Bulletin*, *172*(11), 112778.
28. Salim, H., Jackson, M., Stewart, R. A., & Beal, C. D. (2023). Drivers-pressures-state-impact-response of solid waste management in remote communities: A systematic and critical review. *Cleaner Waste Systems*, *4*(12), 100078. https://doi.org/10.1016/j.clwas.2023.100078.
29. Alamgir, M., Sloan, S., Campbell, M. J., Engert, J., Kiele, R., Porolak, G., Mutton, T., Brenier, A., Ibisch, P. L., & Laurance, W. F. (2019). Infrastructure expansion challenges sustainable development in Papua New Guinea. *PLoS ONE*, *14*(7), 1–20. https://doi.org/10.5061/dryad.3p84s7s.
30. Smith, S. R. (2009). A critical review of the bioavailability and impacts of heavy metals in municipal solid waste composts compared to sewage sludge. *Environment International*, *35*(1), 142–156. https://doi.org/10.1016/j.envint.2008.06.009.
31. Parameswari, K., Padmini, T. K., & Mudgal, B. V. (2015). Assessment of soil contamination around municipal solid waste dumpsite. *Indian Journal of Science and Technology*, *8*(12), 9–10. https://doi.org/10.17485/ijst/2015/v8i36/87437.
32. Kumar, H., Deepak, P., Patra, K., & Acharya, S. (2024). Chromium-induced phytotoxicity and its impact on plant metabolism. *The Acta Physiologiae Plantarum*, *46*(2), 1–23. https://doi.org/10.1007/s11738-023-03646-0.

33. Paul, S., Choudhury, M., Deb, U., Pegu, R., Das, S., & Bhattacharya, S. S. (2019). Assessing the ecological impacts of ageing on hazard potential of solid waste landfills: A green approach through vermitechnology. *Journal of Cleaner Production*, *236*, 117643. https://doi.org/10.1016/j.jclepro.2019.117643
34. Mondal, T., Choudhury, M., Kundu, D., Dutta, D., & Samanta, P. (2023). Landfill: An eclectic review on structure, reactions and remediation approach. *Waste Management*, *164*, 127–142. https://doi.org/10.1016/j.wasman.2023.03.034
35. Choudhury, M., Jyethi, D. S., Dutta, J., Purkayastha, S. P., Deb, D., Das, R., Roy, G., Sen, T., & Bhattacharyya, K. G. (2022). Investigation of groundwater and soil quality near to a municipal waste disposal site in Silchar, Assam, India. *International Journal of Energy and Water Resources*, *6*(1), 37–47. https://doi.org/10.1007/s42108-021-00117-5
36. Choudhury, M., Sharma, A., Pervez, A., Upadhyay, P., & Dutta, J. (2022). Growing menace of microplastics in and around the coastal ecosystem. In *Coastal Ecosystems: Environmental Importance, Current Challenges and Conservation Measures*, pp. 117–137. https://doi.org/10.1007/978-3-030-84255-0_6

CHAPTER 20

Current Trends in Waste Management: Case Studies from the Gulf Countries

ATIN KUMAR[1*], NITISH KARN[1], NANDINI ARYA[1], MOHAMED A. E. ABDEL RAHMAN[2], and ANTONIO SCOPA[3]

[1]*School of Agriculture, Uttaranchal University, Dehradun, Uttarakhand, India*

[2]*Division of Environmental Studies and Land Use, National Authority for Remote Sensing and Space Sciences (NARSS), Cairo, Egypt*

[3]*Scuola di Scienze Agrarie, Forestali, Alimentari ed Ambientali (SAFE) Universitàdegli Studi della Basilicata Viale dell'Ateneo Lucano, Potenza, Italy*

*Corresponding author

ABSTRACT

Waste management is a critical component of environmental protection, public health, resource conservation, and sustainable development. By adopting comprehensive waste management practices that include recycling, waste reduction, and responsible disposal, societies can work toward creating cleaner, healthier, and sustainable environments for current and future generations. This chapter provides a comprehensive analysis of waste management practices in Gulf countries, focusing on case studies from the region to examine trends, challenges, and strategies employed. The study highlights the importance of waste management in promoting environmental sustainability, public health, and resource conservation in the Gulf region. While Gulf countries have made progress in implementing advanced waste management technologies and infrastructure, disparities

Data Analytics and Smart Technologies in Solid Waste Management: Advancing Sustainability.
Moharana Choudhury, Sushobhan Majumdar, and Avnish Chauhan (Eds.)

DOI: 10.1201/9781779640833-24

exist in waste management practices and infrastructure across the region. Challenges such as rapid urbanization, inadequate infrastructure, and limited recycling facilities are identified as key obstacles to effective waste management. The chapter recommends adopting circular economy models, enhancing public awareness, investing in infrastructure, and implementing regulatory frameworks to drive compliance and best practices. By utilizing insights gained from lessons learned, best practices, and innovative solutions, Gulf countries have the opportunity to improve their waste management practices, decrease environmental effects, and progress toward a more sustainable and circular economy approach to waste management.

20.1 INTRODUCTION

Waste management is a multifaceted discipline that encompasses a range of activities aimed at efficiently handling, treating, and disposing of waste materials to minimize environmental impact and promote sustainable practices. The management of waste is crucial for maintaining ecological balance, preserving public health, and conserving natural resources. One of the primary reasons waste management is of paramount importance is its direct impact on environmental sustainability [18]. Inadequate waste disposal can result in the contamination of air, water, and soil, presenting substantial hazards to ecosystems and biodiversity [8]. Through the implementation of efficient waste management techniques like recycling, composting, and waste-to-energy (WTE) technologies, communities can diminish the volume of waste directed to landfills and incinerators [20]. This approach helps alleviate pollution and decrease greenhouse gas emissions [3].

Furthermore, waste management plays a vital role in protecting public health. Inadequate waste disposal practices can create breeding grounds for disease vectors and contaminants, leading to the spread of infectious diseases and posing health risks to communities [10]. Proper waste management, including safe collection, transportation, and treatment of waste, helps prevent the proliferation of harmful pathogens and pollutants, safeguarding the well-being of individuals and communities [17]. Resource conservation is another key aspect of waste management. By recycling and reusing materials, waste management practices contribute to the conservation of natural resources and the reduction of energy consumption [1]. Recycling not only diverts waste from landfills but also minimizes the need for extracting raw materials, thereby promoting a more sustainable use of resources and reducing the environmental footprint of human activities [13].

Economically, effective waste management can generate opportunities for job creation, stimulate innovation in recycling industries, and reduce costs associated with waste disposal. By viewing waste as a potential resource rather than a burden, communities can develop circular economy models that promote the reuse and repurposing of materials, leading to economic benefits and sustainable growth [12].

In the context of Gulf countries, waste management is of particular significance due to factors unique to the region. Gulf countries, such as Qatar, the United Arab Emirates (UAE), Saudi Arabia, Kuwait, Bahrain, and Oman, face challenges related to rapid urbanization, economic growth, and harsh climate conditions [23]. The arid climate in the region poses specific challenges for waste management, as improper disposal practices can lead to environmental degradation and risks to public health [2].

The chapter "Waste Management Current Trend of Gulf Countries as Case Studies" aims to provide a detailed analysis of waste management practices in Gulf countries, focusing on specific case studies from the region. It seeks to examine the existing trends, challenges, and strategies employed by Gulf countries in managing their waste streams, highlighting both successful initiatives and areas requiring improvement [22]. The chapter's scope includes an overview of waste management practices in Gulf countries, encompassing waste collection, transportation, treatment, and disposal methods, as well as the regulatory frameworks and infrastructure supporting solid waste management. It will present case studies from selected Gulf countries to offer insights into their unique waste management approaches, challenges, and outcomes [14].

Overall, the chapter aims to offer a comprehensive examination of waste management practices in Gulf countries through case studies, with the goal of providing valuable insights to policymakers, practitioners, researchers, and stakeholders involved in environmental management and sustainability efforts in the Gulf region [6].

20.2 GENERAL OVERVIEW OF GULF COUNTRIES' WASTE MANAGEMENT

20.2.1 BRIEF INTRODUCTION TO THE GULF COOPERATION COUNCIL

Formed in 1981, the Gulf Cooperation Council (GCC) is a political and economic coalition comprising six Arab nations situated in the

Arabian Peninsula. The member states of the GCC encompass Bahrain, Kuwait, Oman, Qatar, Saudi Arabia, and the UAE. The primary objectives of the GCC are to enhance cooperation and integration among member states in various fields, including economics, politics, security, and culture. The GCC serves as a platform for member countries to collaborate on issues of mutual interest and work toward common goals. Economic cooperation within the GCC is facilitated through initiatives such as the GCC Common Market and the Gulf Customs Union, which aim to promote trade, investment, and economic development in the region [24].

20.2.2 GENERAL STATISTICS ON WASTE GENERATION IN THE GULF REGION

20.2.2.1 PER CAPITA WASTE GENERATION

Gulf nations demonstrate some of the highest per capita waste generation rates worldwide, estimated between 1.5 and 2.5 kg/individual/day. This elevated waste production is shaped by factors like heightened consumption levels, economic affluence, urbanized lifestyles, and a dependence on disposable goods.

20.2.2.2 TOTAL WASTE GENERATION

The considerable volume of waste produced in the Gulf region mirrors the expanding population and urbanization patterns in nations such as Saudi Arabia, the UAE, and Qatar. This escalating waste generation strains the current waste management infrastructure and underscores the need for the establishment of sustainable waste management practices.

20.2.2.3 COMPOSITION OF WASTE

In the Gulf region, waste composition commonly comprises a blend of municipal solid waste (MSW), construction and demolition (C&D) waste, industrial waste, and hazardous waste. MSW, encompassing household, commercial, and institutional waste, represents a substantial segment of the waste stream, emphasizing the necessity for efficient management and disposal tactics.

20.2.2.4 CHALLENGES IN WASTE MANAGEMENT

The Gulf countries face various challenges in managing waste effectively, including inadequate waste collection and disposal infrastructure, limited recycling facilities, improper waste segregation practices, and a lack of awareness about sustainable waste management practices. These challenges contribute to issues such as landfill overcapacity, environmental pollution, and resource depletion [4].

20.2.3 KEY CHALLENGES FACED IN WASTE MANAGEMENT

Waste management encounters various significant challenges on a global scale, including within the Gulf region. Some of the primary challenges in waste management comprise the following.

20.2.3.1 RAPID URBANIZATION

Rapid urbanization leads to increased waste generation due to population growth, changing consumption patterns, and urban lifestyles. Managing the growing volume of waste in urban areas poses challenges for waste collection, transportation, and disposal infrastructure.

20.2.3.2 INADEQUATE INFRASTRUCTURE

The inadequacy of waste management infrastructure, encompassing collection systems, recycling facilities, and disposal sites, obstructs the implementation of effective waste management practices. This deficiency in proper infrastructure can lead to issues such as illegal dumping, open burning, and environmental pollution.

20.2.3.3 LIMITED RECYCLING AND RESOURCE RECOVERY

The limited availability of recycling facilities and resource recovery programs contributes to high levels of waste disposal in landfills. Increasing recycling rates and promoting circular economy principles are essential for reducing waste and conserving resources.

20.2.3.4 WTE CHALLENGES

Implementing WTE technologies faces challenges such as high capital costs, technological barriers, and public acceptance issues. While WTE can help reduce landfill waste and generate energy, overcoming these challenges is essential for its effective implementation.

20.2.3.5 HAZARDOUS WASTE MANAGEMENT

The proper management and disposal of hazardous waste, which includes medical waste, electronic waste, and industrial chemicals and present notable challenges. Insufficient treatment of hazardous waste can result in severe environmental and health consequences.

20.2.3.6 CLIMATE CHANGE IMPACTS

Climate change worsens the difficulties in waste management by causing extreme weather events, increasing sea levels, and altering environmental conditions. Adjusting waste management strategies to alleviate the impacts of climate change and decrease greenhouse gas emissions is essential for sustainable waste management.

20.3 CASE STUDY 1: UAE

20.3.1 INTRODUCTION

The Emirate of Ajman has witnessed a significant increase in waste quantities attributed to population growth and industrial expansion. This has necessitated a strategic approach to waste management to address environmental concerns and promote sustainability [3].

20.3.2 CURRENT WASTE MANAGEMENT PRACTICES

Municipalities and contracted companies are responsible for collecting MSW, with permits issued based on estimated waste generation. The UAE aims for 75% waste recycling, with initiatives like waste segregation using color-coded bins being implemented in cities like Sharjah.

20.3.3 FUTURE DIRECTIONS

Addressing gaps in organic waste recycling infrastructure and enhancing community engagement in waste sorting are key priorities for future waste management strategies. Continued focus on innovative waste management solutions and sustainable practices will be crucial to meeting waste management targets and ensuring environmental sustainability in Ajman and the UAE [3].

20.4 CASE STUDY 2: SAUDI ARABIA

20.4.1 INTRODUCTION

Saudi Arabia, like many countries, faces challenges in managing C&D waste due to rapid urbanization and industrial growth. To address these issues and promote sustainability, innovative waste management strategies are being explored in the country [19].

20.4.2 KEY INITIATIVES

The establishment of WTE facilities can play a crucial role in converting waste into energy, reducing landfill volumes, and generating renewable energy. By utilizing technologies such as incineration or gasification, Saudi Arabia can address waste disposal challenges while contributing to energy sustainability. By utilizing Internet of Things technology and sensors, intelligent waste management systems can streamline waste collection routes, monitor bin fill levels in real-time, and improve overall waste management efficiency. These systems facilitate data-driven decision-making, resulting in more efficient waste collection and disposal practices. Implementing anaerobic digestion (AD) technology can help convert organic waste into biogas and nutrient-rich digestate.

20.5 CASE STUDY 3: QATAR

20.5.1 WASTE GENERATION AND COMPOSITION

Qatar generates around 1.8 kg of waste per person per day, with organic waste accounting for 60% of the total waste produced [16]. Despite the high proportion of degradable and recyclable waste (80%), the predominant waste disposal methods in Qatar are landfilling and composting, contributing to greenhouse gas emissions and climate change.

20.5.2 CURRENT WASTE MANAGEMENT PRACTICES

The recycling sector in Qatar is nascent, with less than 15% of waste being recycled, while over 50% is directed to landfills. Renewable energy sources, including biomass and waste (38%) and solar photovoltaic installations (5%), form a small fraction of Qatar's energy mix, indicating room for growth in utilizing waste for energy generation.

20.5.3 TECHNOLOGICAL INNOVATIONS IN WASTE TREATMENT

Qatar is exploring advanced waste treatment technologies, such as thermal processes like pyrolysis and gasification, to convert waste into value-added products like chars, liquids, and syngas. These technologies offer opportunities for fuel generation, power production, and soil bioremediation, showcasing the potential for sustainable waste management practices.

20.6 CASE STUDY 4: KUWAIT

20.6.1 BACKGROUND

Kuwait, a country known for its rapid development and construction projects, faces challenges in managing C&D waste effectively [5].

20.6.2 STUDY AREA AND SELECTED PROJECTS

A total of seven varied projects were chosen for assessment, encompassing ongoing government initiatives, a commercial project, and residential developments. Among the government projects were constructions for the College of Law and College of Social Sciences at Kuwait University. Additionally, a commercial hotel building and two residential complexes were included in the chapter. Two demolition sites in Kuwait City were also part of the evaluation [15].

20.6.3 RESEARCH DESIGN AND DATA COLLECTION

The research methodology included a blend of primary and secondary data collection techniques. Quantitative data on C&D waste generation were gathered through site visits, documentation, and interviews with relevant authorities. Secondary data from reports, journals, and internet sources complemented the primary data analysis. Data gathering encompassed face-to-face interviews, surveys, and on-site visits to C&D locations [9].

20.6.4 PROPOSED STRATEGIES

Based on the findings, the chapter proposed strategies to minimize C&D waste generation and improve waste management practices in Kuwait. Recommendations included enhancing coordination among stakeholders, implementing effective waste segregation practices, and promoting recycling and reuse of construction materials.

20.7 CASE STUDY 5: OMAN

20.7.1 INTRODUCTION

Oman, a country located in the Middle East, is facing increasing challenges in waste management due to rapid urbanization, population growth, and inadequate infrastructure. The management of waste has become a critical issue for environmental sustainability and public health in the country [18].

20.7.2 WASTE MANAGEMENT STRATEGIES AND INNOVATIONS

Oman is investigating cutting-edge WTE technologies to transform waste into usable energy, effectively tackling both waste management and energy generation requirements. The production of hydrogen gas from landfill leachate using advanced photo-fermentation processes represents a promising avenue for sustainable energy production and waste management in Oman. Embracing the principles of the main 4Rs (rethink, reduce, reuse, and recycle) and implementing measures such as banning nonbiodegradable plastic bags are recommended for enhancing waste management practices in the country [21].

20.8 CASE STUDY 6: BAHRAIN

20.8.1 INTRODUCTION

The generation of MSW poses a notable challenge in developing nations like Bahrain, where a large portion of the waste is disposed of in landfills, leading to environmental and public health hazards [11]. The research endeavors to investigate the viability of AD technology for handling biodegradable waste in Bahrain, a country that produces approximately 2 million tons per year of MSW [1].

20.8.2 CHALLENGES IN WASTE MANAGEMENT

The organic fraction of MSW in Bahrain is high, reaching 60% by weight, similar to other middle- and low-income cities. Improper management of organic waste can lead to increased greenhouse gas emissions, particularly methane, which is 25 times more potent than carbon dioxide. Landfill space in Bahrain is limited, and the existing landfill has already reached its capacity, necessitating the exploration of alternative waste management solutions.

20.9 IDENTIFICATION OF COMMON CHALLENGES AND SUCCESS FACTORS

The rapid urbanization and population growth in the Gulf countries have led to increased waste generation, straining existing waste management infrastructure and services. Many Gulf countries face challenges in developing comprehensive recycling infrastructure, resulting in a high percentage of recyclable materials being disposed of in landfills. Proper handling, treatment, and disposal of hazardous waste, including medical, electronic, and industrial waste, require specialized facilities and expertise, posing challenges in the region.

Continued investment in waste management infrastructure, including modern landfills, recycling facilities, composting plants, and WTE projects, is crucial for improving waste management practices and capacity.

20.10 FUTURE TRENDS AND INNOVATIONS

20.10.1 EMERGING TRENDS IN WASTE MANAGEMENT IN THE GULF REGION

The emerging trends in waste management in the Gulf region are crucial for addressing the increasing challenges of waste generation and environmental impact. These trends include the implementation of advanced waste management technologies, such as smart waste monitoring systems and Internet of Things solutions. These technologies enable real-time monitoring of waste levels, optimize collection routes, and improve overall efficiency in waste management.

20.10.1.1 INCREASED RESEARCH ACTIVITY

Minimal research activity on solid waste in the GCC region, especially using a multidisciplinary approach, indicates a growing recognition of the importance of research in addressing waste management challenges. This increased research activity will contribute to the development of innovative waste management strategies in the Gulf region.

20.10.1.2 FOCUS ON SUSTAINABLE DEVELOPMENT GOALS

Many waste management systems in the Gulf region are now aligning their efforts with the Sustainable Development Goals (SDGs) set by the United Nations. These goals aim to address various aspects of sustainable development, including waste management, and promote sustainable practices in the region. Some specific SDGs that are relevant to waste management include SDG 11, which focuses on ensuring access to safe and affordable housing and sustainable cities, and SDG 12, which emphasizes responsible consumption and production [7].

20.10.1.3 INNOVATIVE WASTE UTILIZATION

Innovative waste utilization refers to finding creative and sustainable ways to utilize waste as a valuable resource. These strategies involve the implementation of WTE projects, where waste is used as a fuel source to generate electricity or heat [18]. Innovative waste utilization strategies have the potential to not only reduce the amount of waste sent to landfills but also generate renewable energy and valuable resources.

20.10.1.4 CONSTRUCTION WASTE MANAGEMENT

Effective management of construction waste is a critical component of sustainable waste practices in the Gulf countries. Given that the construction sector generates a substantial volume of waste materials like concrete, wood, and metals, it is imperative to handle and dispose of this waste properly to mitigate its adverse environmental impact. This entails implementing strategies such as recycling and reusing construction materials, ensuring proper waste segregation at construction sites, and advocating for sustainable construction methods.

20.10.2 TECHNOLOGICAL INNOVATIONS AND THEIR POTENTIAL IMPACT

20.10.2.1 ADVANCED WASTE TREATMENT TECHNOLOGIES

Technologies like pyrolysis and gasification offer the potential to convert waste into value-added products such as chars, liquids, and syngas. These processes can help in efficient waste disposal, energy recovery, and the production of useful by-products, contributing to a circular economy approach.

20.10.2.2 THERMAL PROCESSES FOR ENERGY RECOVERY

Thermal processes, including incineration and gasification, can be utilized for energy recovery from waste. By converting waste into heat or electricity, these technologies can help reduce the reliance on fossil fuels, lower greenhouse gas emissions, and promote renewable energy generation.

20.10.2.3 WATER TREATMENT APPLICATIONS

By utilizing waste by-products like bottom ash and fly ash from MSW incineration for water treatment applications, technological innovations can contribute to sustainable water management practices. These by-products can be used for filtration, adsorption, and other water treatment processes, enhancing water quality and resource efficiency.

20.10.2.4 RENEWABLE ENERGY GENERATION

WTE technologies, such as AD and landfill gas recovery, can enable the generation of renewable energy from organic waste. This not only helps in reducing waste volume but also contributes to the diversification of energy sources and the mitigation of climate change through the reduction of methane emissions from landfills.

20.11 CONCLUSION

Examining waste management practices in Gulf countries through case studies highlights the ongoing efforts and challenges in managing waste streams effectively. The region has made significant strides in implementing innovative solutions, such as WTE facilities, recycling programs, and sustainable waste management infrastructure. However, areas still require improvement, including enhancing public awareness, strengthening regulatory frameworks, and investing in advanced technologies.

Gulf countries can further enhance their waste management strategies by embracing circular economy concepts, promoting the 4Rs, and integrating cost-effective solutions. Policymakers, practitioners, and stakeholders must collaborate in developing comprehensive waste management plans that prioritize environmental conservation, resource efficiency, and sustainable development.

Continuing focus on sustainable practices, community engagement, and technological advancements will be essential in achieving effective waste management practices in Gulf countries. By learning from successful initiatives, addressing existing challenges, and implementing tailored solutions, the region can pave the way for a cleaner, greener, and more sustainable future for future generations.

KEYWORDS

- **waste management**
- **sustainability**
- **recycling**
- **circular economy**
- **urbanization**
- **environmental impact**

REFERENCES

1. Abbas, S. Y. (2020). Feasibility of anaerobic digestion as an option for biodegradable waste management in the Kingdom of Bahrain. *Journal of Research in Environmental Science and Toxicology*, *9*(1), 16–21. https://doi.org/10.14303/jrest.2020.04
2. Abdallah, W. M. (2008). *Accounting, Finance, and Taxation in the Gulf Countries*. In Palgrave Macmillan US eBooks. https://doi.org/10.1057/9780230614543
3. Al-Dabbagh, R. (2021). Waste management strategy and development in Ajman, UAE. *Renewable Energy and Environmental Sustainability*, *6*, 14. https://doi.org/10.1051/rees/2021005
4. Alkhouzaam, A., Rebeeh, Y., Elhafez, O., & Elhadi, E. (2022). Food and water security in Qatar: Current challenges caused by the blockade and proposed solutions. In *Advances in Science, Technology & Innovation*, pp. 773–780. https://doi.org/10.1007/978-3-030-76081-6_98
5. Almusawi, M. B. H., Karim, A. T. B. A., & Ethaib, S. (2022). Evaluation of construction and demolition waste management in Kuwait. *Recycling*, *7*(6), 88. https://doi.org/10.3390/recycling7060088
6. Alsulaili, A., Alshamali, F., & Aldabbous, A. (2024). Toward sustainable waste management: Estimating emissions from alternative practices in Gulf Cooperation

Council countries. *International Journal of Environmental Science and Technology*, 21, 7165–7176.

7. Al-Rumaihi, A., McKay, G., Mackey, H. R., & Al-Ansari, T. (2020). Environmental impact assessment of food waste management using two composting techniques. *Sustainability (Switzerland)*, *12*(4), 1595. https://doi.org/10.3390/su12041595
8. Choudhury, M., Jyethi, D. S., Dutta, J., Purkayastha, S. P., Deb, D., Das, R., Roy, G., Sen, T., & Bhattacharyya, K. G. (2022). Investigation of groundwater and soil quality near to a municipal waste disposal site in Silchar, Assam, India. *International Journal of Energy and Water Resources*, *6*(1), 37–47. https://doi.org/10.1007/s42108-021-00117-5
9. Chowdary, A.V., Saini, N., Kumar, A., Kumar, S., Ballabh, J., Bhatt, S.S., Bhatt, A., Prakash, S., & Patel, A. (2023). Influence of integrated nutrient management on physiological parameters of lentil (Lens culinaris Medik.). *Plant Science Today*, *10*(3), 94–97. https://doi.org/10.14719/pst.2124
10. Clarke, S. F., & Almannai, S. (2022). *Chapter 15 Sustainable Waste Management in Qatar Charting the Emergence of an Integrated Approach to Solid Waste Management*. In Berghahn Books, pp. 367–390. https://doi.org/10.1515/9781782383727-018
11. Fibiger, T. (2022). *Stability or Democracy? The Failed Uprising in Bahrain and the Battle for the International Agenda*. In Berghahn Books, pp. 81–94. https://doi.org/10.1515/9781782384663-006
12. Ibrahim, O., Al-Kindi, G., Qureshi, M. U., & Maghawry, S. Al. (2022). Challenges and construction applications of solid waste management in middle East Arab Countries. In *Processes*. MDPI, Vol. 10, Issue 11. https://doi.org/10.3390/pr10112289
13. Kumar, A., Das, N., Sharma, H., & Singh, R. (2024). Impact of Agrochemicals on the Environment and Ecological Alternatives. In Karmaoui, A. (Ed.), *Water-Soil-Plant-Animal Nexus in the Era of Climate Change*. IGI Global, pp. 317–328. https://doi.org/10.4018/978-1-6684-9838-5.ch015
14. Kumar, A., Kumar, S., Juyal, R., Sharma, H., & Bisht, M. (2023). Nanoscience in agricultural steadiness. In Dar, G. H., Bhat, R. A., & Mehmood, M. A. (Eds.), *Microbiomes for the Management of Agricultural Sustainability*. Springer: Cham, pp. 285–296. https://doi.org/10.1007/978-3-031-32967-8_17
15. Kumar, R., Sachan, S., Kumar, R., Parmar, K., Mishra, A. K., Kumar, A., & Mishra, A. K. (2023). Unveiling use of Information and Communication Technology (ICT) in Western Uttar Pradesh, India: The Farmers' perspective. *Journal of Global Innovations in Agricultural Sciences*, *11*, 521–531. https://doi.org/10.22194/JGIAS/23.1197
16. Mariyam, S., Cochrane, L., Zuhara, S., & McKay, G. (2022). Waste management in Qatar: A systematic literature review and recommendations for system strengthening. In *Sustainability (Switzerland)*. MDPI, Vol. 14, Issue 15. https://doi.org/10.3390/su14158991
17. Mondal, T., Choudhury, M., Kundu, D., Dutta, D., & Samanta, P. (2023). Landfill: An eclectic review on structure, reactions and remediation approach. *Waste Management*, *164*, 127–142. https://doi.org/10.1016/j.wasman.2023.03.034
18. Okedu, K. E., Barghash, H. F., & Al Nadabi, H. A. (2022). Sustainable waste management strategies for effective energy utilization in Oman: A review. *Frontiers in Bioengineering and Biotechnology*, *10*, 825728. https://doi.org/10.3389/fbioe.2022.825728

19. Ouda, O. K. M., Peterson, H. P., Rehan, M., Sadef, Y., Alghazo, J. M., & Nizami, A. S. (2018). A case study of sustainable construction waste management in Saudi Arabia. *Waste and Biomass Valorization*, *9*(12), 2541–2555. https://doi.org/10.1007/s12649-017-0174-9
20. Paul, S., Choudhury, M., Deb, U., Pegu, R., Das, S., & Bhattacharya, S. S. (2019). Assessing the ecological impacts of ageing on hazard potential of solid waste landfills: A green approach through vermitechnology. *Journal of Cleaner Production*, *236*, 117643. https://doi.org/10.1016/j.jclepro.2019.117643
21. Roy, P., & Choudhury, M. (2022). Bioeconomy and green plastic production. In Baskar, C., Ramakrishna, S., & Daniela La Rosa, A. (Eds.), *Encyclopedia of Green Materials*. Springer: Singapore. https://doi.org/10.1007/978-981-16-4921-9_189-1
22. Sharma, H., Kumar, A., Singh, R., & Kumar, S. (2023). Impact of nanopriming and omics-based applications for sustainable agriculture. In Singh, A., Rajput, V., Ghazaryan, K., Gupta, S., & Minkina, T. (Eds.), *Nanopriming Approach to Sustainable Agriculture*. IGI Global, pp. 175–191. https://doi.org/10.4018/978-1-6684-7232-3.ch008
23. Waxin, M., Kumra, S., & Zhao, F. (2020). Workforce localization in the Arab Gulf countries: How do organizations socialize the members of a powerful minority? *Sustainability*, *12*(14), 5847. https://doi.org/10.3390/su12145847
24. Yousaf, S., Ullah, M. I., Sheikh, M. R., & Saboor, A. (2023). Nexus of remittances, poverty and economic growth in Gulf Cooperation (GCC) countries. *iRASD Journal of Economics*, *5*(4), 1087–1095. https://doi.org/10.52131/joe.2023.0504.0181

Index

A

Advanced landfill gas recovery systems, 205
Aedes aegypti, 291
Aerobic digestion, for organic waste, 111
Agricultural waste, 181–182
Air quality degradation, inadequate waste management and, 43–44
Albopictus species, 291
Alcaligenes faecalis, 261
Anaerobic digestion (AD), 16, 110, 151
  future recommendation, 159–160
  of organic waste, 157–158
  sustainable organic waste management strategies, role in, 158–159
Ant Colony Systems (ACS), 219
Anunobi, 291
Artificial intelligence (AI), 218
Artificial neural networks (ANNs), 219
Aspergillus clavatus, 261
Aspergillus flavus, 261
Aspergillus nomius, 261
Aspergillus terreus, 261
Automated recycling gadgets, 237

B

Backtracking Search Algorithms (BSA), 219
Beeswax, 255
Biochemical conversion, 15–16
  anaerobic digestion, 16
  bioethanol production, 16
Biochemical oxygen demand (BOD), 216
Biodiversity
  loss of, inadequate waste management and, 44–45
Bioethanol production, 16
Biomedical waste, 126
Bioremediation, of plastic pollution, 259–261, 260*f*
  by bacteria, 261, 262*t*
  by fungi, 261, 262*t*
  by worms, 262, 262*t*
Blockchain technology, on solid-waste management, 56–59, 219
  carbon credits and emission reduction, 58–59
  data security and privacy, 58
  decentralized waste exchanges, 57
  incentive programs and tokenization, 57
  international collaboration and standards, 58
  public awareness and education, 58
  smart contracts for compliance, 57
  supply chain traceability, 57
  verification of waste management practices, 58
  waste characterization and recycling tracking, 57
Busty garbage runs, 238

C

Catalytic combustion, 184*t*
Central Pollution Control Board (CPCB), 181
Chemical oxygen demand (COD), 216
CHP. *See* Combined heat and power
Circular economy transition, 79
Climate change, and waste management
  capacity building, 282
  circular economy principles, promote, 281
  community engagement, 281
  feedback loops, 275–276
    potential for waste management to mitigate climate change, 275–276
    waste management to GHG emissions, contribution of, 275
  impacts of, 274
  implications for waste collection, treatment, and disposal, 274–275

infrastructure resilience, 281
innovation, 282
integration of waste management into climate change adaptation strategies, 281
intelligent framework for managing and measuring, 278*f*
managing waste with context to, 276
overgeneration of solid waste, 273
overview, 272–273
Paris Agreement of 2015, 272
policy interventions to enhance resilience in waste management, 279–281
investing in research and innovation, 280
promoting public–private partnerships, 280–281
strengthening regulatory frameworks, 279–280
public behavior, understanding, 282
recommendations for future research and action, 282
risk assessment, 281
smart waste management system, 282
technology in, 277–279
"discovery analytics," 278
"predictive analytics," 279
real-time data collection, 277–278
waste composition and generation patterns, changes in, 274
Climate change, waste management
extreme weather events, increased frequency and intensity of, 274
Clostridia perfringens, 216
Clostridium botulinum, 261
Coleoptera, 262
Combined heat and power (CHP), 190
Composting (CP), 151
future recommendation, 159–160
of organic waste, 157–158
sustainable organic waste management strategies, role in, 158–159
Construction and demolition wastes, 123, 167
Contaminated landfills, 141
CPCB. *See* Central Pollution Control Board

D

Data analytics
integration of IoT with, 239
smart waste management in smart city ecosystem, 240–241
in waste sorting and recycling, 203
Data-driven solutions, for waste management, 219–220
artificial neural networks, 219
challenges and barriers to adoption, 220–221
data privacy and security concern, 220
financial constraints, 221
infrastructure and technology requirement, 221
resistance to change from traditional practices, 221
future directions and opportunities, 224–225
emerging trends, 224
potential applications of advanced technologies, 224–225
research priorities for enhancing sustainability, 225
GIS-based smart collection systems, 219
optimizing waste collection routes, 219–220
policy implications and recommendations, 221–224
collaboration between government, industry, and academia, 223–224
education and outreach programs to promote awareness, 223
incentives to adopt innovative practices, 223
regulatory framework to support, 222
real-time monitoring systems, 220
Decentralized waste exchanges, 57
Deposal facilities, managed, 143–144
Desulfotomaculum nigrificans, 261
Disposable plates, 256

E

Eco-friendly agriculture practices, 256–257
Eisenia Andrei, 262
Eisenia fetida, 262

Electronic waste (e-waste), 126
in developing countries (case study), 45–46
in Western Europe, 79
Electrooxidation, 258–259
Energy from waste (EfW). *See also* Waste-to-energy
mechanisms, 185, 186*f*
scope of, 190–191
Engineered landfills, 144
Environmental impacts, of inadequate waste management
air quality degradation, 43–44
loss of biodiversity, 44–45
soil contamination, 41–42
water pollution, 42–43
Environmental Protection Act, 187
E-scrap booths, 238–239
Extended producer responsibility (EPR) schemes, 206
Extracellular polymeric substances (EPS), 257

F

Fixed-bed gasifiers, 185
Fixed grate/direct flame, 184*t*
Fluidized bed incinerators, 183*t*
Fluorescent bulbs and lamps, 167
Food waste
recycling in Seoul, 113–114
in Western Europe, 79
Friendly clothing, 255

G

Galleria mellonella, 145, 262
Garbage disposal or dumping. *See also* Landfills
limiting release of gases, 144–145
minimizing, 144–145
Garbage van loading system, 237–238
Garbage volume detector, 237
Gasification, 14–15, 16*f,* 185
GCC. *See* Gulf Cooperation Council
Global solid-waste management, 53–54
Global trash generation, 122
Global Waste Index, 123, 124*t*
Global waste production, 150, 151, 152*t*
GPS-enabled routing systems, 204–205
Gravity separation, 127
Greenhouse gas (GHG), 172
contributed to global warming, 277*f*
emission, 272*f,* 275
Gulf Cooperation Council (GCC), 319–320
Gulf countries, waste management in, 318–330
case studies
Bahrain, 326–327
Kuwait, 324–325
Oman, 325–326
Qatar, 324
Saudi Arabia, 323
United Arab Emirates, 322–323
challenges faced in waste management, 321–322
climate change impacts, 322
hazardous waste management, 322
inadequate infrastructure, 321
limited recycling and resource recovery, 321
rapid urbanization, 321
WtE challenges, 322
common challenges and success factors, identification of, 326–327
emerging trends in waste management, 327–328
construction waste management, 328
increased research activity, 327
innovative waste utilization, 328
Sustainable Development Goals, focus on, 327–328
Gulf Cooperation Council, 319–320
overview, 318–319
statistics on waste generation, 320–321
challenges in waste management, 321
composition of waste, 320
per capita waste generation, 320
total waste generation, 320
technological innovations
advanced waste treatment technologies, 328
renewable energy generation, 329

thermal processes for energy recovery, 329
water treatment applications, 329

H

Habitat degradation, 44–45
Hazardous and special waste, in Latin America, 166–175
environment and human health, effects on, 170–171
factors affect production of, 170
management of
challenges in, 172
current strategies and policies for, 171–172
sustainable and long-term solutions, 173
overview, 166–168
recommendations for minimization of production of, 173–174
waste crisis in develops inequality, 168–169
Hazardous wastes
environmental risks, 189
household, 167
human health risks, 189
HCW. *See* Healthcare facilities
Healthcare facilities (HCW), 296
Household hazardous wastes, 167

I

Ideonella sakaiensis, 261
Illegal dumping detection
precision waste collection for, 111
satellite imaging for, 111
Inadequate or poor waste management, 38–47
changing consumption patterns, 39
comprehensive solutions for, 46–47
electronic waste in developing countries (case study), 45–46
environmental clean-up liabilities, 296
environmental impacts of, 122
air quality degradation, 43–44
loss of biodiversity, 44–45
soil contamination, 41–42
water pollution, 42–43
impact on biodiversity, 291–292
negative effects on environment and human health, technological remedies for, 296–298
in Oceania, 310
overview, 38
rapid urbanization and industrial growth, 38–39
spread of diseases, 291
in Sub-Saharan Africa (case study), 292–295
Ethiopia, 294
Ghana, 294
Kenya, 294
Nigeria, 294–295
South Africa, 294, 295
Zimbabwe, 295
unseen consequences, 40–41
visible consequences of, 40
waste associated with healthcare, types of, 298*f*
Incineration, 12–14, 151, 183
advantages of, 13
controlling emissions, 188–190
bag filters, 188
catalytic reduction, 190
cyclones, 188
dry sorbent injection, 190
electrostatic precipitators, 189, 189*f*
scrubbers, 189–190
disadvantages of, 14
emission standards and regulations, 187–188
environmental implications of, 186
municipal solid waste, 185
pollutants caused due to, 187*t*
process, 13*f*
reduces waste volume and toxicity, 216
types of, 183*t*–184*t*
India
municipal solid waste in, 108
"Swachh Bharat Mission," 25–26, 80
waste management, 25–26
Western Europe, 79–80
Industrial growth, environmental impacts of, 38–39

Industrial Revolution 4.0 (IR 4.0), 81
Industrial waste, 123, 167, 181
Informal waste sector, 131
Internet of Things (IoT), 81, 218
IoT. *See* Internet of Things
IOT-enabled waste bins, in solid-waste management, 59–61
  cost efficiency, 60
  environmental impact reduction, 60
  integration with smart city initiatives, 60
  optimized collection routes, 59
  predictive maintenance, 59
  real-time monitoring and data collection, 59
  security and privacy considerations, 60–61
  user engagement and education, 60
IR 4.0. *See* Industrial Revolution 4.0

L

Landfills, 29. *See also* Garbage disposal or dumping
  controlled, 143–144
  conventional, 9
  design and regulations of, 143–144
    managed deposal facilities, 143–144
    open garbage pits, 143
  engineered, 144
  gas, power from, 205
  gases releases, 151
  industrial, 141
  leachate, 153
  managing, 140–145
  organic waste breakdown in, 290*f*
  overload, 201
  sanitary, 9, 10*f*, 108
  sustainable, 144
Latin America
  hazardous and special waste in, 166–175
    environment and human health, effects on, 170–171
    factors affect production of, 170
    management of
      challenges in, 172
      current strategies and policies for, 171–172
      sustainable and long-term solutions, 173
    overview, 166–168
    recommendations for minimization of production of, 173–174
  inadequate garbage collection and sanitation, 169
  types of waste generated in, 169*f*
  urban poor, 169
  waste composition in, 168*f*
  waste crisis in develops inequality, 168–169
Leachate, 153
Lepidoptera, 262
Liquid injection incinerators, 184*t*
Long-range atmospheric transport mechanism (LRATM), 167
Loss of biodiversity, inadequate waste management and, 44–45
LRATM. *See* Long-range atmospheric transport mechanism
Lumbricus terrestris, 262

M

Machine learning, in waste sorting and recycling, 203
Macroplastics, 252–253
Managed deposal facilities, 143–144
Manual sorting, 127
Material recovery, in solid waste management, 108–116
  challenges and considerations, 114
  community engagement, 114–115
  contamination issues in, 114
  current challenges in, 116
  emerging technologies in, 110–111
    aerobic digestion for organic waste, 111
    blockchain for waste traceability, 110–111
    drones and remote sensing, 110
    precision waste collection for illegal dumping detection, 111
    robotic waste sorting, 111
    satellite imaging for illegal dumping detection, 111

future trends in solid waste management, 115
overview, 108–109
policy and regulatory frameworks, 114
public awareness, 114–115
successful approaches to (case studies), 112–114
effective source separation in Taipei, 113
innovative food waste recycling in Seoul, 113–114
Oslo's advanced waste-to-energy facilities, 113
San Francisco pioneering zero-waste objectives, 112
technological barriers in, 114
traditional techniques for, 110
unblocking resources
importance of, 109
integration of digital solutions in, 111–112
Mechanical sorting, 127
Medical or infectious wastes, 167
Membrane bioreactors, 257, 258*f*
Microplastics, 253
Mining wastes, 167
Moving grate incinerators, 183*t*
MSW. *See* Municipal solid waste
MSWI. *See* Municipal solid waste incineration
Multiple hearth incinerators, 184*t*
Municipal solid waste incineration (MSWI), 185
Municipal solid waste (MSW), 22–23, 123, 181
collection coverage for cities in Africa, 292*f*
composition in sub-Saharan Africa, 297*f*
global production of, 154
in India, 108
organic matter in, 110
produced by Western Europe, 77
in Southeast Asia
composition of, 95*t*
disposal methods, 95*t*
historical and projected, 93*t*
management in Thailand, 100*f*
waste generation and disposal, disparities in, 99

N

Nonhazardous waste, 289

O

Oceania subcontinent, waste management in, 304–313
case studies, 306–308
Australia, 308
Fiji, 307
Papua New Guinea, 307
Samoa, 307–308
study site, 306*f*
environmental and social challenges, 310–311
improper waste disposal, 310
policy-making, sustainable development, and environmental conservation efforts, 311
social and economic implications of current practices, 310–311
future prospect, 312
implementing technological advancements, 309–310
literature review, 305–306
overview, 304–305
smart waste solutions and WtE initiatives, 309
sustainable waste management in, opportunities for, 311–312
technological innovations, 308–309
Oil and petroleum-related wastes, 167
Open dumping, 153
Open garbage pits, 143
Organic waste (OW), 150
anaerobic digestion of, 157–159
breakdown in landfills, 290*f*
composting of, 157–159
environmental imapcts of, 152–153
management in global and Indian contexts, 153–155
power from, 205
Organosilane, 258
Overflowing bins, 235

P

Pacific Regional Waste and pollution management program, 311
Paris Agreement of 2015, 272
Pesticides and agrochemical wastes, 167
PET, See Polyethylene terephthalate
Philippines
Ecological Solid Waste Management Act, 101
managing plastic waste, struggles with, 94
Plastic pollution
in developing regions, strategies to limit, 263–264
future perspective, 264
impact of, 254, 255*f*
nanomaterial-based solutions, 262–263
sources of, 254
strategies to deal with, 254–262
finding alternatives of plastic, 255–257
preventives strategies, 254
technological strategy, 257–262
in Western Europe, 79
Plastics, 252
in African countries, 297*f*
alternatives of, 255–257, 256*f*
beeswax, 255
disposable plates, 256
eco-friendly agriculture practices, 256–257
friendly clothing, 255
wooden combs and toothbrushes, 255
macroplastics, 252–253
microplastics, 253
single-use, 170
Plastic waste reduction, 52
Pleurotus ostreatus, 113, 261
Polyethylene terephthalate (PET), 261
Precision waste collection, for illegal dumping detection, 111
Pseudomonas alcaligenes, 261
Pseudomonas stutzeri, 261
Public–private partnerships, 67, 280–281
Pyrolysis, 14, 184
process of, 14, 15*f*
steps involved, 15

R

Radioactive wastes, 167
Radio-frequency identification (RFID) technology, 204
Rapid urbanization
inadequate waste management, 38–39
waste generation in Southeast Asia, 93
waste production, increase in, 180
Recycled waste sites, 144
Recycling platforms, 238–239
Recycling (waste)
data analytics and machine learning in, 203
data-driven solutions for, 219–220
food waste (in Seoul), 113–114
tracking, 57
Renewable energy generation, 329
Roadmap for Sustainable Waste Management and Resource Circulation in South Asia, 2019-2030, 27
Robotic waste sorting, 111
Rotary Kiln incinerators, 183*t*
Route optimization algorithms, in solid-waste management, 55–56

S

SAARC (South Asian Association for Regional Cooperation), 31
Samoa Trash Authority (SWA), 307
Sanitary landfill, 10*f*
Satellite imaging, for illegal dumping detection, 111
Single-use plastics, 170
Smart bins and collection systems, 204–205
GPS-enabled routing systems, 204–205
radio-frequency identification technology for waste tracking, 204
Smart city ecosystem, smart waste management in, 239–241
data analytics and reporting, 240–241
energy recovery, 240
IoT sensors, 240
route optimization, 240
waste segregation and recycling, 240
Smart garbage box, 236–237

Smart waste management
case studies and success stories, 241–243
Barcelona's waste management system, 241–242
Dubai's waste-to-energy plant, 243
Indore's waste management system, 243
San Francisco's zero waste initiative, 242
smart city of Singapore, 241
Surat's integrated solid waste management system, 243
waste-to-energy plants in Sweden, 242
challenges and limitations, 245–246
installation of smart sensor, 245
misconceptions of smart sensors operation, 245
recycling, 245
uneven waste leveling in waste bins, 246
unorganized truck routes, 245
emerging technologies, 243–245
advanced waste-to-energy technologies, 244
AI and machine learning, 244
blockchain technology for waste traceability, 244
enhanced waste collection and tracking, 244–245
IoT sensors and smart bins, 244
robotics and automation, 244
future trends, 243
integration of IoT and data analytics, 239
overview, 234
in smart city ecosystem, 239–241
data analytics and reporting, 240–241
energy recovery, 240
IoT sensors, 240
route optimization, 240
waste segregation and recycling, 240
technologies, 236–239
automated recycling gadgets, 237
busty garbage runs, 238
e-scrap booths and recycling platforms, 238–239
garbage van loading system, 237–238
garbage volume detector, 237
smart garbage box, 236–237
solar-powered junk compactors, 238
Soil contamination, inadequate waste management and, 41–42
Solar-powered junk compactors, 238
Solid organic waste (SOW), 150–151
anaerobic digestion, 157
overproduction of, 159
Solid waste
agricultural waste, 181–182
biomedical waste, 126
categories of, 5–8
based on sources, 6–7, 7*t*
based on types, 7–8, 8*t*
classification of, 181–182
composition, 125*f*
construction and demolition waste, 123
electronic waste, 126
factors influencing, 5
global treatment and disposal of, 126*f*
industrial waste, 123, 181
initial steps in management of, 6*f*
municipal. *See* Municipal solid waste
overview, 4–5
sustainable management approach, 9–12
source segregation, 9–10
three Rs (reduce, reuse, and recycle), 11–12
waste-to-energy conversion, 12–16
biochemical conversion, 15–16
gasification, 14–15
incineration, 12–14
pyrolysis, 14
Solid Waste Management Rules in 2016, 155
Solid-waste management (SWM), 52–68, 76
blockchain technology on, 56–59
carbon credits and emission reduction, 58–59
data security and privacy, 58
decentralized waste exchanges, 57

incentive programs and tokenization, 57
international collaboration and standards, 58
public awareness and education, 58
smart contracts for compliance, 57
supply chain traceability, 57
verification of waste management practices, 58
waste characterization and recycling tracking, 57
current challenges in, 62–64
impacts of inadequate waste management, 63–64
waste generation and composition, global trends in, 62–63
future directions and challenges, 64–67
addressing barriers to adoption of progressive practices, 66–67
emerging technologies and innovations, 66
trends in solid-waste generation and management, 65
global, 53–54, 122
governance structures and policy frameworks in Western Europe, 81–85
international collaboration in, 52, 53
IOT-enabled waste bins, 59–61
cost efficiency, 60
environmental impact reduction, 60
integration with smart city initiatives, 60
optimized collection routes, 59
predictive maintenance, 59
real-time monitoring and data collection, 59
security and privacy considerations, 60–61
user engagement and education, 60
material recovery in, 108–116. *See also* Material recovery, in solid waste management
overview, 52
predictive analytics for waste generation, 61
route optimization algorithms in, 55–56
waste composition analysis, 54–55
waste-to-energy solutions, 62
Source-based waste segregation, benefits of, 127–130
behavioral change, 130
community engagement, 130
economic benefits, 129–130
cost savings, 129
job creation, 129–130
revenue generation, 129
environmental benefits, 128–129
energy recovery, 129
pollution prevention, 128–129
resource conservation, 128
waste reduction, 128
public health improvement, 130
social benefits, 130
Source segregation, 9–10, 126
South Asia, waste management in, 22–32
challenges in, identifying, 26–27
description of study area, 23
effective implementation, suggestions for, 28–31
artificial intelligence and cyber-governance, innovations in, 31
environmental awareness, 28
global and regional cooperation, promote, 31
government intervention, 29–30
organizational and personnel capabilities, enhance, 30
strengthening cooperation and coordination, 29–30
technical advice, 28–29
technology system, establish, 30
effective policies to promote, 25–26
in India, 25–26
overview, 22–23
peace building innovations for, 31
processes for, 23–24
sustainable waste management
challenges and opportunities to promote, overcoming, 27
experience in, 24–25
in Maldives, 25
in Nepal, 24
in Pakistan, 24–25

Special wastes, 167
Sustainable Business Council (SBC), 276
Sustainable landfills, 144
Sustainable waste management
anaerobic digestion and composting, role of, 158–159
bottom-up approach to, 77
certifications and verifications, 58
hierarchy, 156
practices, 156–157
techniques, 9–12
source segregation, 9–10
three Rs (reduce, reuse, and recycle), 11–12, 216–217
"Swachh Bharat Mission," 25–26, 80
SWM. *See* Solid-waste management
Sydney Energy Recovery Facility, 308

T

Technological strategy, for plastic pollution, 257–262
bioremediation, 259–261, 260*f*
by bacteria, 261, 262*t*
by fungi, 261, 262*t*
by worms, 262, 262*t*
electrooxidation, 258–259
membrane bioreactors, 257, 258*f*
organosilane, 258, 259*f*
Technology, in waste management, 200–208
challenges and considerations, 203*f*, 207
continued evolution toward circular economy, 206–207
emerging technologies in biomass and bioenergy, 205
ethical and environmental considerations, 207–208
overview, 200
smart bins and collection systems, 204–205
GPS-enabled routing systems, 204–205
radio-frequency identification technology for waste tracking, 204
successful implementations of (case studies), 206
technological solutions, rise of, 202–203
data analytics and machine learning in waste sorting and recycling, 203
incorporating IoT-enabled sensors for monitoring, 202–203
waste-to-energy innovations, 205
Tenebrio molitor, 262
Thailand
community-based initiative, 101
MSW management in, 100*f*
waste containers at Ubon Ratchathani University, 100*f*
Thermal processes in waste management, 182–185
gasification, 185
incineration, 183, 183*t*–184*t*
pyrolysis, 184
Traditional waste management
challenge in, 232–233
current issues in, 234–235
drawbacks of, 235–236
carbon emissions, 235
inefficiencies in fixed schedules and procedures, 236
overflowing bins, 235
traffic congestion, 235
unsanitary environments, 235
economic effect of, 234
environmental effect of, 233, 233*f*
practices and limitations, 215–217
incineration, 216
landfilling, 215–216
reduce, reuse, and recycle, 216–217
Traveling salesman problem (TSP), 56
Trichoderma viride, 261
TSP. *See* Traveling salesman problem

U

UNEP. *See* United Nations Environment Programme
United Nations Environment Programme (UNEP), 276
United Nations Sustainable Development Goals, 53
United States Environmental Protection Agency (USEPA), 9

Urbanization, environmental impacts of, 38–39
USEPA. *See* United States Environmental Protection Agency

V

Volatile organic compounds (VOCs), 190

W

Waste characterization, 57
Waste composition
analysis, 54–55
climate change on, impacts of, 274
global trends in, 62–63
in Latin America, 168*f*
Waste disposal
improper disposal, 288–289
poor, 290
unauthorized dumping, 289
Waste-gas flare, 184*t*
Waste generation
climate change on, impacts of, 274
global trends in, 62–63
population and, 151–152
predictive analytics for, 61
projected, 125*f*
in top ten countries, 124*t*
Western Europe, 79*t*
patterns of waste generation, 76–77
trends in waste generation, 77–78
world scenario of, 123
Waste management
climate change. *See* Climate change, and waste management
and climate change, 272–283
data-driven solutions for, 219–220
data in, role of, 217–218
goals of, 290*f*
in Gulf countries, 318–330
hierarchy, 156*f*
inadequate or poor, 38–47
innovative technologies for, 218–219
in Oceania subcontinent, 304–313
solid. *See* Solid-waste management
in South Asia, 22–32
sustainable. *See* Sustainable waste management
technology in, 200–208
thermal processes in, 182–185
traditional, practices and limitations of, 215–217
incineration, 216
landfilling, 215–216
reduce, reuse, and recycle, 216–217
traditional challenges in, 201–202
growing global concerns and calls for change, 201–202
landfill overload and environmental impacts, 201
limited effectiveness of conventional recycling methods, 201
in Western Europe, 76–87
Waste management system, 182*f*
of Barcelona, 241–242
of Indore, 243
of Surat, 243
Waste Minimization Act of New Zealand, 311
Waste segregation or sorting
data analytics and machine learning in, 203
significance and challenges of, 130–133
financial constraints, 132
inadequate waste segregation, 131
informal waste sector, 131
lack of infrastructure, 131
policy and regulatory gaps, 132
rapid urbanization and population growth, 132–133
sociocultural attitudes and behaviors, 132
source-based, benefits of, 127–130
economic benefits, 129–130
environmental benefits, 128–129
social benefits, 130
techniques, 126–127
biological treatment, 127
gravity separation, 127
manual sorting, 127
mechanical sorting, 127
source segregation, 126
world scenario of, 123

Waste-to-energy conversion, 12–16
biochemical conversion, 15–16
anaerobic digestion, 16
bioethanol production, 16
gasification, 14–15, 16*f*
incineration, 12–14
advantages of, 13
disadvantages of, 14
process, 13*f*
pyrolysis, 14
process of, 14, 15*f*
steps involved, 15
Waste-to-energy (WtE), 185. *See also* Energy from waste
Asia-Pacific area, growth in, 180
challenges in Gulf countries, 322
emission regulations in, 187–188
innovations, 205
in Oceania, 309
solutions, 62
Water pollution, inadequate waste management and, 42–43
Western Europe, waste management in
challenges faced in, 78–79
community engagement, 84–86
future directions and research needs, 87
integrating technology and policy innovation, 81–84
application of IR 4.0 technologies, 81–82
global cities adopted IR 4.0 technologies (case study), 83
implications for policies and practices, 83–84
prospective applications, 82–83
key takesaways for implementation, 80–81
overview, 76–87
policy interventions, 84–86
waste generation, 79*t*
patterns of, 76–77
trends in, 77–78
waste management in India *vs.,* 79–80
Wooden combs and toothbrushes, 255
World Bank, 304
classification of Southeast Asian countries, 91–92, 92*f*

For Product Safety Concerns and Information please contact our EU representative GPSR@taylorandfrancis.com Taylor & Francis Verlag GmbH, Kaufingerstraße 24, 80331 München, Germany